The Principles of Flight
for Pilots

Aerospace Series List

Cooperative Path Planning of Unmanned Aerial Vehicles	Tsourdos et al	November 2010
Principles of Flight for Pilots	Swatton	October 2010
Air Travel and Health: A Systems Perspective	Seabridge et al	September 2010
Design and Analysis of Composite Structures: With Applications to Aerospace Structures	Kassapoglou	September 2010
Unmanned Aircraft Systems: UAVS Design, Development and Deployment	Austin	April 2010
Introduction to Antenna Placement & Installations	Macnamara	April 2010
Principles of Flight Simulation	Allerton	October 2009
Aircraft Fuel Systems	Langton et al	May 2009
The Global Airline Industry	Belobaba	April 2009
Computational Modelling and Simulation of Aircraft and the Environment: Volume 1 – Platform Kinematics and Synthetic Environment	Diston	April 2009
Handbook of Space Technology	Ley, Wittmann Hallmann	April 2009
Aircraft Performance Theory and Practice for Pilots	Swatton	August 2008
Surrogate Modelling in Engineering Design: A Practical Guide	Forrester, Sobester, Keane	August 2008
Aircraft Systems, 3rd Edition	Moir & Seabridge	March 2008
Introduction to Aircraft Aeroelasticity And Loads	Wright & Cooper	December 2007
Stability and Control of Aircraft Systems	Langton	September 2006
Military Avionics Systems	Moir & Seabridge	February 2006
Design and Development of Aircraft Systems	Moir & Seabridge	June 2004
Aircraft Loading and Structural Layout	Howe	May 2004
Aircraft Display Systems	Jukes	December 2003
Civil Avionics Systems	Moir & Seabridge	December 2002

The Principles of Flight
for Pilots

P. J. Swatton

A John Wiley and Sons, Ltd., Publication

This edition first published 2011
© 2011 John Wiley & Sons Ltd

Registered office
John Wiley & Sons Ltd, The Atrium, Southern Gate, Chichester, West Sussex, PO19 8SQ, United Kingdom

For details of our global editorial offices, for customer services and for information about how to apply for permission to reuse the copyright material in this book please see our website at www.wiley.com.

Library of Congress Cataloging-in-Publication Data

Swatton, P. J. (Peter J.)
   The principles of flight for pilots / P. J. Swatton.
      p. cm.
Includes index.
ISBN 978-0-470-71073-9 (pbk.)
1. Airplanes–Piloting.   2. Aerodynamics.   3. Flight.   I. Title.
TL710.S774   2010
629.132–dc22   2010014529

A catalogue record for this book is available from the British Library.

Print ISBN: 9780470710739
ePDF ISBN: 9780470710937
oBook ISBN: 9780470710944

Set in 9/11 Times by Aptara Inc., New Delhi, India.

Contents

Series Preface

The field of aerospace is wide ranging and covers a variety of products, disciplines and domains, not merely in engineering but in many supporting activities. These combine to enable the aerospace industry to produce exciting and technologically challenging products. A wealth of knowledge is contained by practitioners and professionals in the industry in the aerospace fields that is of benefit to other practitioners in the industry, and to those entering the industry from University or other fields.

The Aerospace Series aims to be a practical and topical series of books aimed at engineering professionals, operators and users and allied professions such as commercial and legal executives in the aerospace industry. The range of topics spans design and development, manufacture, operation and support of the aircraft as well as infrastructure operations, and developments in research and technology. The intention is to provide a source of relevant information that will be of interest and benefit to all those people working in aerospace.

The other books in the Aerospace Series concentrate very much on the technical aspects of Airframe, Structure and Systems - providing technical descriptions that are of use to engineers and designers. In most of these books the Human Machine interface is described, especially in Aircraft Display Systems.

Aircraft Performance, Theory and Practice for Pilots by P. J. Swatton extended the Series from the Design phase of the life-cycle into the operate phase by introducing aspects of the aircraft that are essential to the pilot.

In this book, Principles of Flight for Pilots, the author takes this a step further by introducing principles of flight in a comprehensive and easy to use compendium of knowledge complemented by self-assessment exercises. The book is packed with information from basic aerodynamics and stability through aerodynamic principles for level flight, manoeuvre and high speed flight. Even though this book is aimed squarely at pilots wishing to study for the EASA ATPL and CPL examinations, it should also be considered as essential reading for students wishing to enter the field of aero engineering and for practitioners in systems engineering, design, aerodynamics and testing.

Allan Seabridge

Preface

Since the Wright brothers' triumphant production of a flying machine in 1903, followed by Bleriot's successful navigation of the Channel in 1909, the mysteries of how an aeroplane flies have fascinated almost everyone. Although aerodynamics is a complicated subject it is essential that all aviators have a basic understanding of the principles of flight for the safety of themselves and those on the ground, without the prerequisite of comprehending all of the mathematics involved. This is the prime objective of the syllabus formulated by the JAA and now adopted by EASA. Although the knowledge and manipulation of some formulae is required, the syllabus limits it to those necessary to safely execute the duties of a pilot.

The aim of this book is to provide a trustworthy work of reference for pilots. It is collated and presented in such a manner that it will not only help student pilots to pass the examination but will also enable experienced personnel to gain a deeper understanding of the Principles of Flight and related subjects. **It is not intended to be a comprehensive study of aerodynamics.**

An examination in Principles of Flight is set by the Flight Crew Licensing Department of the Civil Aviation Authority (CAA) acting as an agent for EASA. To validate a licence, together with other requirements, a candidate must attain a mark of at least 75% in the examination.

Principles of Flight for Pilots

The Complete Manual. This manual has been written in a manner for easy learning primarily for trainee pilots wishing to study for the EASA ATPL and CPL licence examinations. It is also a useful reference book for qualified transport aeroplane pilots and has been comprehensively indexed for easy use.

The manual is divided into seven parts. Each part contains the necessary number of chapters to explain the appropriate topic in detail. After each chapter is a set of self-assessed questions that have been gleaned from the feedback of previous candidates in the Principles of Flight examination over the past nine years. The calculations and explanations to the correct solutions are those of the author are given in Chapter 19.

Part 1 – The Preliminaries. This part of the manual is devoted to an introduction to that area of basic physics applicable to the principles of flight and to the definitions that are used in the subsequent chapters.

Part 2 – Basic Aerodynamics. Theoretical aspects of aeroplane control and lift generation are confined to this part of the manual.

Part 3 – Level Flight Aerodynamics. This part is devoted to lift analysis, lift augmentation, drag, stalling and the thrust and power essential to maintain level flight.

Part 4 – Stability. This part examines in detail the complex topics of aeroplane static and dynamic stability.

Part 5 – Manoeuvre Aerodynamics. Level-flight manoeuvres such as turns and dives together with the aerodynamics of climbs and descents are the main topics of this part of the manual.

Part 6 – Other Aerodynamic Considerations. High-speed flight, including supersonic flight, is explained in detail because of the EASA syllabus requirements; despite the fact that there are no supersonic transport aeroplanes any longer. CPL examination candidates should ignore Chapter 15 – High Speed Flight.

Part 7 – Conclusion. This part includes a summary of the major components of the Principles of Flight syllabus and the solutions to all of the self-assessed exercises

The author would like to stress that, although *The Principles of Flight for Pilots* is directed towards explaining basic theory of flight, the explanations, advice and interpretations given are his alone, and not necessarily shared by EASA or any other legislative body. It does not seek to replace any of the works mentioned in the bibliography, but should be used in conjunction with them. References quoted in the text of the manual were current in May 2010.

Every effort has been made to ensure that the information contained in *The Principles of Flight for Pilots* was up-to-date at the time of publication; but readers are reminded that every document listed in the bibliography on which this book is based is subject to amendment. It is true that major changes of policy are not implemented without adequate warning and publicity; but minor alterations could escape notice and every reader is advised to pay careful attention to any amendment list issued by the CAA and EASA. No responsibility is accepted for any errors or discrepancy.

P. J. Swatton

Acknowledgements

My grateful thanks once again go to David Webb who has willingly given his expert advice and contributed in no small part by drawing all of the illustrations using his computer.

The Principles of Flight Examination

This manual contains the information required to cover the ATPL (A) and CPL (A) Learning Objectives for the EASA subject 081 - Principles of Flight. The examination in this subject is from 0930 to 1030 on the first day of the examinations for ATPL candidates and contains 40 questions. For CPL candidates the examination is from 0900 to 0945 on the first day of the examinations and contains 34 questions.

The main reference documents for the Principles of Flight examination are:
(1) EU-OPS1
(2) AMC Definitions
(3) CS-23 Normal and Commuter Aeroplanes
(4) CS-25 Large Aeroplanes
(5) Civil Aviation Aeronautical Information Circulars

List of Abbreviations

a	Acceleration
A	Cross-Sectional Area
A/F	Airfield
A and AEE	The Aeroplane and Armament Experimental Establishment
aal	above aerodrome level
AC	Aerodynamic Centre
AFM	Aeroplane Flight Manual
agl	above ground level
AIC	Aeronautical Information Circular
AIP	Aeronautical Information Package
amsl	above mean sea level
AoA	Angle of Attack
AR	Aspect Ratio
ASD	Accelerate/Stop Distance
ASDR	Accelerate/Stop Distance Required
ASIR	Airspeed Indicator Reading
ATM	Aerodynamic Twisting Moment
AUM	All-Up Mass
AUW	All-Up Weight
BHP	Brake Horsepower
BRP	Brake Release Point
C of A	Certificate of Airworthiness
CP	Centre of Pressure
CAA	Civil Aviation Authority
CAP	Civil Aviation Publication
CAS	Calibrated Airspeed
C_D	Coefficient of Drag
C_{DI}	Coefficient of induced drag
C_{DP}	Coefficient of parasite drag
C_{DA}	Mean Coefficient of drag in the air
C_{DG}	Mean Coefficient of drag on the ground
CF	Centrifugal Force
CG	Centre of Gravity
C_L	Coefficient of Lift
C_{Lmax}	Maximum Coefficient of Lift
C_n	Yawing Moment Coefficient
C_M	Pitching Moment
C_{M0}	Pitching Moment at the Zero Lift value

CP	Critical Point
CS	Certification Standards Document
CSU	Constant Speed Unit
CTM	Centrifugal Twisting Moment
DA	Density Altitude
EAS	Equivalent Airspeed
EASA	European Aviation Safety Agency
F	Force
FAA	Federal Aviation Administration
FAR	Federal Aviation Regulations
FLL	Field-length-limited
g	Acceleration due to gravity
GE	Ground Effect
G/S	Groundspeed
IAS	Indicated Airspeed
IAT	Indicated Air Temperature
ICAO	International Civil Aviation Organisation
ISA	International Standard Atmosphere
JAA	Joint Aviation Authority
JAR	Joint Aviation Requirements
JSA	Jet Standard Atmosphere
kg	kilogram(s)
km	kilometre(s)
kt	nautical miles per hour (knots)
KE	Kinetic Energy
L	Rolling moment
LD	Landing Distance
LE	Leading Edge
LER	Leading Edge Radius
LSS	Local Speed of Sound
m	Mass
M	Mach Number
M/S	Mass per unit area of a wing (wing loading)
MAC	Mean Aerodynamic Chord
M$_{CDR}$	Critical Drag Rise Mach Number
M$_{CRIT}$	Critical Mach Number
M$_{DET}$	Detachment Mach Number
M$_{FS}$	The True Mach Number of an aeroplane
M$_L$	The Local Mach Number
M$_{MO}$	Maximum Operating Mach Number
n	Load Factor
N	Newton
NP	Neutral Point
OAT	Outside Air Temperature
PCU	Propeller Control Unit
PIO	Pilot-Induced Oscillation
ps	Static Pressure
pt	Total Pressure
q	Dynamic Pressure
RAF	Relative Airflow
RAS	Rectified Airspeed
Re	Reynold's Number

ROC	Rate of Climb
ROD	Rate of Descent
RPM	Revolutions per Minute
S	Wing Area
SG	Specific Gravity
SM	Static Margin
SP	Stagnation Point
STOL	Short-field take-off and landing
TAS	True Airspeed
TAT	Total Air Temperature
TE	Trailing Edge
THS	Trimmable Horizontal Stabilizer
TOD	Take-off Distance
TOM	Take-Off Mass
TOR	Take-off Run
TOW	Take-Off Weight
TP	Trim Point
V_A	Design Manoeuvring Speed
V_b	Basic Stalling Speed
V_B	Design Speed for maximum gust intensity
V_C	Design Cruising Speed
V_{CLmax}	CAS of the maximum CL.
V_D	Design Diving Speed
V_{DD}	Design Drag Devices speed
V_{EF}	The assumed speed of engine failure during the take-off ground run
V_F	Design Flap Speed
V_{FE}	Maximum speed for flying with flaps extended
V_{FO}	Maximum speed for operating the flaps
V_{IMD}	The velocity of minimum drag
V_{IMP}	The velocity of minimum power
V_{LE}	The maximum speed with the undercarriage (landing gear) extended
V_{LO}	The maximum speed at which the undercarriage (landing gear) may be operated
V_M	Manoeuvre Stalling Speed.
V_{MC}	The minimum control speed with the critical power unit inoperative
V_{MCG}	The minimum control speed on the ground with the critical power unit inoperative
V_{MCL}	The minimum control speed on the approach to land
$V_{MCL(1out)}$	The minimum control speed on the approach to land with one engine inoperative
V_{MCL-2}	The minimum control speed on the approach to land with two engines inoperative
V_{IMD}	Velocity of minimum drag IAS
V_{IMP}	Velocity of minimum power IAS
V_{MD}	Velocity of minimum drag TAS
V_{MO}	The maximum operating speed
V_{MP}	Velocity of minimum power TAS
V_{MS}	The minimum stalling speed
V_{MS0}	The minimum stalling speed with the flaps in the landing setting
V_{MS1}	The minimum stalling speed for the case under consideration
V_{MU}	The minimum unstick speed
V_{NE}	Never exceed speed
V_{NO}	Maximum normal operating speed.
V_O	The speed of the freestream airflow over an aerofoil surface
V_{RA}	The rough-air or turbulence speed
V_{REF}	The reference landing speed

V_S	Stalling speed CAS
V_{S0}	The stalling speed CAS with the flaps at the landing setting
V_{S1}	The stalling speed CAS for the configuration under consideration
V_{S1g}	Stalling speed CAS at 1g
V_{SR}	Reference stalling speed CAS
V_{SR0}	Reference stalling speed CAS in the landing configuration
V_{SR1}	Reference stalling speed CAS for the configuration under consideration
V_{SW}	The speed at which the onset of the natural or artificial stall warning activates
V_X	The speed at which the maximum gradient of climb will be achieved
V_Y	The speed at which the maximum rate of climb will be achieved
WC	Wind Component
WED	Water-Equivalent Depth

Weight and Mass

Before starting any calculations it is necessary to explain the difference between a Newton (N), which is a unit of force, a kilogram (kg), which is a unit of mass and weight, which is the force acting on a body by gravity. Most of us know what we mean when we use the term weight and become confused when the term mass is used in its place. In all of its documents the JAA consistently use the term mass whereas the majority of aviation documents produced by the manufacturers use the term weight. The following are the definitions of each of the terms and should help clarify the situation:

Mass The quantity of matter in a body as measured by its inertia is referred to as its mass. It determines the force exerted on that body by gravity, which is directly proportional to the mass. Gravity varies from place to place and also decreases with increased altitude above mean sea level.

Weight The force exerted on a body by gravity is known as its weight and is dependent on the mass of the body and the strength of the gravitational force for its value. Weight = mass in kg × gravity in Newtons. Thus, the weight of a body varies with its location and elevation above mean sea level but the mass does not change for the same body.

　　The change of weight of an object due to its changed location is extremely small, even at 50 000 ft above mean sea level, however, it is technically incorrect and the term mass should be used. *For the purposes of this manual the terms weight and mass are interchangeable.* In the questions asked in the JAA examinations the word mass is used most of the time. *IEM OPS 1.605.*

The Newton A Newton is a unit of force, which equals mass × acceleration.

　　1 Newton = 1 kg × 1 m/s^2. At the surface of the Earth the acceleration due to gravity equals 9.81 m/s^2. Thus, the force acting on 1 kg at the Earth's surface is 9.81 Newtons. To simplify calculations in the examination the acceleration due to gravity is given as 10 m/s^2 therefore 1 kg is equal to 10 Newtons.

Part 1
The Preliminaries

1 Basic Principles

1.1 The Atmosphere

The Earth's atmosphere is the layer of air that surrounds the planet and extends five hundred miles upwards from the surface. It consists of four concentric gaseous layers, the lowest of which is the troposphere in which all normal aviation activities take place. The upper boundary of the troposphere is the tropopause, which separates it from the next gaseous layer, the stratosphere. The next layer above that is the mesosphere and above that is the thermosphere.

The height of the tropopause above the surface of the earth varies with latitude and with the season of the year. It is lowest at the poles being approximately 25 000 feet above the surface of the Earth and 54 000 feet at the Equator. These heights are modified by the season, being higher in the summer hemisphere and lower in the winter hemisphere.

Above the tropopause the stratosphere extends to a height of approximately one hundred thousand feet. Although these layers of the atmosphere are important for radio-communication purposes, because of the ionised layers present, they are of no importance to the theory of flight.

Since air is compressible the troposphere contains the major part of the mass of the atmosphere. The weight of a column of air causes the atmospheric pressure and density of the column to be greatest at the surface of the Earth. Thus, air density and air pressure decrease with increasing height above the surface. Air temperature also decreases with increased height above the surface until the tropopause is reached above which the temperature remains constant through the stratosphere.

1.2 The Composition of Air

Air is a mixture of gases the main components of which are shown in Table 1.1.

Water vapour in varying quantities is found in the atmosphere up to a height of approximately 30 000 ft. The amount in any given air mass is dependent on the air temperature and the passage of the air mass in relationship to large areas of water. The higher the air temperature the greater the amount of water vapour it can hold.

1.2.1 The Measurement of Temperature

Centigrade Scale. The Centigrade scale is normally used for measuring the air temperature and for the temperature of aero-engines and their associated equipment. On this temperature scale water freezes at 0° and boils at 100° at mean sea level.

The Principles of Flight for Pilots   P. J. Swatton
© 2011 John Wiley & Sons, Ltd

Table 1.1   Gas Components of the Air.

Element	Volume	Weight
Nitrogen	78.09%	75.5%
Oxygen	20.95%	23.1%
Argon	0.93%	1.3%
Carbon Dioxide	0.03%	0.05%

Note: For all practical purposes the atmosphere is considered
to contain 21% oxygen and 79% nitrogen.

Kelvin Scale. Often for scientific purposes temperatures relative to absolute zero are used in formulae regarding atmospheric density and pressure. Temperatures relative to absolute zero are measured in Kelvin. A body is said to have no heat at absolute zero and this occurs at a temperature of −273.15 °C.

1.2.2   Air Density

Air density is mass per unit volume. The unit of air density is either kg per m^3 or gm^{-3} and the symbol used is 'ρ'. The relationship of air density to air temperature and air pressure is given by the formula:

$$\frac{p}{T\rho} = \text{constant}$$

where ρ is the density, p is pressure in hPa and T is the absolute temperature.

The Effect of Air Pressure on Air Density. If air is compressed the amount of air that can occupy a given volume increases. Therefore, both the mass and the density are increased. For the same volume if the pressure is decreased then the reverse is true. From the formula above if the air temperature remains constant then the air density is directly proportional to the air pressure. If the air pressure is doubled so is the air density.

The Effect of Air Temperature on Air Density. When air is heated it expands so that a smaller mass will occupy a given volume and provided that the air pressure remains constant then the air density will decrease. Thus, the density of the air is inversely proportional to the absolute temperature. The rapid decrease of air pressure with increased altitude has a far greater effect on the air density than does the increase of density caused by the decrease in temperature for the same increased altitude. Thus, the overall effect is for the air density to diminish with increased altitude.

The Effect of Humidity on Air Density. Until now it has been assumed that the air is perfectly dry; such is not the case. In the atmosphere there is always some water vapour present, albeit under certain conditions a miniscule amount. However, in some conditions the amount of water vapour present is an important factor when determining the performance of an aeroplane. For a given volume the amount of air occupying that volume decreases as the amount of water vapour contained in the air increases. In other words, air density decreases with increased water-vapour content. It is most dense in perfectly dry air.

1.3   The International Standard Atmosphere

The basis for all performance calculations is the International Standard Atmosphere (ISA) which is defined as a perfect dry gas, having a mean sea level temperature of +15 °C, which decreases at the rate of 1.98 °C for every 1000 ft increase of altitude up the tropopause which is at an altitude of 36 090 ft above which the temperature is assumed to remain constant at −56.5 °C. The mean sea level (MSL) atmospheric pressure is assumed 1013.2 hPa (29.92 in. Hg). See Table 1.2.

Table 1.2 International Standard Atmosphere (Dry Air).

Pressure	Temperature		Density	Height		Thickness of 1 hPa layer	
hPa	°C	°F	gm^{-3}	m	ft	m	ft
1013.2	15.0	59.0	1225	0	0	8.3	27
1000	14.3	57.7	1212	111	364	8.4	28
950	11.5	52.7	1163	540	1773	8.8	29
900	8.6	47.4	1113	988	3243	9.2	30
850	5.5	41.9	1063	1457	4781	9.6	31
800	2.3	36.2	1012	1949	6394	10.1	33
750	−1.0	30.1	960	2466	8091	10.6	35
700	−4.6	23.8	908	3012	9882	11.2	37
650	−8.3	17.0	855	3591	11 780	11.9	39
600	−12.3	9.8	802	4206	13 801	12.7	42
550	−16.6	2.1	747	4865	15 962	13.7	45
500	−21.2	−6.2	692	5574	18 289	14.7	48
450	−26.2	−15.2	635	6344	20 812	16.1	53
400	−31.7	−25.1	577	7185	23 574	17.7	58
350	−37.7	−36.0	518	8117	26 631	19.7	65
300	−44.5	−48.2	457	9164	30 065	22.3	75
250	−52.3	−62.2	395	10 363	33 999	25.8	85
200	−56.5	−69.7	322	11 784	38 662	31.7	104
150	−56.5	−69.7	241	13 608	44 647	42.3	139
100	−56.5	−69.7	161	16 180	53 083	63.4	208

1.3.1 ISA Deviation

It is essential to present performance data at temperatures other than the ISA temperature for all flight levels within the performance-spectrum envelope. If this were to be attempted for the actual or forecast temperatures, it would usually be impracticable and in some instances impossible.

To overcome the presentation difficulty and retain the coverage or range required, it is necessary to use ISA deviation. This is simply the algebraic difference between the actual (or forecast) temperature and the ISA temperature for the flight level under consideration. It is calculated by subtracting the ISA temperature from the actual (or forecast) temperature for that particular altitude. In other words:

ISA Deviation = Ambient temperature − Standard Temperature

Usually, 5 °C bands of temperature deviation are used for data presentation in Flight Manuals to reduce the size of the document or to prevent any graph becoming overcrowded and unreadable.

1.3.2 JSA Deviation

As an alternative to ISA deviation some aircraft manuals use the Jet Standard Atmosphere (JSA) Deviation that assumes a temperature lapse rate of 2°/1000 ft and that the atmosphere has no tropopause, the temperature is, therefore, assumed to continue decreasing at this rate beyond 36 090 ft.

1.3.3 Height and Altitude

Three parameters are used for vertical referencing of position in aviation. They are the airfield surface level, mean sea level (MSL) and the standard pressure level of 1013.2 hPa. It would be convenient if the performance data could be related to the aerodrome elevation because this is fixed and published in the Aeronautical Information Publication.

However, this is impractical because of the vast range that would have to be covered. Mean sea level and pressure altitude are the only permissible references for assessing altitude for the purposes of aircraft performance calculations, provided that the one selected by the manufacturers for the Flight Manual is used consistently throughout the manual. Alternatively, any combination of them may be used in a conservative manner.

Table 1.3 ISA Height in Feet above the Standard Pressure Level.*

Pressure (hPa)	0	2	4	6	8
1030	−456	−509	−563	−616	−670
1020	−185	−240	−294	−348	−402
1010	+88	+33	−22	−76	−131
1000	+363	+308	+253	+198	+143
990	640	584	529	473	418
980	919	863	807	751	695
970	1200	1143	1087	1031	975
960	1484	1427	1370	1313	1256
950	1770	1713	1655	1598	1541
940	2059	2001	1943	1885	1828
930	2351	2293	2235	2176	2117
920	2645	2587	2528	2469	2410
910	2941	2882	2823	2763	2704
900	3240	3180	3120	3060	3001
890	3542	3482	3421	3361	3300
880	3846	3785	3724	3663	3603
870	4153	4091	4029	3968	3907
860	4463	4401	4339	4277	4215
850	4777	4714	4651	4588	4526
840	5093	5029	4966	4903	4840
830	5412	5348	5284	5220	5157
820	5735	5670	5606	5541	5476
810	6061	5996	5930	5865	5800
800	6390	6324	6258	6192	6127
790	6722	6656	6589	6523	6456

*Enter with QFE to read Aerodrome Pressure Altitude. Enter with QNH to read the correction to apply to Aerodrome/Obstacle Pressure Altitude.

Using MSL avoids the problem of the range of heights and would be ideal from a safety viewpoint; but again this would be too variable because of the temperature and pressure range that would be required.

The only practical datum to which aircraft performance can be related is the standard pressure level of 1013.2 hPa. See Table 1.3.

1.3.4 Pressure Altitude

In Aeroplane Flight Manuals (AFMs) the word **altitude** refers strictly to **pressure altitude**, which can be defined as the vertical distance from the 1013.2 hPa pressure level. Therefore, aerodrome and obstacle elevations must be converted to pressure altitude before they can be used in performance graphs. Many large aerodromes provide the aerodrome pressure altitude as part of their hourly weather reports.

To correct an aerodrome elevation to become a pressure altitude if Table 1.3 is not available use the following formulae:

A/F Pressure Altitude = Aerodrome elevation in ft + [(1013.2 hPa − QNH) × 27 ft]

Aerodrome Pressure Altitude = (1013.2 hPa − QFE) × 27 ft

To correct an altitude for the temperature errors of the altimeter use the following formula:

Altitude Correction = 4 × ISA Deviation × Indicated Altitude ÷ 1000

1.3.5 Density Altitude

The performance data for small piston/propeller-driven aeroplanes is calculated using *density altitude*, which is pressure altitude corrected for nonstandard temperature. It is the altitude in the standard atmosphere at which the prevailing density occurs and can be calculated by using the formula:

Density Altitude = Pressure Altitude + (118.8 × ISA Deviation)

1.4 The Physical Properties of Air

Air is a compressible fluid. It can therefore, flow or change its shape when subjected to very small outside forces because there is little cohesion of the molecules. If there was no cohesion between the molecules and therefore no internal friction then it would be an 'ideal' fluid, but unfortunately such is not the case.

1.4.1 Fluid Pressure

The pressure in a fluid at any point is the same in all directions. Any body, irrespective of shape or position, when immersed in a stationary fluid is subject to the fluid pressure applied at right angles to the surface of that body at that point.

1.4.2 Static Pressure

The pressure of a stationary column of air at a particular altitude is that which results from the mass of air in the column above that altitude and acts in all directions at that point. The static pressure decreases with increased altitude as shown in Table 1.3. The abbreviation for the static pressure at any altitude is P.

1.4.3 Dynamic Pressure

Air in motion has energy because it possesses density (mass per unit volume) that exerts pressure on any object in its path. This is dynamic pressure, which is signified by the notation (q) and is proportional to the air density and the square of the speed of the air. A body moving through the air has a similar force exerted on it that is proportional to the rate of movement of the body. The energy due to this movement is kinetic energy (KE), which is equal to half the product of the mass and the square of the speed. Bernoulli's equation for incompressible airflow states that the kinetic energy of one cubic metre of air travelling at a given speed can be calculated from the following formula:

$$KE = \frac{1}{2}\rho V^2 \text{ joules}$$

Where ρ is the air density in kg per m^3 and V is the airspeed in metres per second.

Note: 1. A joule is the work done when the point of application of a force of one Newton is displaced by one metre in the direction of the force.

Note: 2. A Newton is that force that when applied to a mass of one kg produces an acceleration of 1 metre per second per second.

If a volume of air is trapped and brought to rest in an open ended tube the total energy remains constant. If such is the case then KE becomes pressure energy (PE), which for practical purposes is equal to $\frac{1}{2}\rho V^2$ Newtons per m^2. If the area of the tube is S square metres then:

$$S \text{ (dynamic + static pressure)} = \frac{1}{2}\rho V^2 \text{ Newtons}$$

The term $\frac{1}{2}\rho V^2$ is common to all aerodynamic forces and determines the load imposed on an object moving through the air. It is often modified to include a correction factor or a coefficient. The term is used to describe the dynamic pressure imposed by the air of a certain density moving at a given speed and that is brought completely to rest. Therefore, $q = \frac{1}{2}\rho V^2$.

Note: Dynamic pressure cannot be measured on its own because static pressure is always present. This total pressure is known as pitot pressure.

1.5 Newton's Laws of Motion

1.5.1 Definitions

It is essential to remember the following definitions regarding the motion of a body:

a. **Force**. That which changes a body's state of rest or of uniform motion in a straight line is a force, the most familiar of which is a push or a pull.
b. **Inertia**. The tendency of a body to remain at rest or, if moving, to continue in a straight line at a constant speed is inertia.
c. **Momentum**. The product of mass and velocity is momentum.

1.5.2 First Law

Newton's first law of motion, the law of inertia, states that every body remains in a state of rest or uniform motion unless compelled to change its state by an applied force. Bodies at rest or in a state of steady motion are said to be in equilibrium and have the property of inertia. Where motion results from an applied force, the force exerted is the product of mass and acceleration. Mathematically:

$$F = ma$$

where F = Force; m = mass; a = acceleration.

1.5.3 Second Law

Being the product of mass and velocity, momentum is a vector quantity that involves motion in the direction of the velocity. If the body is in equilibrium then there is no change to the momentum, however, if the forces are not in equilibrium Newton's second law of motion states that the rate of change of momentum, the acceleration, is proportional to the applied force in the direction in which that force acts and inversely proportional to the mass of the object. By transposing the formula for Newton's First Law then:

$$a = F/m$$

1.5.4 Third Law

Newton's third law of motion states that for every action there is an equal and opposite reaction. A body at rest on a surface applies a force to that surface and an opposite force is applied by the surface to the body. A free falling body is acted on by gravity the force (F) is measured in Newtons and is calculated as:

$$F = mg$$

where F = Force in Newtons; m = mass in kg; g = acceleration due to gravity of 9.81 m/s^2.

1.6 Constant-Acceleration Formulae

A constant acceleration is when the velocity of a body is changing at a constant rate. Four formulae can be derived using the abbreviations v = the final velocity; u = the initial velocity; a = the acceleration; t = the time interval over which the acceleration took place; s = the distance travelled during the period of motion.

a. **The Final Velocity (1)**. The final velocity of a body is equal to the initial velocity plus the acceleration made during the time interval during which the acceleration took place and can be determined at any time by the formula:

$$v = u + at$$

b. **The Distance Travelled (1)**. The distance travelled during the period of motion is equal to the mean velocity multiplied by the time over which the acceleration took place and can be calculated by using the formula:

$$s = \tfrac{1}{2}(u + v)t$$

c. **The Distance Travelled (2)**. By substitution in the formulae a and b a second method of calculating the distance travelled can be derived as follows:

$$s = ut + \tfrac{1}{2}at^2$$

d. **The Final Velocity (2)**. Similarly, a formula for the final velocity can be derived by substitution as follows:

$$v^2 = u^2 + 2as$$

$$\text{or } v = \sqrt{(u^2 + 2as)}$$

1.7 The Equation of Impulse

The impulse of a force is the change in momentum (final momentum – initial momentum) of force acting on a body and is usually identified by the initial **J**. It is the product of that force and time. The SI unit of impulse is the Newton second (N s) **NOT** Newtons per second (N/s).

Using substitution in the formula for Newton's first law of motion and the formula for the final velocity (1) the following formula can be derived:

a. $F = ma$
b. By dividing by m then $a = F/m$.
c. $v = u + at$
d. By transposition becomes $v = u + t(F/m)$
e. By multiplication by m and transposition becomes: $Ft = mv - mu$
f. Therefore, the value of the impulse of the force $\mathbf{J} = \mathbf{mv} - \mathbf{mu}$ **or**

$$\mathbf{J = m(v - u)} \text{ or } \mathbf{Impulse = mass \times speed\ change}.$$

1.8 The Basic Gas Laws

There are three basic gas laws regarding the relationship between pressure ($\mathbf{P}$), volume ($\mathbf{V}$) and temperature ($\mathbf{T}$) of the gas, which were formulated in the past. They are:

1.8.1 Boyles Law

This law states that if the mass of gas is fixed and the temperature of the gas remains constant then the volume of the gas is inversely proportional to the pressure. In other words, if the volume of a given mass is halved then its pressure will be doubled provided the temperature does not change. Mathematically then:

$$\mathbf{P_1 V_1 = P_2 V_2} \text{ or } \mathbf{PV = constant}.$$

1.8.2 Charles' Law

This law states that if a fixed mass of gas is at a constant pressure its volume will increase by 1/273 of its volume at 0 °C for every 1 °C rise in temperature. Alternatively it can be stated that the volume of a fixed mass of gas is directly proportional to its absolute temperature provided the pressure remains constant. Mathematically then:

$$\frac{\mathbf{V_1}}{\mathbf{T_1}} = \frac{\mathbf{V_2}}{\mathbf{T_2}} \text{ or } \frac{\mathbf{V}}{\mathbf{T}} = \mathbf{constant}$$

1.8.3 Pressure Law

This law states that for a fixed mass of gas at constant volume the pressure increases by 1/273 of its volume at 0 °C for every 1 °C rise in temperature. Thus, provided the volume of a mass of gas does not change the pressure is directly proportional to its temperature. Mathematically then:

$$\frac{\mathbf{P_1}}{\mathbf{T_1}} = \frac{\mathbf{P_2}}{\mathbf{T_2}} \text{ or } \frac{\mathbf{P}}{\mathbf{T}} = \mathbf{constant}$$

1.8.4 The Ideal Gas Equation

If the formulae in each of the gas laws a, b and c above are combined into a single equation it provides the equation for the ideal gas. Mathematically then:

$$\frac{\mathbf{P_1 V_1}}{\mathbf{T_1}} = \frac{\mathbf{P_2 V_2}}{\mathbf{T_2}} \text{ or } \frac{\mathbf{PV}}{\mathbf{T}} = \mathbf{constant}$$

1.9   The Conservation Laws

Bernoulli's principle is for the conservation of energy and the continuity equation is for the conservation of mass; both have equations that are directly related to the conservation laws.

1.10   Bernoulli's Theorem

Bernoulli stated that moving gas has four types of energy:

a. potential energy due to its height
b. kinetic energy due to movement
c. heat energy due to its temperature
d. pressure energy due to its compression

Bernoulli's theorem states that that the sum of the energies within an ideal gas in a streamline flow remains constant. It is the same principle as that of the conservation of energy. However, it does not account viscosity, heat transfer or compressibility effects. Below 10 000 ft and 250 kt airspeed compressibility effects can be safely ignored and the flow density is assumed to be constant.

Conventionally, airflow at speeds less than 0.4 Mach are considered to be an ideal gas because it is not compressed and the friction forces are small in comparison with the inertial forces, this is referred to as an inviscid flow. The sum of all forms of mechanical energy, that is the sum of kinetic energy and potential energy remains constant along a streamline flow and is the same at any point in the streamline flow. *This principle does not apply to the boundary layer* because mechanical and thermal energy is lost due to the skin friction, which is an effect of viscosity.

Bernoulli's theorem when applied to the airflow past an aerofoil at less than Mach 0.4 shows that the total pressure is equal to the sum of the dynamic pressure (the pressure caused by the movement of the air) and the static pressure (the pressure of the air not associated with its movement) and can be expressed as the following equation:

Total Pressure (pt) = Dynamic Pressure (q) + Static Pressure (ps)

The total pressure is constant along a streamline flow. Therefore, if static pressure decreases then dynamic pressure increases and vice versa. Air that is flowing horizontally flows from high pressure to low pressure. The highest speed occurs where the pressure is lowest and the lowest speed is where the pressure is highest.

In a freestream airflow, a *favourable pressure gradient* is one in which the static pressure decreases with distance downstream. An *adverse pressure gradient* is one in which the static pressure increases with distance downstream. A freestream airflow will accelerate in a favourable pressure gradient and decelerate in an adverse pressure gradient.

It is this principle that is utilised in the construction of the airspeed indicator. The dynamic pressure is the difference between the stagnation pressure and the static pressure. The airspeed indicator is calibrated to display the indicated airspeed appropriate to the dynamic pressure.

The flow speed can be measured in a pipe in which the tube diameter is restricted. The reduction in diameter increases the speed of flow and simultaneously decreases the pressure of the flow. This is referred to as the 'Venturi effect.'

1.10.1   Viscosity

Viscosity is a measure of the degree to which a fluid resists flow under an applied force. A fluid or liquid that is highly viscous flows less readily than a fluid or liquid that has low viscosity. The internal friction of the gas or liquid determines its ability to flow or its fluidity. Viscosity for air is the resistance of one layer of air to the movement over a neighbouring layer. When considering the effects of scale in wind-tunnel tests this fact is of great importance in the determination of aerofoil surface friction. The greater the friction or viscosity of a gas or liquid the greater is its resistance to flow. Temperature affects the viscosity but unlike liquids, air becomes more viscous as ambient temperature increases and is less able to flow readily. The viscosity of air is not affected by changes of air density that are not caused by temperature.

1.11   The Equation of Continuity

Mass cannot be created or destroyed. Air mass flow is steady and continual. The equation of continuity states that for an incompressible fluid flowing in a cylinder the rate at which the fluid flows past any given point is the same everywhere in the cylinder. In other words, the flow rate is equal to the mass flowing past divided by the time interval. Because the air density remains constant then the flow rate is constant throughout the cylinder. The product of the air density, the velocity and the cross-sectional area, the mass flow, is always constant.

$$\text{Mass Flow} = \rho AV = \text{constant}$$

where ρ = air density; A = cross-sectional area of the cylinder; V = flow velocity.

The equation of continuity applies equally to a cylinder that has a variable diameter know as a Venturi tube. If the inside diameter of the cylinder decreases it causes the inside of the cylinder to have a neck. Because the flow rate is constant anywhere in a cylinder through which there is streamline airflow then for the equation of continuity to remain valid, because the air pressure decreases and the air density remains constant, then the speed of the airflow must increase to ensure the same quantity of air passes any given point in the neck of the Venturi. Because the speed of the air increases and the static pressure decreases the streamlines move closer together and vice versa.

Therefore, it is true to say that as the diameter of a stream tube decreases the stream velocity increases and to maintain a constant mass flow the static pressure decreases, the dynamic pressure increases but the total pressure remains constant. The air density is the same after the change has occurred as it was before the event.

If the temperature of streamline airflow changes, the air density will alter. Increased temperature will decrease the air density and if the stream speed remains constant then the mass flow past any point decreases and vice versa. The drag experienced in a stream tube is directly proportional to the air density of the flow. If the air density is halved then the drag is also halved.

1.12   Reynolds Number

Reynolds, a 19th-century physicist, discovered from experiments that a sphere placed in a streamline flow of fluid caused the flow to change from a smooth flow to a turbulent flow. Furthermore, he found that the transition point from smooth to turbulent flow in all cases occurred at the maximum thickness of the body relative to the flow. It was also ascertained that the speed at which the transition from smooth to turbulent flow occurs is when the flow velocity reaches a value that is inversely proportional to the diameter of the sphere. Thus, turbulence occurs at a low speed over a large sphere and at a high speed over a small sphere.

Since then it has been established that the airflow pattern over an exact model of an aeroplane is precisely the same as the airflow over the actual aeroplane. Thus, the laws of aerodynamics are true and there is no error due to scale effect when based on Reynolds principles. These state that the value of velocity × size must be the same for both the model and the full-size aeroplane.

It has been determined that the similarity of flow pattern is the same for both the model and the full size aeroplane if the value of the following formula remains constant:

$$\frac{\text{density} \times \text{velocity} \times \text{size}}{\text{viscosity}}$$

For every wind-tunnel test there is one Reynolds number (R), which is always published in the results of any test. The ratio between the inertial and viscous (friction) forces is the Reynolds number. It is used to identify and predict different flow regimes such as laminar or turbulent flow. It can be calculated by the formula:

$$\text{Reynolds Number (Re)} = \frac{\rho VL}{\mu} = \frac{\text{Inertial Forces}}{\text{Viscous Forces}}$$

where ρ is density in kg per m^3; V is the velocity in metres per second; L is the chord length and μ is the viscosity of the fluid.

A *small Reynolds number* is one in which the *viscous force is predominant* and indicates a steady flow and **smooth** fluid motion. A *large Reynolds number* is one in which the *inertial force is paramount* and indicates random eddies and **turbulent** flow.

1.12.1   Critical Reynolds Number (Re$_{crit}$)

The change from smooth laminar flow to turbulent flow is gradual. There is a range during which the transition from smooth to turbulent flow takes place the Critical Reynolds number occurs approximately half way through this range. Its value is determined experimentally and is dependent on the exact flow configuration.

To determine the airflow around an aeroplane a scale model is tested in a wind tunnel using the same Reynolds number as the actual aeroplane to determine the airflow behaviour. The results are directly proportional to the size of the model in relation to the size of the actual aeroplane. For example, the flow velocity and behaviour of a quarter size model has to be increased fourfold for use with the actual aeroplane. This is called 'dynamic similarity' and is significant when determining the drag characteristics of an aeroplane.

1.13   Units of Measurement

Item	S.I. Units	Description
Wing Loading M/S (mass per unit area)	N/m^2	Force in Newtons divided by the wing area in square metres.
Dynamic Pressure (q)	N/m^2	Force in Newtons divided by the area in square metres.
Power	N m/s	This is the measure of the work done in a specific time period. It is the force in Newtons multiplied by speed in metres per second.
Air Density	g/m^3	Mass in kilograms divided by volume in cubic metres.
Force (Newton)	$kg\ m/s^2$	Mass multiplied by acceleration in metres per second per second.
Pressure	N/m^2	Force in Newtons divided by the area in square metres. In the USA they use psi as a measure of pressure (pounds per square inch).

Self-Assessment Exercise 1

Q1.1 Which formula or equation describes the relationship between force (F), acceleration (a) and mass (m)?
(a) $a = F.m$
(b) $F = m/a$
(c) $F = m.a$
(d) $m = F.a$

Q1.2 Bernoulli's equation can be written as:
(pt $-$ total pressure, ps $=$ static pressure, q $=$ dynamic pressure)
(a) $pt = q - ps$
(b) $pt = q + ps$
(c) $pt = ps - q$
(d) $pt + ps = q$

Q1.3 If the continuity equation is applicable to an incompressible airflow in a tube at low subsonic speed, if the diameter of the tube changes the air density after the change:
(a) will be greater than it was before
(b) will be less than it was before
(c) is dependent on the change
(d) will be the same as before the change

Q1.4 If the continuity equation is applicable to an incompressible airflow in a tube at low subsonic speed, if the diameter of the tube increases the speed of the flow:
(a) increases
(b) becomes sonic
(c) decreases
(d) remains the same

Q1.5 In the SI system kg. m/s^2 is expressed as a:
(a) Joule
(b) Watt
(c) Newton
(d) Pascal

Q1.6 The unit of wing loading (i) M/S and (ii) dynamic pressure q are:
(a) (i) N/m; (ii) kg
(b) (i) N/m^2; (ii) N/m^2
(c) (i) N/m^3; (ii) kg/m^3
(d) (i) kg/m; (ii) N/m^2

Q1.7 The total pressure is:
(a) static pressure plus dynamic pressure
(b) static pressure minus dynamic pressure
(c) $^1/_2 \rho V^2$
(d) measured parallel to the local stream

Q1.8 The static pressure of the flow in a tube:
(a) decreases when the diameter decreases
(b) is total pressure plus dynamic pressure
(c) is the pressure at the point at which the velocity is zero
(d) increases when the diameter decreases

Q1.9 Bernoulli's equation can be written as:
(pt $=$ total pressure; ps $=$ static pressure; q $=$ dynamic pressure)
(a) $pt = ps/q$
(b) $pt = ps + q$
(c) $pt = ps - q$
(d) $pt = q - ps$

Q1.10 The unit of power measurement is:
  (a) kg m/s^2
  (b) PA/m^3
  (c) N/m
  (d) N m/s
Q1.11 Static pressure acts:
  (a) perpendicular to the direction of flow
  (b) in the direction of the total pressure
  (c) in all directions
  (d) in the direction of the flow
Q1.12 Which of the following statements regarding Bernoulli's theorem is correct?
  (a) The total pressure is zero when the stream velocity is zero
  (b) The dynamic pressure is maximized at the stagnation point
  (c) The dynamic pressure increases as the static pressure decreases
  (d) The dynamic pressure decreases as static pressure decreases
Q1.13 The units of measurement used for air density (i)...........and its force (ii)............are:
  (a) (i) N/ kg; (ii) kg
  (b) (i) kg/m^3; (ii) N
  (c) (i) kg/m^3; (ii) kg
  (d) (i) N/m^3; (ii) N
Q1.14 The units used to measure air density are:
  (a) kg/cm^3
  (b) Bar
  (c) kg/m^3
  (d) psi
Q1.15 The unit of measurement used for pressure is:
  (a) lb/gal
  (b) kg/dm^2
  (c) psi
  (d) kg/m^3
Q1.16 Which of the following formulae is correct?
  (a) a = M ÷ F
  (b) F = M × a
  (c) a = F × M
  (d) M = F × a
Q1.17 The product kg and m/s^2 is:
  (a) the Newton
  (b) psi
  (c) the Joule
  (d) the Watt
Q1.18 The units used to measure wing loading (i) and dynamic pressure (ii) are:
  (a) (i) N/m^2; (ii) N/m^2
  (b) (i) N m; (ii) N m
  (c) (i) N; (ii) N/m^2
  (d) (i) N/m^2; (ii) Joules
Q1.19 Bernoulli's theorem states:
  (a) dynamic pressure increases as static pressure increases
  (b) dynamic pressure increases as static pressure decreases
  (c) dynamic pressure is greatest at the stagnation point
  (d) dynamic pressure is zero when the total pressure is greatest

Q1.20 Bernoulli's theorem states:
  (a) The sum of all the energies present is constant in a supersonic flow.
  (b) The sum of the pressure and the kinetic energy is a constant in a low subsonic streamline flow.
  (c) Air pressure is directly proportional to the speed of the flow.
  (d) Dynamic pressure plus pitot pressure is a constant.

Q1.21 If the temperature of the air in a uniform flow at velocity V in a stream tube is increased the mass flow:
  (a) remains constant and the velocity increases
  (b) increases
  (c) remains constant and the velocity decreases
  (d) decreases

Q1.22 If the continuity equation is applicable to an incompressible airflow in a tube at low subsonic speed, if the diameter of the tube decreases the speed of the flow:
  (a) increases
  (b) becomes supersonic
  (c) decreases
  (d) remains the same

Q1.23 The equation for power is:
  (a) N/m
  (b) N m/s
  (c) Pa/s^2
  (d) $kg/m/s^2$

Q1.24 If the temperature of streamline airflow in a tube at constant speed is increased it will:
  (a) increase the mass flow
  (b) not affect the mass flow
  (c) increase the mass flow if the tube is divergent
  (d) decrease the mass flow

Q1.25 If the air density in a stream tube flow is halved, the drag is decreased by a factor of . . . .
  (a) 8
  (b) 4
  (c) 6
  (d) 2

Q1.26 Regarding a Venturi in a subsonic airflow, which of the following statements is correct?
  (i) The dynamic pressure in the undisturbed airflow and the airflow in the throat are equal.
  (ii) the total pressure in the undisturbed airflow and in the throat are equal.
    (a) (i) correct; (ii) correct
    (b) (i) correct; (ii) incorrect
    (c) (i) incorrect; (ii) incorrect
    (d) (i) incorrect; (ii) correct

Q1.27 If the velocity in a stream tube is increased, the streamlines:
  (a) remain the same
  (b) move further apart
  (c) move closer together
  (d) are not affected by the velocity

Q1.28 As subsonic air flows through a convergent duct, static pressure (i) . . . . and velocity (ii) . . . .
  (a) (i) increases; (ii) decreases
  (b) (i) increases; (ii) increases
  (c) (i) decreases; (ii) decreases
  (d) (i) decreases; (ii) increases

2 Basic Aerodynamic Definitions

2.1 Aerofoil Profile

The definitions used with reference to an aerofoil's shape are shown in Figure 2.1 and are as follows:

a. **Camber.** The curvature of the profile view of an aerofoil is its camber. The amount of camber and its distribution along the chord length is dependent on the performance requirements of the aerofoil. Generally, low-speed aerofoils require a greater amount of camber than high-speed aerofoils. See Figure 2.1(a).

b. **Chordline.** The straight line joining the centre of curvature of the leading-edge radius and the trailing edge of an aerofoil is the chordline. See Figure 2.1(a).

c. **Chord.** The distance between the leading edge and the trailing edge of an aerofoil measured along the chordline is the chord. See Figure 2.1(a).

d. **Fineness Ratio.** The fineness ratio is the ratio of the length of a streamlined body to its maximum width or diameter. A low fineness ratio has a short and fat shape, whereas a high fineness ratio describes a long thin object. See Figure 4.2.

e. **Leading Edge Radius.** The radius of a circle, centred on a line tangential to the curve of the leading edge of an aerofoil, and joining the curvatures of the upper and lower surfaces of the aerofoil is the leading-edge radius or nose radius of an aerofoil. See Figure 2.1(a).

f. **Maximum Thickness.** The maximum depth between the upper and lower surfaces of an aerofoil is its maximum thickness. See Figure 2.1(a).

g. **Mean Aerodynamic Chord.** The chordline that passes through the geometric centre of the plan area (the centroid) of an aerofoil is the Mean Aerodynamic Chord (MAC) and for any planform is equal to the chord of a rectangular wing having the same wing span and the same pitching moment and lift. It is usually located at approximately 33% of the semi-span from the wing root. The MAC is often used to reference the location of the centre of gravity particularly with American-built aeroplanes. It is the primary reference for longitudinal stability.

h. **Mean Camber Line.** The line joining points that are equidistant from the upper and lower surfaces of an aerofoil is the mean camber line. Maximum camber occurs at the point where there is the greatest difference between the mean camber line and the chordline. When the mean camber line is above the chordline the aerofoil has positive camber. See Figure 2.1(a).

The Principles of Flight for Pilots   P. J. Swatton
© 2011 John Wiley & Sons, Ltd

i. **Mean Geometric Chord.** This is the average chord length of an aerofoil is the mean geometric chord or alternatively it is defined as the wing area divided by the wingspan.

j. **Quarter Chordline.** A line joining the quarter chord points along the length of a wing is the quarter chordline, i.e. a line joining all of the points 25% of the chord measured from the leading edge of the aerofoil.

k. **Root Chord.** The line joining the leading edge and trailing edge of a wing at the centreline of the wing is the root chord. See Figure 2.1(b).

l. **Thickness/Chord Ratio.** The ratio of the maximum thickness of an aerofoil to the chord length expressed as a percentage is the thickness/chord ratio. This is usually between 10% and 12%.

m. **Washout.** A reduction in the angle of incidence of an aerofoil from the wing root to the wing tip is the washout. This is also known as the geometric twist of an aerofoil. See Figure 2.1(b).

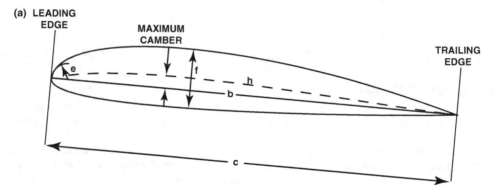

Figure 2.1(a)   The Aerofoil Shape.

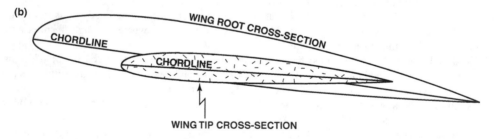

Figure 2.1(b)   Aerofoil Washout.

2.2   Aerofoil Attitude

The definitions used with reference to an aerofoil's attitude shown in Figure 2.2 are as follows:

a. **Angle of Attack.** The angle subtended between the chordline of an aerofoil and the oncoming airflow is the angle of attack.

b. **Angle of Incidence.** The angle subtended between the chordline and the longitudinal axis is the angle of incidence. It is usual for the chordline of a commercial transport aeroplane to be at an angle of incidence of 4° when compared with the longitudinal axis because in level flight, at relatively high speeds, it produces the greatest amount of lift for the smallest drag penalty. If such is the case, the angle of attack and the pitch angle will be 4° apart, no matter what the attitude of the aeroplane.

c. **Critical Angle of Attack.** The angle of attack at which the maximum lift is produced is the critical angle of attack often referred to as the stalling angle of attack. A fixed-wing aeroplane is stalled at or above the critical angle of attack.

d. **Climb or Descent Angle.** The angle subtended between the flight path of an aeroplane and the horizontal plane is the climb angle or descent angle.

e. **Longitudinal Dihedral.** The difference between the angle of incidence of the mainplane and the tailplane is the longitudinal dihedral. See Figure 2.2(b).

f. **Pitch Angle.** The angle subtended between the longitudinal axis and the horizontal plane is the pitch angle.

g. **Pitching Moment.** Certain combinations of the forces acting on an aeroplane cause it to change its pitch angle. The length of the arm multiplied by the pitching force is referred to as a pitching moment and is counteracted by the use of the tailplane. A pitching moment is positive if it moves the aircraft nose-upward in flight and negative if it moves it downward.

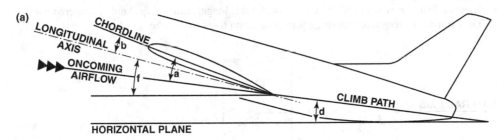

Figure 2.2(a)   The Aerofoil Attitude.

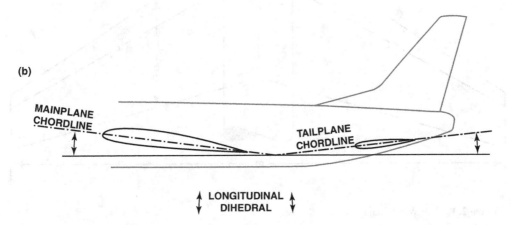

Figure 2.2(b)   Longitudinal Dihedral.

2.3   Wing Shape

The definitions used with reference to the shape of an aeroplane's wing are shown in Figure 2.3 and are as follows:

a. **Aspect Ratio.** The ratio of the span of an aerofoil to the mean geometric chord is the aspect ratio, which is sometimes expressed as the square of the span divided by the wing area.

$$\text{Aspect Ratio} = \frac{\text{span}}{\text{chord}} = \frac{\text{span}^2}{\text{wing area}}$$

b. **Mean Aerodynamic Chord.** The chordline that passes through the geometric centre of the plan area (the centroid) of an aerofoil is the Mean Aerodynamic Chord (MAC).
c. **Quarter Chordline.** The line joining all points at 25% of the chord length from the leading edge of an aerofoil is the quarter chordline.
d. **Root Chord.** The straight line joining the centres of curvature of the leading and trailing edges of an aerofoil at the root is the root chord.
e. **Sweep Angle.** The angle subtended between the leading edge of an aerofoil and the lateral axis of an aeroplane is the sweep angle.
f. **Taper Ratio.** The ratio of the length of the tip chord expressed as a percentage of the length of the root chord is the taper ratio.
g. **Tapered Wing.** Any wing on which the root chord is longer than the tip chord is a tapered wing.
h. **Tip Chord.** The length of the wing chord at the wing tip is the tip chord.

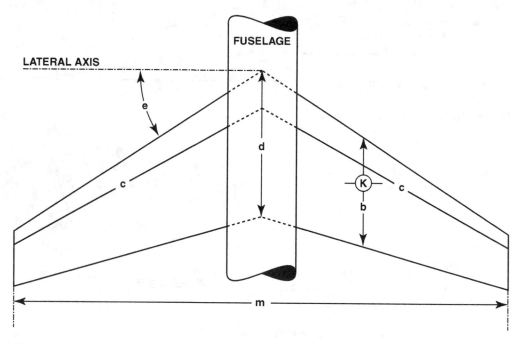

Figure 2.3   The Wing Shape.

i. **Wing Area.** The surface area of the planform of a wing is the wing area.
j. **Swept Wing.** Any wing of which the quarter chordline is not parallel to the lateral axis is a swept wing.
k. **Wing Centroid.** The geometric centre on a wing plan area is the wing centroid.
l. **Mean Geometric Chord.** The average chord length of an aerofoil is the mean geometric chord, or alternatively it is defined as the wing area divided by the wingspan.
m. **Wingspan.** The shortest distance measured between the wing tips of an aeroplane is its wingspan.

The angle of inclination subtended between the front view of a wing 25% chordline (the wing plane) and the lateral horizontal axis is shown in Figure 2.4 and referred to as:

a. **dihedral** if the inclination is upward.
b. **anhedral** if the inclination is downward.

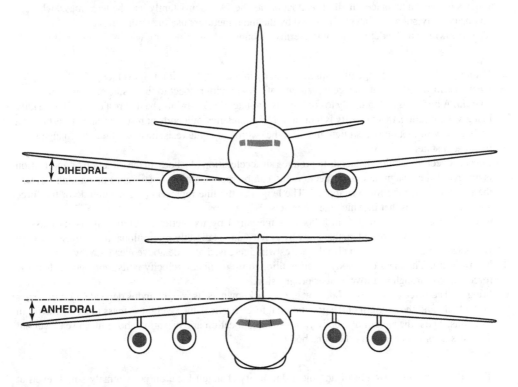

Figure 2.4 Dihedral and Anhedral.

2.4   Wing Loading

The forces acting on a wing are defined below:

a. **Centre of Pressure (CP).** The point on the chordline through which the resultant of all of the aerodynamic forces acting on the wing is considered to act is the centre of pressure.
b. **Coefficient.** A numerical measure of a physical property that is constant for a system under specified conditions is a coefficient, e.g. C_L is the coefficient of lift.
c. **Load Factor (n).** The total lift of an aerofoil divided by the total mass is the load factor. i.e.

$$n = \frac{\text{total lift}}{\text{mass}}$$

d. **Wing Loading.** The mass per unit area of a wing is the wing loading. i.e.

$$\textbf{Wing Loading} = \frac{\text{mass}}{\text{wing area}} = \frac{M}{S}$$

2.5   Weight and Mass

Before starting any calculations it is necessary to explain the difference between a Newton (N), which is a unit of force, a kilogram (kg), which is a unit of mass and weight which is the force acting on a body by gravity. Most of us know what we mean when we use the term weight and become confused when the term mass is used in its place. In all of its documents the JAA consistently use the term mass, whereas the majority of aviation documents produced by the manufacturers use the term weight.

The following are the definitions of the terms encountered when dealing with weight or mass:

a. **Velocity.** The rate of change of position of a body in a given direction is its velocity.
b. **Acceleration.** The rate of change of velocity of a body with respect to time is its acceleration.
c. **Inertia.** A body has the tendency to continue in its state of equilibrium, be it at rest or moving steadily in a given direction, this property is inertia. It is the tendency of a body at rest to remain at rest or of a body to stay in motion in a straight line. In other words, it is the resistance of a body to change from its present state.
d. **Force.** A vector quantity that tends to produce an acceleration of a body in the direction of application is a force. It has magnitude and direction and can be represented by a straight line passing through the point at which the force is applied. The length of the line represents the magnitude of the force and the direction is that in which the force acts.
e. **Mass.** The quantity of matter in a body as measured by its inertia is referred to as its mass. It determines the force exerted on that body by gravity, and is directly proportional to the mass. Gravity varies from place to place and also decreases with increased altitude above mean sea level.
f. **Momentum.** The mass of a body when multiplied by its linear velocity is its momentum. It is the force gained through its movement or progression.
g. **Weight.** The force exerted on a body by gravity is known as its weight and is dependent on the mass of the body and the strength of the gravitational force for its value. Weight = mass in kg × gravity in Newtons. Thus, the weight of a body varies with its location and elevation above mean sea level but the mass does not change for the same body.

The alteration to the weight of an object due to its changed location is extremely small, even at 50 000 ft above mean sea level, however, it is technically incorrect and the term mass should be used. *For the purposes of this manual the terms weight and mass are interchangeable. IEM OPS 1.605.*

2.5.1   The Newton

A Newton is a unit of force, which equals mass × acceleration.

1 Newton = 1 kg × 1 m/s^2. At the surface of the Earth the acceleration due to gravity equals 9.80665 m/s^2. Thus, the force acting on 1 kg at the Earth's surface is 9.80665 Newtons. To simplify calculations in the examination the acceleration due to gravity is given as 10 m/s^2 therefore 1 kg is equal to 10 Newtons.

2.6   Airspeeds

2.6.1   Airspeed Indicator Reading (ASIR)

The reading of the airspeed indicator without correction is the ASIR. The correction made to allow for inaccuracies in the construction of the instrument, which are permissible within its normal working accuracy, are referred to as instrument error. The ASIR becomes the indicated airspeed (IAS) when corrected for instrument error.

2.6.2 Indicated Airspeed (IAS)

The speed of an aeroplane as shown on its pitot/static airspeed indicator calibrated to reflect standard atmosphere adiabatic compressible flow at mean sea level *uncorrected* for airspeed system errors, but corrected for instrument error is the indicated airspeed. The abbreviation V_I is the prefix used for any speed that is an indicated airspeed, e.g. V_{IMD}.

The dynamic pressure formula for IAS is $q = \frac{1}{2}\rho V^2$ and is based on Bernoulli's equation that assumes air is incompressible, which is almost true for speeds below 300 kt True Airspeed (TAS). As altitude increases in a standard atmosphere the value of the true airspeed for a given IAS increases. *CS Definitions page 15.*

2.6.3 Calibrated Airspeed (CAS)

The IAS corrected for position (or system) error is calibrated airspeed (CAS), which is equal to equivalent airspeed and true airspeed in a standard atmosphere at mean sea level. The abbreviation used for this speed is V_C. *CS Definitions page 19.*

The CAS in a standard atmosphere can be approximately converted to a true airspeed (TAS) by calculating the percentage increase to apply to the CAS, which is equal to the density altitude in 1000s of feet multiplied by 1.5. Thus, the formula to roughly determine the TAS is:

TAS = CAS + (Density Altitude in 1 000s × 1.5)% correction.

e.g. CAS 80 kt at 7000 ft Density Altitude $= 80 + (7 \times 1.5)\% = 80 + (10.5\%$ of 80$) =$ TAS 88.4 kt.

2.6.4 Rectified Airspeed (RAS)

The airspeed value that is obtained when ASIR has been corrected for both instrument and position error is rectified airspeed. This term is rarely used in civil-aviation performance, preference being given to the term calibrated airspeed that has the same value.

2.6.5 Equivalent Airspeed (EAS)

Below 300 kt TAS the compression of the air when it is brought to rest as in a pitot tube is insignificant. For all practical purposes compressibility is ignored below 300 kt TAS but becomes increasingly significant as the speed increases above that speed. The CAS when corrected for adiabatic compressible flow at a particular altitude becomes equivalent airspeed (EAS). In a standard atmosphere at mean sea level the values of CAS and EAS are the same. The abbreviation used for EAS is V_E. *CS Definitions page 8.*

2.6.6 True Airspeed (TAS)

The actual speed of the aeroplane relative to the undisturbed air surrounding the aeroplane is referred to as the true airspeed. *True airspeed is equal to the EAS divided by the square root of the relative density.* It therefore takes account of the density of the atmosphere and can be calculated by the formula:

$$\text{TAS} = \frac{\text{EAS}}{\sqrt{\sigma}}$$

where $\sigma =$ relative density. The abbreviation used for this speed is V, e.g. V_{MD}, V_{MP}. *CS Definitions page 18.*

2.6.7 Mach Number

A Mach number is the result of the actual TAS of an aeroplane when divided by the TAS of the local speed of sound expressed as a number to two decimal places for a specified altitude and ambient temperature. It is of particular significance to jet aeroplanes at high altitude because it is likely to limit the maximum operating speed. Mach number is directly proportional to the static air temperature expressed in Kelvin and enables the effect of compressibility to be predicted. *CS Definitions page 16.*

2.7 Speed Summary

To summarize, then, if the IAS is corrected for position and instrument error in a standard atmosphere at MSL it is equal to CAS, EAS and TAS. Rarely do the atmospheric conditions exactly match those of the standard atmosphere. Either the surface temperature or the density altitude or both do not conform to these conditions, in which case the corrections must be applied in the correct sequence.

The corrections required for position and instrument error are normally combined and tabulated on a card that is positioned close to the ASI. Corrections are shown for the whole range of indicated airspeeds at which it is possible for that type of aeroplane to fly from zero to the never-exceed speed, V_{NE}.

If there is only one pitot/static system in the aeroplane then the same correction card will apply to all the airspeed indicators fed by that system. However, if more than one pitot/static system is fitted to the aeroplane then separate correction cards must be provided for each system and used for the appropriate airspeed indicators fed by each of the systems.

Usually, the correction card provided is for the flaps retracted configuration because this is the only configuration that can be used for the whole range of indicated airspeeds possible. The mass used for the card is normally the maximum certificated mass of the aeroplane. If altitude and mass make a significant difference to the corrections for this configuration then there will be a graph in the aeroplane flight manual (AFM) from which it is possible to interpolate for other masses and/or altitudes. There will also be graphs available for other configurations.

For take-off and landing, it may be possible to use a variety of flap settings. If such is the case, then a series of graphs will be provide in the AFM, a separate graph for each flap setting to enable the pilot to determine the correction to apply to the IAS. The application of this correction, no matter whether from a card or a graph, is a simple addition or subtraction applied to the indicated airspeed.

Compressibility of the air causes an error to the reading given by the airspeed indicator, which becomes significant when the TAS exceeds 300 kt. The value of the correction for compressibility varies with speed and altitude and can be determined from a graph or by means of a navigation computer. Application of the correction to CAS produces EAS. For take-off and landing the effect of compressible flow is negligible irrespective of the aerodrome elevation. Therefore, for these phases of flight CAS and EAS may be considered the same. That is why some manufacturers specify the performance speeds in EAS and others in CAS.

Irrespective of which speeds are used for take-off and landing due allowance must be made for density error if the TAS is to be computed. Density error is that which is caused by the difference between the ambient air density and that, which was used to calibrate the instrument, the standard atmosphere density. Air density is affected by both air temperature and altitude. Therefore, these two factors must be used to determine the correction to be applied. It may be applied either graphically, by calculation or by means of the navigation computer to obtain TAS.

Because the manufacturers specify the speeds for take-off and landing in terms of CAS or EAS the pilot must apply the corrections for instrument and position error in the opposite sense to that shown on the correction card to determine what the target speed should be indicated on the airspeed indicator.

2.8 The Effect of Altitude on Airspeeds

The effect that increasing altitude at a constant speed has on other speeds is shown in Figure 2.5 for climbs below the tropopause and above the tropopause. The tropopause in a standard atmosphere occurs at 36 090 feet. The effects differ because of the isothermal layer above the tropopause.

2.8.1 a. Below the Tropopause

1. A constant IAS/CAS climb will result in an increasing TAS and Mach number as the climb progresses as shown in the first diagram.
2. A constant TAS climb will cause the Mach number to increase as the climb progresses but the IAS/CAS will decrease as illustrated in the second diagram.
3. Climbing at a constant Mach number will result in both the IAS/CAS and the TAS decreasing as the climb progresses, as shown in the third diagram.

2.8.2 b. Above the Tropopause

1. A constant IAS/CAS climb will cause both the TAS and Mach number to increase with increasing altitude.
2. Climbing at either a constant TAS or Mach number will result in the IAS/CAS decreasing as the climb progresses.

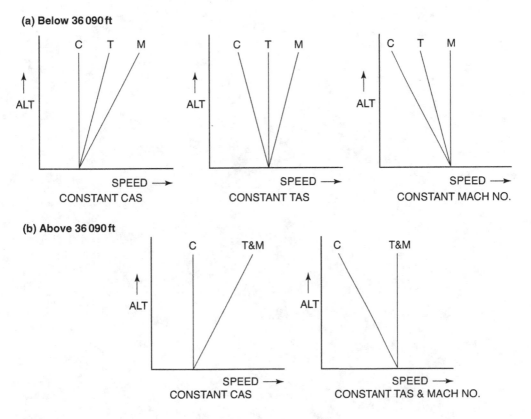

Figure 2.5   The Effect of Altitude on Speeds.

Pilots use the airspeed indicator to determine the forward speed of the aeroplane in the lower atmosphere; because of this the aircraft manufacturers provide the recommended speeds for various phases of flight in terms of IAS. Modern instruments are manufactured and installed with no instrument error therefore the airspeed indicator reading (ASIR) is the indicated airspeed (IAS).

The pressure error correction on modern aeroplanes is extremely small, therefore it may be assumed that IAS is equal to calibrated airspeed (CAS) and this is equal to equivalent airspeed (EAS) at mean sea level in a standard atmosphere.

The difference between IAS and EAS is compressibility, which is small at low altitude and becomes increasingly significant as altitude increases. If the IAS remains constant with increase of altitude and compressibility is accounted for then EAS will decrease.

If compressibility is ignored then IAS can be assumed to be equal to EAS at all altitudes. EAS is the speed used for the aerodynamic theoretical diagrams in Part 3 of this Manual. Although the assumption that IAS is equal to EAS is technically incorrect, for the purposes of Part 3 of this Manual IAS and EAS are interchangeable.

Self-Assessment Exercise 2

Q2.1 The relative thickness of an aerofoil is expressed in:
(a) centimetres
(b) metres
(c) degrees of cross section
(d) % chord

Q2.2 The line joining the leading and trailing edges of an aerofoil that is precisely midway between the upper and lower surfaces is the:
(a) mean aerodynamic chord
(b) average camber line
(c) mean camber line
(d) chordline

Q2.3 The angle of attack of an aerofoil is the angle between the:
(a) longitudinal axis and the chordline
(b) chordline and the undisturbed airflow
(c) longitudinal axis and the horizontal
(d) longitudinal axis and the undisturbed airflow

Q2.4 With respect to increasing altitude, which of the following statements is correct?
(a) At a constant Mach number the IAS increases.
(b) At a constant TAS the Mach number decreases.
(c) At a constant IAS the Mach number increases.
(d) At a constant IAS the TAS decreases.

Q2.5 The angle subtended between the longitudinal axis of an aeroplane and the chordline of the wing is the:
(a) angle of incidence
(b) glide-path angle
(c) angle of attack
(d) climb-path angle

Q2.6 Lateral wing dihedral is the angle between the 0.25 chordline and the:
(a) longitudinal axis
(b) vertical axis
(c) longitudinal horizontal axis
(d) lateral horizontal axis

Q2.7 In a constant Mach number climb in a standard atmosphere to FL 350 the TAS will:
(a) remain constant
(b) decrease
(c) increase to FL 100, then decrease above FL100
(d) increase

Q2.8 The angle of attack of a wing is the angle between the (i) …….. and the (ii) ………
(a) (i) longitudinal axis; (ii) free stream direction
(b) (i) chordline; (ii) camber line
(c) (i) chordline; (ii) free stream direction
(d) (i) longitudinal axis; (ii) chordline

Q2.9 Angle of attack of a wing is defined as:
(a) the angle between the aeroplane longitudinal axis and the wing chordline
(b) the angle to obtain the maximum lift/drag ratio
(c) the angle between the wing chordline and the direction of the relative airflow
(d) the angle between the climb path and the horizontal

Q2.10 The difference between IAS and TAS will:
   (a) increase with temperature decrease
   (b) increase with increasing air density
   (c) decrease at high speeds
   (d) decrease with decreasing altitude

Q2.11 In a constant TAS climb the Mach number will:
   (a) decrease
   (b) remain constant
   (c) decrease to FL100 and then increase
   (d) increase

Q2.12 The aspect ratio of a wing is the ratio between the (i) . . . .and the (ii) . . . .
   (a) (i) mean chord; (ii) the root chord
   (b) (i) wing span; (ii) the mean geometric chord.
   (c) (i) wing span; (ii) the root chord
   (d) (i) tip chord; (ii) the wing span

Q2.13 The angle between the airflow and the chordline is the:
   (a) glide path
   (b) climb path
   (c) angle of incidence
   (d) angle of attack

Q2.14 Descending at a constant Mach number, the effect on the TAS is that it:
   (a) decreases as air pressure increases
   (b) decreases as altitude decreases
   (c) remains constant
   (d) increases

Q2.15 To predict the effect of compressibility it is necessary to determine the:
   (a) Mach number
   (b) EAS
   (c) TAS
   (d) IAS

Q2.16 The angle of attack of a wing is defined as the angle between:
   (a) the upper surface airflow and the chordline
   (b) the undisturbed airflow and the mean camber line
   (c) the undisturbed airflow and the chordline
   (d) the upper surface airflow and the mean camber line

Q2.17 True airspeed is:
   (a) higher than the speed of the undisturbed airstream
   (b) equal to IAS multiplied by the air density at mean sea level
   (c) lower than the speed of the undisturbed airflow
   (d) lower than the IAS at ISA conditions at altitudes below mean sea level

Q2.18 Assuming ISA conditions, and all other conditions are constant then if an aeroplane flies at the same angle of attack in straight and level flight at two different altitudes then the TAS will be:
   (a) lower at the higher altitude.
   (b) higher at the higher altitude.
   (c) the same at both altitudes.
   (d) impossible to determine.

Q2.19 The MAC of a given wing planform is the:
   (a) chord of a large rectangular wing
   (b) average chord of the actual aeroplane
   (c) wing area divided by the wing span
   (d) chord of a rectangular wing with same moment and lift

Q2.20 Load factor is:
(a) 1/Bank angle
(b) wing loading
(c) Lift/Mass
(d) Mass/Lift

Q2.21 Flying at a constant CAS below mean sea level in ISA conditions the TAS:
(a) is less than it would be at mean sea level
(b) equal to the TAS at mean sea level
(c) is greater than it would be at mean sea level
(d) equal to the TAS at mean sea level but the IAS is higher

Q2.22 Which of the following is the correct method for measuring lateral dihedral? It is the angle between:
(a) the 25% chordline and the horizontal longitudinal axis
(b) the 25% chordline and the horizontal lateral axis
(c) the wing plane and the horizontal
(d) the 25% chordline and the longitudinal axis

Q2.23 A line joining points that are equidistant from the upper and lower surfaces of an aerofoil and joining the centres of curvature of the leading and trailing edge is the:
(a) average camber line
(b) mean aerodynamic chord
(c) mean chordline
(d) mean camber line

Q2.24 The angle of attack is the angle between the:
(a) undisturbed airflow and the chordline
(b) undisturbed airflow and the mean camber line
(c) local airflow and the chordline
(d) local airflow and the mean camber line

Q2.25 Aspect ratio is defined as:
(a) wingspan divided by the wing tip chord
(b) the wing tip chord divided by the wingspan
(c) the wingspan divided by the mean chord
(d) the 25% chord divided by the wingspan

Q2.26 The angle of the attack is defined as the angle subtended between the chordline of:
(a) wing to the relative free-stream flow
(b) wing to the fuselage datum
(c) tailplane to the wing chordline
(d) tailplane to the fuselage

Q2.27 The measured thickness of an aerofoil section is usually expressed as:
(a) a percentage of the wingspan
(b) related to the camber
(c) a percentage of the chord
(d) metres

Q2.28 Assuming ISA conditions, which of the following statements regarding a climb is correct?
(a) At a constant TAS the Mach number decreases with increased altitude.
(b) At a constant Mach number the IAS increases with increased altitude.
(c) At a constant IAS the TAS decreases with increased altitude.
(d) At a constant IAS the Mach number increases with increased altitude.

Q2.29 For two aeroplanes flying in a standard atmosphere at the same mass and the same IAS at different altitudes the relationship of their TAS is:
  (a) greater at the lower altitude
  (b) greater at the higher altitude
  (c) they are equal
  (d) less at the higher altitude

Q2.30 The mean aerodynamic chord of a wing is defined as:
  (a) the same as the mean chord of a rectangular wing of the same span
  (b) the 25% chord of a swept-back wing
  (c) the wing area divided by the wing span
  (d) the wing chord at 66% of the semi-wing span

Q2.31 In a constant Mach number climb the effect of increased altitude on the value of TAS is that it is:
  (a) decreased
  (b) increased
  (c) unaffected
  (d) increased to the tropopause and then decreased

Part 2
Basic Aerodynamics

3 Basic Control

3.1 Aeroplane Axes and Planes of Rotation

The three axes of an aeroplane intersect at the centre of gravity (CG) and are those about which an aeroplane moves in flight. They are the longitudinal, the lateral and the normal axes and are shown in Figure 3.1. If the forces about any axis are not balanced (i.e. the total moments about the axis is not zero) there will be an angular acceleration and the aeroplane will rotate about that axis.

3.1.1 The Longitudinal or Roll Axis

The horizontal straight line joining the most forward point of the fuselage of the aeroplane, the centre of the nose, to the centre of the most rearward point of the aeroplane's fuselage is the longitudinal axis. It is often referred to as the roll axis because the aeroplane, in response to a control input by the pilot or as the result of an external disturbance, drops one wing and raises the other. Thus, it rotates or rolls about the longitudinal axis.

3.1.2 The Lateral or Pitch Axis

The lateral axis is a horizontal straight line passing from one side of the aeroplane to the other, through the CG, at right angles to the longitudinal axis. It is the axis about which the aeroplane moves, in response to a control input by the pilot or as the result of an external disturbance, by raising or lowering the nose of the aeroplane. Thus, it pitches or rotates about the lateral axis. See Figure 3.1.

3.1.3 The Normal or Yaw Axis

The normal axis of an aeroplane is the vertical straight line passing through the CG. It is perpendicular to the other two axes and is the axis about which the aeroplane in response to a control input by the pilot or as the result of an external disturbance moves its nose horizontally or yaws.

3.2 The Flight Controls

The flight controls enable the pilot to rotate the aeroplane about its three axes. The ailerons cause the aeroplane to roll about its longitudinal axis, the elevators enable the aeroplane to pitch around its lateral

The Principles of Flight for Pilots   P. J. Swatton
© 2011 John Wiley & Sons, Ltd

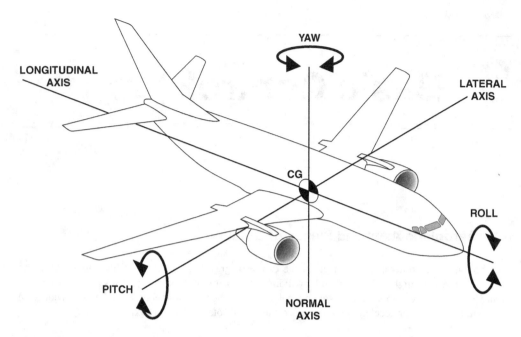

Figure 3.1 The Aeroplane's Axes and Planes of Rotation.

axis and the rudder allows the aeroplane to yaw around its normal axis. The control surfaces are illustrated in Figure 3.2 and produce aerodynamic forces that result in the aeroplane rotating about the appropriate axis.

Each control is normally a small aerofoil section hinged and mounted on the trailing edge of the main aerofoil. It is positioned as far as possible from its rotational axis to maximize the length of its force arm so that it produces the largest moment for the smallest amount of force. Movement of the flight control changes the effective angle of attack of the main aerofoil on which it is sited and produces an aerodynamic force to generate the required reaction.

The tailplane, often called the horizontal stabiliser, is a small version of the wing usually positioned at the rear of the fuselage, which may be fixed or adjustable on which the elevator is hinged. Its main purpose is to counterbalance the nose-down pitching moment, caused by the lift generated by the wings acting through the CP. It does this by producing a tail-down pitching moment of equal magnitude to the pitching moment to maintain the equilibrium. This enables the aeroplane to fly at a combination of different speeds and CG positions. Without a tailplane a conventional aeroplane would be restricted to one combination of speed and CG for it to be longitudinally stable.

Only a small amount of downward lift or downforce is required to produce the required counterbalancing moment because of the great length of the arm of the stabiliser surface CP from the CG. This reduces the pitching moment to zero and maintains the equilibrium for a range of speeds and CG positions. The magnitude of the downforce is dependent on the strength of the airflow over the tailplane. If this airflow suddenly weakens then the tailplane will stall, but in a negative sense, causing a sudden reduction of the downward aerodynamic force, this is sometimes called a negative tail stall.

The tailplane may be fixed or all-moving. If it is fixed the downward force is produced by a combination of the lift produced by the stabiliser and the elevators. Some aeroplanes are designed with all-moving tailplanes; these are stabilators and can have the angle of attack adjusted in flight to produce the required downforce.

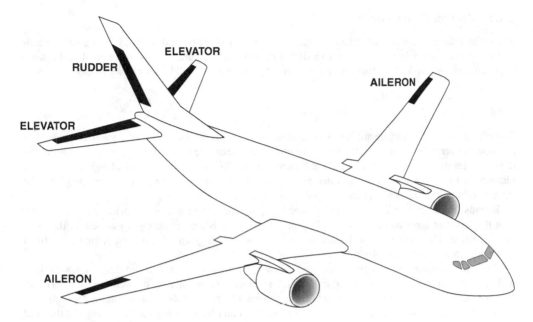

Figure 3.2   The Control Surfaces.

3.3   The Elevators

Control of the aeroplane about its lateral axis, or pitch control, is accomplished by using the elevators. See Figure 3.2. The elevators are hinged control surfaces located on the horizontal tailplane usually at the rear of the aeroplane. The elevators occupy a considerable proportion of the tailplane surface area and normally comprise a single control surface on each tailplane that will act in unison. A forward movement of the control column (stick) by the pilot will cause the elevators to go down and a backward movement will cause them to go up.

The effect produced by the elevators is to increase the camber of the tailplane when the stick is pushed forward, which causes an increase of tailplane lift and an upward force on the underside of the control surface that results in the aeroplane rotating about its lateral axis through the CG and the nose being depressed. The total lift of the aeroplane is decreased by the amount of tailplane lift so caused. The opposite is true when the stick is pulled back, the camber is negative and the lift generated is downward which when trimmed becomes the counterbalancing force required for longitudinal equilibrium.

3.4   Pitch Control

The magnitude of the pitch control exercised by the elevators is dependent on:

a. The area of the control surface.
b. The degree of control-surface deflection.
c. The speed of the aeroplane and the distance between the control surface CP and the CG of the aeroplane (the moment arm).
d. The mainplane angle of attack.

3.4.1   Control Surface Area

The greater the elevator surface area the larger is the aerodynamic force developed for a given control-surface deflection angle at any given IAS. If the control-surface deflection is maintained at the same angle and the IAS is doubled then the aerodynamic force produced will be four times as large.

3.4.1.1   Control Surface Angular Deflection

The amount of control surface angular deflection necessary is greater for an aeroplane with a forward CG position to overcome the longitudinal stability of such an aeroplane which during take-off, rotation will require extra stick force. Irrespective of the location of the CG, the amount of control angular deflection required is always greater at low indicated airspeeds to produce the same amount of tailplane lift for the same conditions than it is at a higher airspeed.

Because the stabilising effect of the tailplane is greater during a manoeuvre than in steady level flight then the inherent longitudinal static stability of an aeroplane is greater during a manoeuvre than it is in level flight. Thus, the elevator deflection angle required for a given CG position is greater during a manoeuvre than it is in level flight.

For a jet aeroplane with engines mounted below the wing if the thrust is suddenly increased there will be a nose-up pitching moment about the CG. To maintain a zero pitching moment it requires a down elevator deflection to counter the moment, this is achieved by pushing the control column forward. One of the advantages of siting engines on the rear fuselage of an aeroplane is a sudden change of thrust of this nature has less effect on the longitudinal control.

3.4.2   The Moment Arm

The effectiveness of any control surface is dependent not only on the degree of angular deflection but also on the speed of the aeroplane. The greater the speed of the aeroplane the smaller is the angular deflection necessary to achieve the same attitude change that would be attained at a lower airspeed with a larger angular deflection. *The angular deflection of the control surfaces is inversely proportional to the square of the equivalent airspeed.*

The location of the CG also affects the measure of authority attained by the elevators because it affects the length of the moment arm and consequently the pitching moment of the elevators, or the force applied by the elevators to change the attitude of the aeroplane. A forward CG position increases the length of the moment arm and increases the applied force for a given elevator angular deflection above that of an aft CG position with the same angular deflection because it decreases the length of the moment arm.

If the CG is too far forward, at the low airspeeds experienced during take-off and landing, it is possible that the pilot will need an angular deflection of the elevators greater than the maximum physically possible one. This is extremely dangerous at this critical stage of flight. To prevent such an occurrence the CG must remain within the safe limits of the CG envelope at all times in flight.

Large aeroplanes must take account of the movement of the CG as fuel is burnt and ensure that it remains in the CG envelope at all times. Trim drag can be reduced to a minimum if the CG is kept at the aft limit of the envelope; this reduces the thrust requirement and the fuel consumption, thus increasing the maximum range and the maximum endurance.

3.4.3   Angle of Attack

The downwash of the airstream over the upper surface of the mainplane directly influences the effectiveness of the elevators of a fuselage-mounted tailplane. A change of downwash alters the effective angle of attack of the tailplane and the lift that it generates. This changes the pitching moment of the tailplane and requires a deflection of the elevators to maintain a given attitude.

3.5 Alternative Pitch Controls

Some aeroplanes have the horizontal stabiliser positioned on top of the vertical fin as a "T" tail to improve its aerodynamic efficiency. There are three alternative means of controlling the pitch of an aeroplane other than the elevators; they are the variable incidence (VI) tailplane, the stabilator and the elevons.

3.5.1 Variable Incidence Tailplane

Instead of using elevators to change the angle of attack and camber of the tailplane, the whole of the tailplane lifting surface is constructed in such a manner that it has a mutable incidence or changeable angle of attack. This is the 'variable incidence' (VI) tailplane or 'adjustable horizontal stabiliser.' The advantages of this tailplane over a conventional fixed tailplane, with an elevator and trim tab, is that the trim is more powerful, it creates less drag and retains full elevator movement at extreme trim angles. See Figure 3.3. This is also known as the trimmable horizontal stabiliser (THS) and is able to cope with a larger range of CG positions than traditional elevator trim tabs.

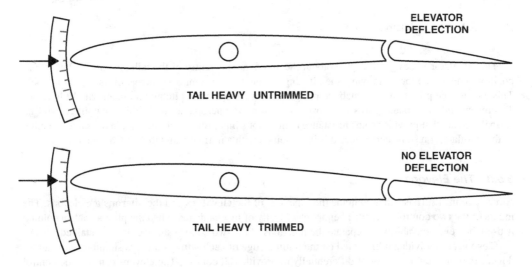

Figure 3.3 The Variable Incidence Tailplane.

The angle of incidence of an adjustable stabiliser is controlled by the aeroplane's trim system, which alters the jack datum setting. The position of a properly trimmed variable-incidence tailplane depends, most importantly on, the CG position, the airspeed and the thrust setting, which does not affect the stick-neutral position. When correctly trimmed the elevator angle relative to the horizontal stabiliser is neutral and the stabilised speed is referred to as the 'trim speed.' Thus, the trim may be used to set the desired speed without having to hold the elevator out of its trimmed position. Aircraft having a fixed stabilator have a trim tab on the trailing edge of the elevators for this purpose.

A tail-heavy aeroplane would require the elevator to be deflected downward, which when the variable-incidence tailplane is trimmed has no elevator deflection; it is in the neutral position and the leading edge of the adjustable stabiliser is up. If it becomes jammed in the cruise position, when approaching to land select a higher landing speed and/or use a lower flap setting. When trimmed for a forward CG position (nose heavy) the stabiliser leading edge is lower than the trailing edge.

3.5.2   The Stabilator

In transonic flight shockwaves generated by the tailplane render the elevator unusable. To overcome this disadvantage transonic and supersonic aeroplanes are fitted with an 'all-moving' or 'all-flying' tailplane to maintain manoeuvrability when flying at speed greater than MCRIT and to counteract 'Mach tuck' (See Chapter 15). However, there are light-aeroplane examples of its use such as the Cessna 177 and the Piper Cherokee.

The stabilator is a tailplane that has no elevators but of which the whole tailplane angle of incidence or angle of attack can be altered in flight to provide the required pitch and trim control by moving the control column. The word stabilator has been derived from the words **stabiliser** and **elevator** because it combines the functions of both. See Figure 3.4.

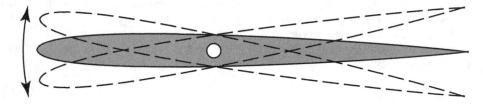

Figure 3.4   The Stabilator.

The stabilator is designed to pivot about the aerodynamic centre of the tailplane at the 25% chord position where, like the main wing, the pitching moment is constant irrespective of its angle of attack. This enables the pilot to use the stabilator to generate a considerable pitching moment with little effort. To prevent light-aeroplane pilots from over controlling the aeroplane, the stabilator is fitted with an antibalance tab that provides some resistance to the pilot's input force by deflecting in the same direction as the stabilator, on some aeroplanes this is combined with a normal trim tab. See Chapter 10.

3.5.3   The Elevons

Aircraft control surfaces that combine the functions of the **elevators** and the **ailerons** are elevons. The inputs of the two controls are mixed either electrically or mechanically when the pilot's control column is used, however, the ability to separate the functions of the two control surfaces is still retained.

Elevons are individually mounted on the trailing edge of each wing and move simultaneously up or down to provide pitch control or differentially to provide roll control. The elevons can provide control in both planes at the same time when the control column is appropriately positioned.

3.6   The Rudder

The control of an aeroplane about its normal or yaw axis is accomplished by using the rudder. The rudder is a hinged control mounted vertically on a post at the rear of the aeroplane known as the fin or vertical stabiliser. Usually, there is a small fillet mounted at the forward base of the fin; this is the dorsal fin which is fitted to prevent the force acting on a fully deflected rudder in a sideslip from suddenly reversing; this undesirable event is known as 'rudder lock.'

The rudder occupies a surface area equal to or greater than the surface area of the fixed vertical fin on which the vertical post is mounted and its deflection is controlled by the movement of a horizontal bar by the pilot's feet. To yaw the aeroplane to the left the left foot is pushed forward and this causes the rudder surface to move to the left against which the airflow produces a force that yaws the aeroplane about its normal axis, acting through the CG, to the left. The opposite is true if the right foot is moved forward.

3.7 Yaw Control

The magnitude of the yaw control exercised by the rudder is dependent on:

a. The area of the control surface;
b. The degree of control-surface deflection;
c. The speed of the aeroplane;
d. The distance between the control surface CP and the CG of the aeroplane (the moment arm).

3.7.1 Control-Surface Area

The greater the rudder surface area the larger is the aerodynamic force developed for a given control-surface deflection angle at any given IAS. The greatest amount of authority exercised by the rudder improves with increased surface area/and or increased moment arm from the CG and/or increased IAS.

3.7.1.1 Control-Surface Deflection

The amount of control surface angular deflection necessary for a given rate of change of direction is less for an aeroplane with a forward CG position than it is for the same rate of change of direction for an aeroplane with an aft CG. Irrespective of the location of the CG, the amount of deflection required is always greater at low indicated airspeeds to produce the same rate of turn, for the same conditions, than it is at a higher airspeed. Thus, the effectiveness of the rudder increases with increasing airspeed. *The angular deflection of the rudder is inversely proportional to the square of the equivalent airspeed.* A rudder deflection causes a yaw rotation about the vertical axis and a secondary effect is roll rotation about the longitudinal axis the same way.

On some high-speed aeroplanes the rudder is fitted with a rudder/pedal ratio changer that automatically limits the maximum amount of rudder deflection possible as the airspeed increases and so prevents the possibility of overstressing the airframe of the aeroplane. The deflection decrease may be progressive or in steps associated with the flap-retraction speeds. The method employed on many aeroplanes restricts the amount of rudder pedal movement available to match the restricted maximum rudder deflection.

3.7.2 The Moment Arm

The location of the CG also affects the measure of authority attained by the rudder because it affects the length of the moment arm and consequently the turning moment of the rudder, or the aerodynamic force applied by the rudder to change the direction of the aeroplane. A forward CG increases the length of the moment arm and increases the aerodynamic force for a given rudder angular deflection to more than that of an aeroplane with an aft CG for the same angular deflection because it decreases the length of the moment arm.

3.7.2.1 Engine-Induced Yaw

The yawing moments directly caused by the force acting normal to the propeller plane or to the jet engine intake is engine-induced yaw. The location of the engines relative to the position of the CG affects the magnitude of the yawing moment. The further forward the engines the greater is the destabilising effect due to the thrust.

Propeller-driven aeroplanes have additional indirect thrust induced yawing moments that are caused by the influence of the propeller on the speed and direction of the airflow over the vertical tail resulting in significant directional trim changes. (See Chapter 16).

3.8   Asymmetric Engine Yawing Moment

In the event of an engine failure of a multi-engined aeroplane the strength of the asymmetric yawing moment caused by its failure is dependent on the distance of the failed engine from the CG, measured at right angles to the longitudinal axis. The further outboard the failed engine is located the greater is the induced yawing moment.

Because the engines of a propeller-driven aeroplane have to be mounted on the wing sufficiently far from the fuselage to account the necessary propeller clearance then these aeroplanes always experience a large yawing moment when an engine fails that may be difficult to control. This is not so with a jet-propelled aeroplane because the wing-mounted engines on these aeroplanes can be mounted relatively close to the fuselage thus creating a relatively short moment arm, which produces a small yawing moment that is easier to control. In such a situation the propeller-driven aeroplane also has a greater tendency to roll than the jet-propelled aeroplane.

On suffering an engine failure the unbalanced thrust and drag produces a yawing moment that must be counteracted by the deflection of the rudder. The side force generated by the rudder deflection when multiplied by the moment arm (the distance that the centre of pressure of the rudder is from the CG) has to produce a counteracting moment equal to and opposite to the yawing moment.

The yawing effect of an engine failure is most difficult to control during take-off and the take-off climb path shortly after becoming airborne because during this phase of flight the engines are generating maximum take-off thrust. The loss of the thrust from the failed engine is therefore of greater significance than it would be at the cruise thrust setting and results in a large yawing moment.

At this stage of flight the indicated airspeed is low, consequently the rudder has little authority and is unable to generate much force. It therefore requires a large angular deflection of the rudder to generate sufficient strength to produce a large enough moment to counteract the yawing moment caused by the engine failure.

The distance between the thrust line of the operating engine and the CG is the asymmetric thrust moment arm and the thrust of the operating engine provides the force to generate the moment. The distance from the CP of the fin and the CG of the aeroplane is the moment arm of the fin side force, which is derived from the impingement of the airflow on the deflected rudder and when multiplied together create the opposing moment.

The lowest IAS at which full rudder deflection provides sufficient force to create a balancing yawing moment is the minimum directional control speed. It is this requirement that determines the value of V_{MCG}, the minimum control speed with the critical engine inoperative on the ground, and V_{MC}, the minimum control speed with the critical engine inoperative in the air. The maximum that V_{MC} may be is $1.13V_{SR}$. *CS 25.149(c).*

The side force on the fin requires a small amount of bank to be applied towards the operative engine to provide a horizontal component of the lift to balance the side force. It is important not to overbank because of the increased stalling speed. Up to 15° of bank is of little significance because the increase is only 2% but at angles of bank greater than this the stalling speed increases dramatically (See page 174).

With rudder deflection at or close to maximum, the fin angle of attack is close to stalling and any decrease of airspeed could lead to a sideslip, fin stall and loss of directional control. Ailerons must be carefully used because excessive use of bank will cause the secondary effect of roll, which is that of a yaw towards the failed engine, more likely to occur.

3.8.1   *Critical Power Unit*

For a multi-engined aeroplane the largest asymmetric thrust moment occurs when an outboard engine fails. That engine failure which most adversely affects the performance is deemed the critical power unit and is used to determine the limiting control speeds that will ensure adequate directional control in such an event.

During a crosswind take-off an aeroplane tends to weathercock into wind. Because of this the failure of the outboard engine on the windward side of the aeroplane is the most difficult to counteract and is, therefore, the critical engine for jet-propelled aircraft. For propeller-driven aeroplanes the direction of propeller rotation determines the critical engine. Viewed from behind, a clockwise rotation dictates that the outboard left engine is critical.

EASA has decreed that the manufacturers should determine VMCG and VMC using the following conditions:

a. **VMCG During the take-off ground run** only primary aerodynamic means may be used to recover control, which means that the assistance gained from the use of nose-wheel steering may not be accounted, because in practice it may not be available due to surface contamination. Furthermore, only sufficient aileron may be used to keep the wings level and the maximum rudder force used is limited to 667 Newtons. The recovery must enable the aeroplane to continue on a path parallel to the original path but separated by no more than 30 ft. *CS 25.149(e).*

b. **VMC During the take-off climb** the maximum bank angle permitted to be accounted is 5° and using that maximum the aeroplane must be able to recover control to within 20° of its original heading. During the recovery the aeroplane must not assume a dangerous attitude nor must it require exceptional piloting skill, alertness or strength to effect the recovery. *CS 25.149(c).*

By plotting the yawing moment coefficient (Cn), due to asymmetric thrust, against airspeed and the counteracting yawing moment coefficient (Cn), due to the maximum rudder deflection, the manufacturer can determine the value of VMCG and VMC from the intersection of the graph lines. See Figure 3.5.

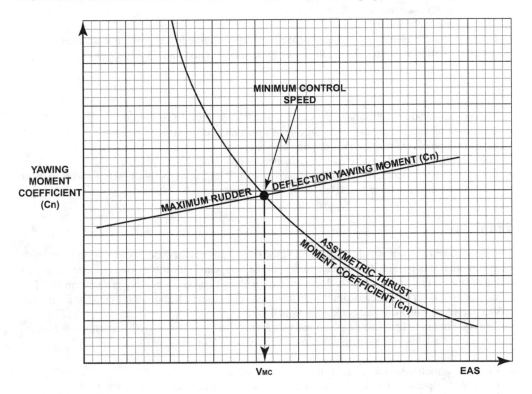

Figure 3.5   VMC.

3.9   Asymmetric Rolling Moment

In addition to the yawing moments the asymmetric thrust causes rolling moments, particularly for twin-engined propeller-driven aeroplanes. The imbalance between the torque produced by the operative engine and the lack of torque from the inoperative engine produces a rolling moment about the longitudinal axis. The direction of propeller rotation determines the direction of the roll.

The imbalance of lift produced by the operative engine and the inoperative engine will induce a sideslip, which produces a rolling moment toward the live engine. This effect is exacerbated by:

a. Extending trailing-edge flaps.
b. A high aeroplane mass.
c. A large dihedral angle.
d. A large sweepback angle.

3.10   Minimum Control Speeds

The configuration used to determine the minimum control speeds is specified in CS 25 as:

3.10.0.1   For Take-off

a. The most unfavourable position of the CG, i.e. the aft limit of the CG envelope.
b. The maximum sea-level take-off mass, i.e. the maximum structural take-off mass.
c. The maximum available take-off power or thrust.
d. The aeroplane trimmed for take-off.
e. For propeller aeroplanes, the propeller of the inoperative engine in the position it achieves without pilot intervention assuming that it fails during the take-off ground run. *CS 25.149(c).*

3.10.0.2   For Landing

a. The most unfavourable position of the CG, i.e. the aft limit of the CG envelope.
b. The most unfavourable mass, i.e. the maximum structural landing mass.
c. Go-around power or thrust on the operating engines.
d. The aeroplane trimmed for approach with all engines operating.
e. For propeller aeroplanes, the propeller of the inoperative engine in the position it achieves without pilot intervention assuming that it fails during a 3° approach. *CS 25.149(f).*

3.10.1 V_{MC}

V_{MC} is the lowest CAS, at which, in the event of the critical power unit suddenly becoming inoperative when airborne, it is possible to maintain control of the aeroplane with that engine inoperative, and to maintain straight flight using no more than 5° of bank without a change of heading greater than 20°. The maximum limit is set at 5° because there is an increasing risk of fin stall above this limit. The magnitude of V_{MC} increases with increased bank angle from 0° to 5°.

The equilibrium of forces about the normal axis is provided by rudder deflection and about the lateral axis by either bank angle or sideslip or a combination of both. V_{MC} may not exceed:

a. Class 'B' aeroplanes – $1.2V_{S1}$ at the maximum take-off mass. *CS 23.149(b).*
b. Class 'A' aeroplanes – $1.13V_{SR}$ at the maximum take-off mass at MSL.*CS 25.149(c).*

3.10.2 V_{MCG}

V_{MCG} is the minimum control speed on the ground. It is the CAS, at which, when the critical engine of a multiengined aeroplane fails during the take-off run and with its propeller, if applicable, in the position it automatically takes, it is possible to maintain control with the use of the primary aerodynamic controls alone *(without the use of nose-wheel steering)* to enable the take-off to be safely continued using normal piloting skill.

V$_{MC}$ must be equal to or less than V$_{LOF}$, because if V$_{EF}$ is equal to V$_{MCG}$ the aeroplane must be able to continue the take-off safely. The lateral control may be used only to keep the wings level. *CS 25.149(e)*.

Recognition of the engine failure by the pilot will be provided by a distinct change of directional tracking by the aeroplane or by the pilot seeing a directional divergence of the aeroplane with respect to the view outside the aeroplane. *AMC 25.149(e)*.

The path of any multi-engined aeroplane from the point of engine failure to the point at which recovery to a direction parallel to the centreline of the runway is attained may not deviate by more than 30 ft laterally from the centreline at any point. *CS 25.149(e)*

The reason that the use of nose-wheel steering is not accounted for in the value of V$_{MCG}$ is because the speed must be able to be used with wet and/or slippery runway surfaces when nose-wheel steering is not available. In practice, if an engine did fail during the take-off ground run on a normal runway surface, nose-wheel steering would be used to assist the recovery, but V$_{MCG}$ ensures that if it is not available because of ice or surface contamination then a safe recovery can still be made. In the event of an engine failure after attaining V$_{MCG}$, if it is less than V$_{1}$ then the take-off must be abandoned.

3.10.2.1   The Effect of the Variables on V$_{MCG}$ and V$_{MC}$

Because V$_{MC}$ and V$_{MCG}$ are both control speeds the only variables that influence their values are air temperature and pressure altitude. *Mass has no effect.*

In conditions of high air density, the engines produce a high thrust output and the loss of the output from one engine during take-off is a proportionate loss of a high output. This will require a considerable amount of force to counteract the tendency to yaw.

Because the rudder and fuselage are of fixed dimensions, sufficient force can be generated only if the aircraft has a high CAS. The loss of an engine in low air density, whilst being proportionately the same, can be counteracted at a lower CAS.

Thus, *both control speeds are decreased in conditions of low air density and increased when the air density is high because of the thrust generated by the engines. No other variable affects V$_{MCG}$ or V$_{MC}$.* Air density will be lowest when the aerodrome pressure altitude is highest, the air temperature is highest and the water-vapour content of the air is highest.

3.10.3   V$_{MCL}$

V$_{MCL}$ is the minimum control speed during the approach and landing with all engines operating in the landing configuration. It is the lowest CAS at which, with the aeroplane in its most critical configuration *on the approach to land*, i.e. the landing configuration, in the event of an engine failure, it is possible to maintain control:

a. With the critical engine inoperative, and the propeller of the failed engine feathered, if applicable, and the operating engine(s) set to go-around power or thrust:
   (1) To maintain straight flight using no more than 5° of bank. *CS 25.149(f)*.
   (2) Roll through 20° from straight flight away from the operative engine in five seconds. *CS 25.149(h)*.
b. Assuming the engine fails while at the power or thrust set to maintain a 3° approach path. *CS 25.149(f)*.

3.10.4   V$_{MCL}$(1out)

V$_{MCL}$(1out) is the one-engine-inoperative landing minimum control speed. The manufacturer may provide this speed for use instead of V$_{MCL}$. *AMC 25.149(f)*.

It is determined for the conditions appropriate to the approach and landing with one engine having failed *before* the start of the approach. The propeller of the inoperative engine, if applicable, may be feathered throughout. *AMC 25.149(f)*.

3.10.5   V$_{MCL-2}$

V$_{MCL-2}$ is the minimum control speed during the approach to land with the critical engine inoperative; it is the lowest CAS at which it is possible when a second critical engine becomes inoperative, and its propeller feathered, if applicable and the remaining engine(s) producing go-around power or thrust to:

a. Maintain straight flight using no more than 5° of bank.
b. Roll through 20° from straight flight, away from the operative engine in five seconds. *AMC 25.149(g)*.
c. Assuming the power or thrust set on the operating engine(s) is that necessary to maintain a 3° approach path when one critical engine is inoperative. *CS 25.149(g)(5)*.
d. The power or thrust on the operating engine(s) is rapidly changed, immediately the second critical engine fails, to the go-around power or thrust setting. *CS 25.149(g)(7)*.

3.10.5.1   *The Effect of the Variables on V$_{MCL}$*

The value of the minimum control speed during approach and landing is influenced by the aeroplane mass and by the density of the air and then only to a minor extent. A high mass requires a high speed and low air density requires a low speed and vice versa. Generally, most manufacturers specify a single speed for all masses and air densities.

3.11   The Ailerons

The control of an aeroplane about its longitudinal or roll axis is accomplished by using the ailerons. The ailerons are hinged control surfaces located at the rearward side of the wing usually at the wing tips occupying a relatively small area of the wing surface. A sideways movement of the control stick or a rotation of the control wheel by the pilot causes one aileron to go down and the aileron on the opposite wing to go up.

The downgoing aileron increases the camber of the wing surface causing an increase of lift and an upward force on the underside of the control surface at that point, which results in that wing being raised. Simultaneously the aileron on the other wing goes up resulting in a downward force on the upper surface of the control that causes that wing to go down. Together, the forces produced by the ailerons rotate the aeroplane about its longitudinal axis.

3.12   Roll Control

Moving the control wheel or control column to the left causes the left aileron to go up and the right aileron to go down together they create a roll of the aeroplane to the left around the CG. Moving the control wheel or control column to the right has the opposite affect. The angular displacement of the ailerons is directly proportional to the amount of control wheel or control column movement. The greater the deflection the greater is the rate of roll, i.e. the bank angular change per unit of time.

A disturbance in the rolling plane causes the angle of attack of the upgoing wing to decrease and the downgoing wing to increase. Provided the aeroplane is not flying close to the stalling speed then the upgoing wing produces less lift than it did before the disturbance and the downgoing wing will produce more lift than it did. Together, these changes result in a rolling moment in opposition to the initial disturbance and have a 'roll damping' effect. Thus, a steady rate of roll is established for any given aileron angle of deflection.

The rate of roll achieved for a given aileron deflection is dependent on:

a. **The surface area of the ailerons.** The size of the ailerons is principally limited by the torsional stiffness of the wings and the induced drag that they cause.
b. **The airspeed of the aeroplane.** The higher the indicated airspeed the greater is the rate of roll attained for a given aileron angular deflection. The force generated by the ailerons is proportional to the square of the EAS. At high airspeeds the force can be large enough to cause the supporting wing to twist that can minimize or reverse the effectiveness of the ailerons often referred to as 'flutter.' See Figure 3.6.
c. **The distance of the ailerons from the longitudinal axis.** To achieve a high rate of roll at low airspeeds when taking off or landing the ailerons are mounted at the end of the trailing edge of the wing. This provides a long moment arm and enables the required rate of roll moment to be attained.

3.12.1 The Flaperon

A control surface that combines the functions of the **flaps** and the **ailerons** is the flaperon. The control surfaces govern the roll or bank of the aeroplane in the same manner as the ailerons but can be lowered simultaneously to perform the function of the flaps. If they can be raised upward they are spoilerons and perform the same function as spoilers.

The input of the two controls are mixed either electrically or mechanically when the pilot's control column is used, however, the ability to separate the functions of the two control surfaces is still retained. The control surfaces are individually mounted on the trailing edge of each wing raised from the surface of the wing so that they are positioned in the undisturbed airflow outside of the boundary layer, so as to maintain total controllability.

3.13 Wing Twist

A desirable characteristic to be included as an aeroplane design feature is wing twist because it ensures that at high angles of attack the wing root will stall before the wing tip and that induced drag is minimized. This enables the pilot to retain a greater degree of roll control when approaching the critical angle of attack during the stall than would normally be the case because the ailerons that are usually located close to the wing tip still have a large degree of authority. There are two types of wing twist that are geometric twist and aerodynamic twist, and both achieve the same result but in a different manner.

3.14 Geometric Twist

Geometric twist is the actual change of the angle of incidence of the chordline from the wing root to the wing tip introduced during construction. The wing cross-section at each point along the wingspan is geometrically similar to the root cross-section but the chordline angle of incidence gradually decreases from the wing root to the wing tip. Thus, the angle of attack of the wing tip is less than that of the wing root. This feature is known as 'washout.' (See page 20).

3.15 Aerodynamic Twist

Aerodynamic twist is the difference between the zero-lift angle at the wing root and the zero-lift angle at the wing tip. The zero-lift line is increasingly rotated downward from the wing root towards the wing tip and this *is achieved by gradually changing the camber of the aerofoil cross-section.* The wing tip cross-section has less camber than the wing-root cross-section.

The wing planform that has the lowest amount of induced drag is elliptical but this is expensive to produce and technically more difficult to construct than a rectangular wing. It is therefore necessary to employ other wing planform shapes, however, the induced drag of any wing planform can be equated to that of an elliptical wing shape if it is designed with the correct amount of wing twist. It has been found that untwisted tapered wings having a tip to root chord ratio of 0.4 produce less induced drag than rectangular wings having the same planform area and aspect ratio developing the same lift. Correct twist application to the design can decrease the induced drag by as much as 15%, which will significantly reduce the total drag. See Figure 8.5.

3.15.1 Twisterons

To avoid the limitations imposed by utilising the wing twist to minimize induced drag, it is possible to implement the twist distribution required by including full-span trailing-edge flaps in the design. These flaps are twisted along their length to produce a spanwise variation of wing twist and have an upward twist at the wing tip end of the flap. These are twisterons, which when extended symmetrically act as normal trailing-edge flaps or when deflected asymmetrically as ailerons to generate high lift and provide roll control.

Twisterons have the advantage that the twist can be varied to minimize the induced drag over a wide range of operating conditions. At low airspeed as twisteron deflection is increased they decrease the amount of elevator deflection required to trim the aeroplane and increase the nose-up pitching moment. By properly linking the elevator deflection to that of the twisteron, accounting the aeroplane mass, air density and airspeed, they reduce the induced drag to the smallest amount possible.

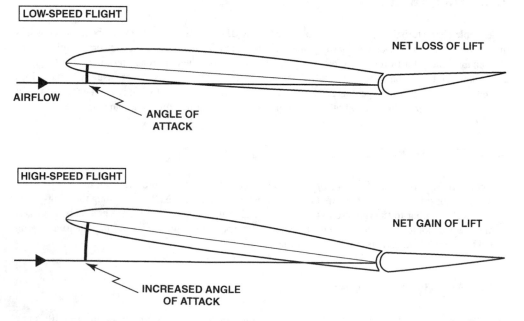

Figure 3.6 High-speed Wing Twist.

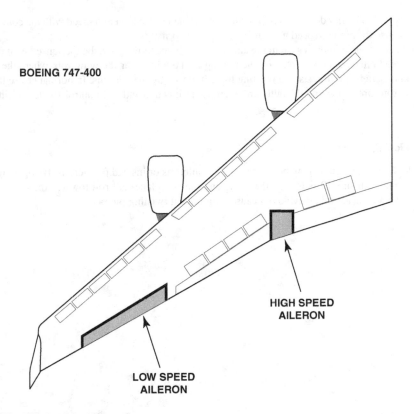

Figure 3.7 The Double-Aileron System.

3.16 High-Speed Twist

The pressure applied to the wing structure by the airflow passing over it has a tendency to twist the wing by an amount, which is in direct proportion to its forward speed. The effect of wing twist at high airspeeds is that the angle of attack of the twisted wing is greater that that of the same wing at a lower airspeed when it is untwisted. The twisted wing develops more lift as a result of the increased angle of attack that may diminish or even reverse the effectiveness of the ailerons, which will be preceded by reduced roll response. See Figure 3.6. To overcome this problem some aeroplanes are fitted with two sets of ailerons; one set for low-speed flight and the second set for high-speed flight. See Figure 3.7.

3.16.1 Low-Speed Ailerons

The low-speed ailerons are conventional ailerons fitted to the outboard trailing edge of the wings. The moment arm is therefore maximized and will give a high rate of roll at low airspeeds. This set of ailerons is locked in the neutral position during high-speed flight and the other set is utilised.

3.16.2 High-Speed Ailerons

The high-speed ailerons have a smaller surface area than that of conventional ailerons and are fitted inboard of the conventional ailerons. The moment arm is therefore less than that of the conventional ailerons and, despite the surface area being less than the conventional aileron, the rate of roll attained for a given aileron angular deflection is equal to that attained at the lower airspeed with the conventional ailerons because the high airspeed gives them greater authority.

By using the double-aileron system the amount of high-speed wing twist experienced is minimized. The high-speed ailerons are mounted on the trailing of the wing near the wing root where the wing is thicker and is better able to resist the twisting force imposed by the ailerons. In normal cruise flight the inboard ailerons are active and the outboard ailerons are inactive and roll-control spoilers if fitted may be active.

3.16.3   Roll Spoilers

Roll spoilers (See Chapter 7) may be used as well as ailerons or instead of ailerons. By opening one of them by a small amount lift is lost on that wing and the aeroplane will roll towards the raised spoiler. They have the advantage that they do not cause a significant twisting moment.

Self-Assessment Exercise 3

Q3.1 If the total sum of the moments about one axis of an aeroplane is not zero it would:
   (a) have control difficulties
   (b) fly a curved path
   (c) fly a normal straight path
   (d) have an angular acceleration about that axis

Q3.2 Why is the effect of nose-wheel steering not accounted for in V_{MCG}?
   (a) Because it has no affect on the value of V_{MCG}.
   (b) Because V_{MCG} must apply to wet and/or slippery runways.
   (c) Because the nose-wheel steering may be inoperative after an engine failure.
   (d) Because it must be possible to abandon a take-off after the nose-wheel leaves the ground.

Q3.3 Regarding the CG and the adjustable stabiliser position, which the following statements is correct?
   (a) The leading edge of the stabiliser is lower for a nose-heavy aeroplane than it is for a tail-heavy aeroplane.
   (b) Stabilizer adjustment for take-off is dependent on flap position only.
   (c) The leading edge of the stabiliser is higher for a nose-heavy aeroplane than it is for a tail-heavy aeroplane.
   (d) At the CG forward limit the stabiliser trim is adjusted to maximum nose-down to obtain maximum elevator authority on rotation at take-off.

Q3.4 Rolling is the rotation of the aeroplane about the:
   (a) wing axis
   (b) longitudinal axis
   (c) vertical axis
   (d) lateral axis

Q3.5 Which of the following statements is correct?
   (i) V_{MCL} is the minimum control speed in the landing configuration.
   (ii) V_{MCL} can be limited by the maximum roll rate.
      (a) (i) incorrect; (ii) incorrect
      (b) (i) correct; (ii) incorrect
      (c) (i) incorrect; (ii) correct
      (d) (i) correct; (ii) correct

Q3.6 For a twinjet low-wing aeroplane with the engines mounted below the wings, if the thrust is suddenly increased the elevator deflection required to maintain a zero pitching moment is:
   (a) none because the thrust line remains the same
   (b) dependent on the location of the CG
   (c) down
   (d) up

Q3.7 Two similar aeroplanes with wing-mounted engines one jet and the other propeller driven, in the event of an engine failure:
   (a) the propeller aeroplane has the greater roll tendency
   (b) they will both have the same yaw tendency
   (c) they will both have the same roll tendency
   (d) the propeller aeroplane will have the least roll tendency

Q3.8 Flaperons are controls that simultaneously use:
   (a) flaps and speed brakes
   (b) flaps and elevator
   (c) ailerons and flaps
   (d) ailerons and elevator

Q3.9 If the stabiliser is jammed in the flight cruise position what action should be taken for landing?
  (a) Select a lower landing speed.
  (b) Relocate as many passengers as possible forward.
  (c) Use the Mach trimmer.
  (d) Select a higher landing speed and/or use a lower flap setting.

Q3.10 During take-off with the CG at the forward limit, if the trimmable horizontal stabiliser is at the maximum setting **and** nose-down position, which of the following statements is correct?
  (a) It will be a normal take-off.
  (b) Rotation will require extra stick force.
  (c) The take-off warning system will be activated.
  (d) The nose-wheel will rise very sharply at rotation.

Q3.11 One of the advantages of siting the engines on the rear fuselage compared with beneath the wings is:
  (a) lighter wing construction
  (b) changes of thrust have less influence on the longitudinal control
  (c) easier engine maintenance
  (d) the wing is less sensitive to flutter

Q3.12 Which of the following statements is correct?
  (i) When the critical engine fails during take-off V_{MCL} can be limiting.
  (ii) V_{MCL} is always limited by the maximum rudder deflection.
      (a) (i) incorrect; (ii) correct
      (b) (i) incorrect; (ii) incorrect
      (c) (i) correct; (ii) correct
      (d) (i) correct; (ii) incorrect

Q3.13 One of the advantages of a movable stabiliser system compared with a fixed stabiliser system is that:
  (a) it leads to greater stability in flight
  (b) the system's complexity is reduced
  (c) it is a more powerful means of trimming
  (d) the structure weighs less

Q3.14 Which of the following statements is correct?
  (i) When the critical engine fails during take-off V_{MCL} can be limiting.
  (ii) V_{MCL} can be limited by the maximum roll rate.
      (a) (i) correct; (ii) incorrect
      (b) (i) incorrect; (ii) correct
      (c) (i) correct; (ii) correct
      (d) (i) incorrect; (ii) incorrect

Q3.15 The motion about the longitudinal axis is;
  (a) pitching
  (b) yawing
  (c) rolling
  (d) slipping

Q3.16 The advantage of a variable-incidence tailplane over a fixed tailplane fitted with elevator and trim tab is:
  (a) flight stability is greater
  (b) the trim is more powerful
  (c) decreased mass penalty
  (d) less complex operating mechanism

Q3.17 The reason the horizontal stabiliser is positioned on top of the fin ('T' tail) on some aeroplanes is:
   (a) to improve the aerodynamic efficiency of the wing
   (b) to decrease the tendency for the aeroplane to superstall
   (c) to improve the aerodynamic efficiency of the vertical tail
   (d) to ensure that the stabiliser is out of the ground effect on take-off

Q3.18 The elevon is a control that simultaneously operates the:
   (a) flaps and elevators
   (b) flaps and ailerons
   (c) elevators and ailerons
   (d) flaps and speed brakes

Q3.19 An aeroplane fitted with a variable-incidence trimming tailplane, when it is properly trimmed the position of the tailplane is . . . .
   (a) dependent on the CG position, speed and thrust
   (b) up
   (c) down
   (d) neutral

Q3.20 When an aeroplane is trimmed correctly the elevator angle relative to the adjustable horizontal stabiliser is:
   (a) up if CG is forward
   (b) down if CG is aft
   (c) is dependent on the speed
   (d) neutral

Q3.21 To maintain a constant pitching moment with an increase of thrust, for an aeroplane with engines mounted below the wing, the elevator deflection is:
   (a) up
   (b) down
   (c) is dependent on the CG location
   (d) remains constant

Q3.22 Rotation about the lateral axis is:
   (a) yawing
   (b) slipping
   (c) pitching
   (d) rolling

Q3.23 If the sum of the moments of an aeroplane in flight is not zero, then the aeroplane will rotate about the:
   (a) CG
   (b) neutral point
   (c) aerodynamic centre
   (d) CP

4 Lift Generation

4.1 Turbulent Flow

An object placed in a steady constant flow of air causes the air to flow around the object. Downstream of the object the airflow becomes turbulent. If the object is a sphere then the airflow around it has certain properties, which were observed by Bernoulli and shown in Figure 4.1 as follows:

a. The transition from a smooth streamlined airflow to a turbulent flow begins at the maximum diameter of the sphere relative to the airflow and the turbulence extends downstream of the sphere.
b. The airflow speed at which the transition to turbulent flow commences is dependent on the maximum diameter of the sphere. For a large sphere it occurs at a relatively low airspeed but for a small sphere it happens at a higher speed.
c. The turbulence downstream of the sphere has a greater depth and extends further downstream for the larger sphere than for a small sphere. In other words, the volume of turbulence is directly proportional to the diameter of the sphere. The *drag is directly proportional to the size of the frontal area.*
d. The stagnation point, at which the airstream velocity is zero and the air pressure is greatest, is at the point furthest into the airflow, i.e. the leading edge of the sphere relative to the airflow.
e. The airflow velocity is greatest and the airflow pressure is least at the maximum diameter of the sphere relative to the airflow where the streamlines converge.

4.2 Streamline Flow

To eliminate or reduce the amount of turbulence experienced downstream of a sphere it is necessary to fill the volume in which the turbulence occurs with a solid object. To do this then the shape has to be altered from a sphere to become streamlined downstream. Its shape then becomes a cone with a hemispherical leading edge. Although this eliminates the turbulence, the shape has little practical value in aviation. However, it is the two-dimensional profile or cross-section of this shape that is used in the design of aerofoils because it produces a streamline flow.

The profile of the object so produced is a symmetrical aerofoil having a rounded leading edge and smooth trailing upper and lower surfaces that gradually taper to meet at a point at the trailing edge. Thus, the airflow is smoothed downstream of the aerofoil to become a smooth steady flow again. This is the 'classical linear flow.' Once again the maximum speed of the airflow is indicated by the closeness of the streamlines. See Figure 4.2.

The Principles of Flight for Pilots   P. J. Swatton
© 2011 John Wiley & Sons, Ltd

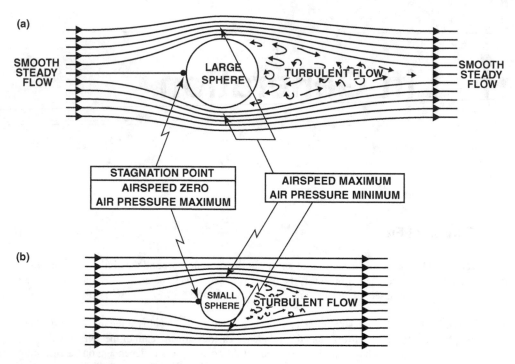

Figure 4.1   Turbulent Airflow.

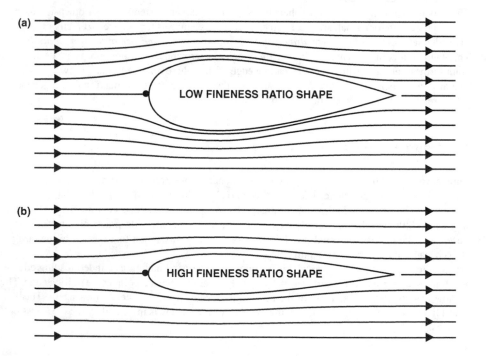

Figure 4.2   Classical Linear Flow. (a) Low Speed Aerofoil. (b) High Speed Aerofoil.

The aerofoil shape that results from a large sphere is thick and can only be used at lower airspeeds. But the aerofoil resulting from the small sphere is slim and is suitable for use at high airspeeds. Both have the same properties as the spheres but without the turbulence downstream.

The thickness/chord ratio is the ratio of the maximum thickness of an aerofoil to its chord length expressed as a percentage. The fineness ratio is the ratio of the chord length to the maximum thickness and is expressed as a nondimensional number. It applies to all streamlined bodies, not just to aerofoils.

4.3   The Boundary Layer

There is a layer of air that clings to the surface of an object and resists movement. This characteristic of the air is its viscosity, which although very small, is enough for the molecules of the air in contact with a surface to stick to the surface. It is static and therefore causes the laminar streamline flow adjacent to this layer to slow down, which in turn produces turbulence. The speed of the airflow changes from zero at the surface of the aerofoil to the full speed of the freeflow airstream a few millimetres away from the surface. This thin layer of sluggish air that clings to the surface of the object and in which the speed change of the airflow takes place is the boundary layer. Bernoulli's principle that the sum of all forms of mechanical energy remains constant does not apply to the boundary layer, because mechanical energy and thermal energy are lost due to skin friction.

The depth of the boundary layer is dependent on the Reynold's number. (See Chapter 1). The Reynold's number accounts for the following factors:

a. the speed of the free airflow;
b. the density of the air;
c. the viscosity of the air;
d. the compressibility of the air.

In aerodynamics the boundary layer is important when considering wing stall, skin friction drag and heat transfer in high-speed flight. Additional factors that affect these considerations are the cleanliness of the aerofoil surface and the condition of the aerofoil surface. Boundary layers may be laminar (layered) or turbulent (disordered). The effect of the boundary layer on lift is accounted for in the coefficient of lift and the effect on drag is accounted for in the coefficient of drag. See Figure 4.3.

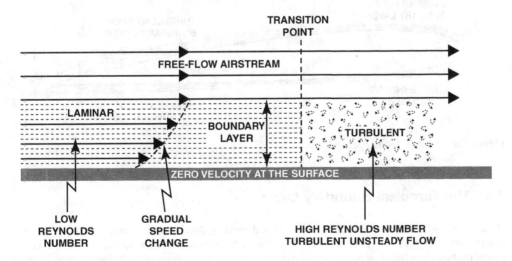

Figure 4.3   The Boundary Layer.

4.4   The Laminar Boundary Layer

A low Reynold's number indicates that the boundary layer is laminar, that there is no velocity component normal to the aerofoil surface and, furthermore, that the airspeed changes gradually with increasing distance from the surface of the aerofoil. At the outer edge of the laminar boundary layer the airflow speed is 99% of Vo, the speed of the freestream airflow. It also has less surface friction than the turbulent boundary layer and because it has very little kinetic energy it has a greater tendency to separate from the aerofoil surface over which it is flowing. If the boundary layer does lift off or 'separate' from the aerofoil surface it creates an effective shape different from the physical shape of the surface.

4.4.1   The Transition Point

The point at which the change from laminar flow to turbulent flow is complete is the transition point. At low angles of attack this point is close to the trailing edge of the aerofoil and gradually moves forward as the angle of attack increases until at approximately 14° it is almost at the leading edge of the aerofoil. Aft of the transition point the mean speed of the airflow increases and the surface friction increases.

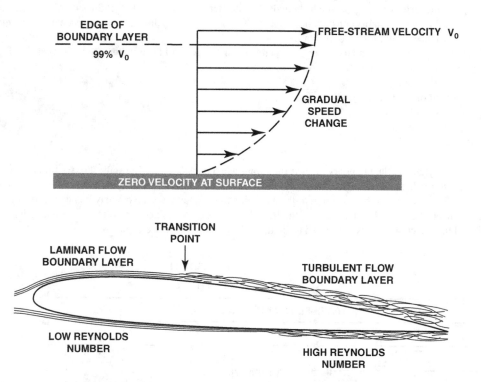

Figure 4.4   An Aerofoil Boundary Layer.

4.5   The Turbulent Boundary Layer

Higher Reynold's numbers indicates a turbulent and swirling flow within the boundary layer that has more kinetic energy than the laminar flow. This turbulent flow occurs aft of the transition point and is that part of the boundary layer that has the greatest change of velocity and the greatest surface friction. Despite this it has less tendency to separate from the aerofoil surface over which it is flowing than the laminar boundary layer because it is better able to withstand the positive pressure gradient. See Figure 4.4.

4.5.1 Leading-Edge Separation

'Controlled separated flow' or 'leading-edge separation' is half-way between steady laminar streamline airflow and turbulent unsteady airflow. A sharp leading edge to an aerofoil causes the laminar boundary layer to separate from the aerofoil surface. Beneath the separated layer a stationary vortex forms that is relatively small at the wing root and increases in size towards the wing tip. This is sometimes referred to as a 'bubble.' The shape of the aerofoil determines the size and shape of the bubble aft of the leading edge. Short bubbles have no effect on the pressure distribution; long bubbles do affect the lift generated even at low angles of attack.

The flow aft of the vortex reattaches itself to the wing surface, as a turbulent boundary layer, a short distance behind the vortex at the wing root but within an increasing distance behind the vortex towards the wing tip. Close to the wing tip it does not reattach to the upper surface but adds to the strength and size of the tip vortex. The leading-edge separation and vortex are shown in Figures 4.5 and 8.3.

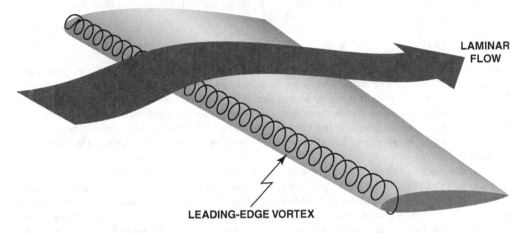

LAMINAR FLOW

LEADING-EDGE VORTEX

Figure 4.5 Leading-Edge Separation.

4.6   Boundary-Layer Control

The boundary layer is the very thin layer of air lying on the surface of an aerofoil that tends to adhere to the surface of the aerofoil. As the wing moves forward the boundary-layer air at first moves slowly aft in the same direction as the streamlined flow, this is laminar or streamlined flow. As the air approaches the centre of the wing chord the surface friction causes the airflow to slow down and become thicker and more turbulent. The point at which the airflow becomes turbulent is the transition point. As either the speed or the angle of attack increases then the separation point moves forward.

If the boundary layer can be controlled so as to remain laminar and unseparated then the performance of the aeroplane would be greatly enhanced because the parasite (profile) drag, caused by surface friction and the shape of the aerofoil, is greatly reduced, whilst the usable angles of attack are greatly increased and lift is dramatically improved at slow speeds.

There are three main methods that are currently used to control the boundary layer so that it remains attached to the aerofoil surface as long as possible. They all depend on the principle of adding kinetic energy to the lower layers of the boundary layer. They are blowing, suction and vortex generators.

4.6.1 Blowing

This type of control of the boundary layer is confined to jet-propelled fighter aircraft that have a high load factor on a small overall wing surface area. Although they were popular in the 1960s they lost their appeal because of the difficulties encountered during manufacture and maintenance. Furthermore, the emphasis during design changed to satisfy a requirement for greater speed, higher load capacity and more manoeuvrability, and because of this the blown system generally disappeared.

The operating method of this system is that a small amount of compressed air is bled off from the compressor stage of the engine and fed through pipes to the channels separated at intervals along the trailing edge of the wing and exhausted through slots in the flaps when the flap reached a specific angle. The air is forced out of the channel downward over the upper surface of the flap and follows the flap profile re-energizing the boundary layer, thus preventing it from stagnating until further along the flap. The overall effect is that the amount of lift produced is increased.

At low airspeeds when the amount of flap deployed is large, the amount of air delivered by this system is a significant portion of the overall airflow. This importantly increases C_{Lmax}, almost doubling the value of that attained without the use of the blown system of augmentation. This also enables the landing speed to be safely decreased by a large amount. Using the same method simultaneously with leading-edge slots further enhances the maximum lift and decreased landing-speed benefits.

4.6.2 Suction

Another method used to force the boundary layer of air to stay in contact with the upper surface of the wing is suction. If the whole surface area were to be covered in a series of small slots through which the entire boundary layer of air could be sucked then it could be completely replaced by undisturbed air, however, the power required to perform such a function is enormous and there is no overall benefit. Nevertheless, replacing the boundary layer over a small portion of the wing is feasible and beneficial to the performance of the aeroplane. The effect it has prevents separation by stabilising and strengthening the boundary layer.

A vacuum sucks air through a porous surface area or through a series of small slots running from the wing root out to the wing tip, at approximately the 50% point of the chordline where the adverse pressure gradient is marked and not where it is just starting. The air is then exhausted rearward through ducts or channels. In this manner the thickness of the boundary layer is decreased and permits the air flowing over it to travel faster. The use of a porous surface area is generally more effective than the use of slots.

4.6.3 Vortex Generators

The boundary layer breaks away from the surface when it is no longer able to overcome the adverse pressure gradient over the upper surface of the wing. The principle of operation of vortex generators is that they create small vortices that cause the faster moving air above the boundary layer to be deflected from the top of the boundary layer to the slower moving air near the aerofoil surface. This transfers momentum and re-energizes the lower layer of the boundary air, delaying the separation point to further aft along the aerofoil surface. Their purpose is to limit the spanwise flow of the boundary layer to prevent tip stalling.

Vortex generators are either small jets of air that issue from the upper surface of the aerofoil at the normal to the surface or small metal plates approximately 25 mm deep standing vertically in a row spanwise along the wing. There are many shapes that may be used for the metal plates, they are small rectangular plates or miniature aerofoil sections and their positioning depends on what the designer wished to achieve. Usually, they are positioned at a small angle of attack to the oncoming airstream.

The selection of the shape and position of the vortex generators is made to ensure that the increase of lift obtained exceeds the amount of extra drag that they incur. The advantage of the air-jet type of vortex generator is that they can be turned off when not required and therefore avoid the drag penalty, whereas the metal plates cannot.

For aeroplanes travelling at high subsonic or transonic speeds, the airflow over the upper surface of the wing is accelerated to such an extent that it becomes sonic and a shockwave is produced. This worsens the situation because it means that the boundary sublayer is subsonic and the high pressure at the shockwave can be transmitted forward. (See Chapter 15).

Vortex generators weaken the shockwave and decrease shock drag but have no diminishing effect on wing-tip vortices. Nevertheless, the vortices produced by the vortex generators diminish shock buffet and are beneficial to the aeroplane's overall performance.

4.7 Two-Dimensional Flow

Although an aerofoil is a three-dimensional object it is necessary at first to consider the airflow around the aerofoil in only two dimensions, the third dimension will be considered later. The two dimensions to be considered are those in a cross-section of an aerofoil in which the airflow motion is restricted to that plane parallel to the freestream airflow.

As air flows around an aeroplane its speed and pressure changes. The spacing of the streamlines indicates the change of speed; streamlines close together indicate a high airspeed and those further apart a low airspeed. The 'equation of continuity' states that mass can neither be created nor destroyed; air mass flow is constant. From Bernoulli's theorem it can be deduced that an increase of airspeed around an aerofoil results in a decrease of air pressure and vice versa.

The lift and drag forces of a wing cross-section are dependent on the pressure distribution around the wing. The pressure distribution surrounding an aerofoil, found by wind-tunnel tests revealed that the two-dimensional flow is dependent on the attitude of the aerofoil with respect to the freestream airflow. In other words, the pressure distribution around an aerofoil is the resultant of the angle of the aerofoil chordline to the oncoming airflow, the angle of attack. The pressure distribution varies with the angle of attack. However, where the streamlines converge in a two-dimensional flow pattern the static pressure will decrease. (See Figure 4.6).

4.8 The Stagnation Point

The stagnation point is that point on an aerofoil at which the approaching airflow becomes stationary. It is located on the leading edge of the aerofoil just below the maximum point of curvature and moves downward and aft along the lower surface of the aerofoil profile as the angle of attack increases. The stagnation pressure, the pressure necessary to stop the airflow at the stagnation point, is equal to the dynamic pressure plus the static pressure.

4.8.1 Aerofoil Upper-Surface Airflow

The airflow from the stagnation point rapidly accelerates upward, the upwash, and rearward over the leading edge and increases in magnitude as the angle of attack increases. The air pressure decreases in value from the stagnation point until it reaches a maximum negative value at the point of maximum curvature of the surface over which it is flowing. At this point the airflow speed is equal to that of the streamline flow and from this point the air pressure increases until by the time it reaches the trailing edge of the aerofoil it attains a small positive value.

The acceleration of the airflow over the upper surface of the wing changes the direction of the streamlines downward and this induces a decreased angle of attack over the tailplane.

4.8.2 Aerofoil Lower-Surface Airflow

The acceleration of the airflow from the stagnation point under the lower surface of an aerofoil is less rapid than that over the upper surface. The air pressure decreases more slowly to a less-negative maximum

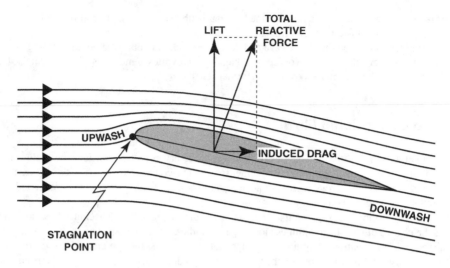

Figure 4.6   Aerofoil Streamline Laminar Flow.

value at the point of maximum curvature. From this point the pressure increases until it attains a small positive value at the trailing edge of the aerofoil. At high angles of attack, the slight concavity of the lower surface towards the trailing edge can cause the air pressure to retain a small positive value along the entire lower surface of the aerofoil. Although this does increase the lifting properties of the aerofoil it also produces undesirable pitching and drag qualities. See Figure 4.7.

4.9 Lift Production

4.9.1 Symmetrical Aerofoils

The curvature of the upper and lower surfaces of a symmetrical aerofoil is the same, consequently at an angle of attack of 0° the negative pressure generated by the acceleration of the airflow over the upper surface is equal to the negative pressure below the lower surface. The centres of pressure of both surface negative-pressure areas are exactly opposite each other at the aerodynamic centre (AC) (described later in this chapter). Therefore, the total pressure for each surface is equal and opposite and the resultant force is zero. Consequently, there is no lift generated and the pitching moment is zero. Therefore, there is no induced drag but there will still be a small amount of parasite drag.

If the angle of attack of the aerofoil is positively increased the effective camber of the upper surface increases and that of the lower surface decreases. As a result, the magnitude of the negative pressure over the upper surface increases and the magnitude of that below the lower surface decreases. The difference between the sizes of the two pressure areas or the combined pressure is upward lift. The greater the angle of attack the greater is the lift generated up to the stalling angle.

Because the centres of pressure of a symmetrical aerofoil remain exactly opposite each other, no matter what the angle of attack, and act through the same point at the aerodynamic centre (defined later in this chapter) then the total lift acts through this point and there is no change to the pitching moment. It remains at zero, which is one of the advantages of this type of aerofoil.

4.9.2 Cambered Aerofoils

The curvature or camber of the upper surface of a cambered aerofoil is greater than that of the lower surface. As a result, the negative pressure generated by the acceleration of the airstream over the upper

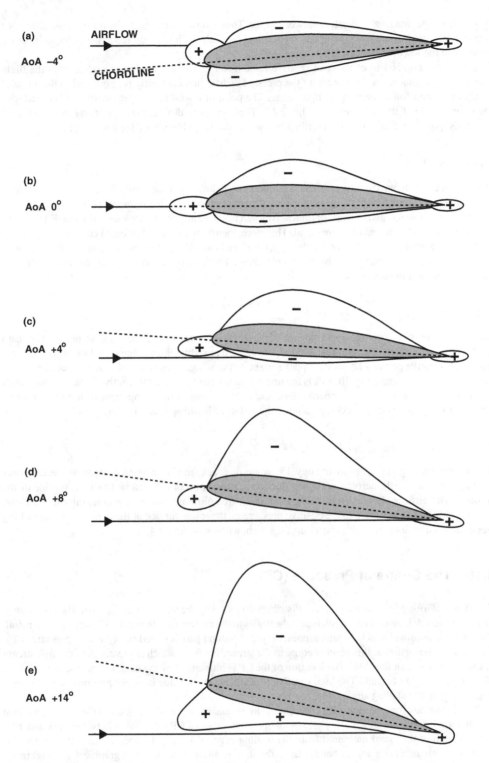

Figure 4.7   Cambered Aerofoil Pressure Distribution.

surface is greater than that beneath the lower surface. The total reactive force is the result of the difference between the air pressure over the upper surface and beneath the lower surface assisted by the positive pressure at the lower leading edge of the aerofoil.

Lift is that component of the total reactive force (see Figure 4.6) that is perpendicular to the flight path of the aeroplane. The magnitude of the pressure distribution is directly proportional to the angle of attack of the aerofoil in the normal flight range. The point of lowest static pressure moves forward with increasing angle of attack as shown in Figure 4.7. There are three different groups of angles of attack for which the pressure distribution is described below and shown in Figure 4.7 for level flight.

4.9.2.1   a. Negative Angles of Attack

Because of the different curvatures of the upper and lower surfaces of the aerofoil even when the angle of attack is zero the aerofoil will still generate a small amount of total lift. *To produce no lift at all a cambered aerofoil must have a negative angle of attack.* At small negative angles of attack the pressure distributions over both surfaces of the aerofoil are equal. Therefore, there is no reactive force and consequently no lift. However, the total pressure vector for the upper surface is aft of the total pressure vector for the lower surface and the AC is exactly midway between them. Thus, a nose-down pitching moment is created about the AC as **a** in Figure 4.7.

4.9.2.2   b. Small Positive Angles of Attack

The negative pressure over the upper surface of the aerofoil is greater than the negative pressure beneath the lower surface. Thus, the total reactive force is upward at right angles to the chordline. It is this large excess of negative pressure above the upper surface of the wing, often referred to as the suction, that is the major factor in generating lift. Lift is the upward component of the total reactive force at right angles to the airflow passing over the upper surface and induced drag is the component of the total reactive force that is in a rearward direction parallel to the airflow. Shown as **c** and **d** in Figure 4.7.

4.9.2.3   c. Large Positive Angles of Attack

Beyond the stalling angle, approximately 15° angle of attack, the large area of negative pressure over the upper surface of the aerofoil collapses due to the separation of the airflow from the surface of the aerofoil. The airflow changes from being a laminar, streamline flow to an unstable, turbulent airflow. The only lift remaining is due to the positive pressure on the lower surface of the aerofoil. At the stalling angle the lift and drag are both maximum. This is shown as **e** in Figure 4.7.

4.10   The Centre of Pressure (CP)

The resultant lifting force of the pressure distribution caused by the airflow over an aerofoil can be shown as a single aerodynamic force, which is at right angles to the airflow over the upper surface of the aerofoil. This force is assumed to act through a point on the chordline that is called the centre of pressure (CP); the speed of the airflow at this point is equal to the streamline flow, which is greater than the free stream or undisturbed streamline flow. The location of the CP is influenced by the camber, the coefficient of lift (CL) and the angle of attack. The total reactive force increases in value in direct proportion to the angle of attack up to the stalling angle.

The CP moves forward along the chordline with increasing angle of attack up to the stalling angle at which point it suddenly moves aft. The normal operating range of CP locations is between 30% and 40% of the length of the chord measured from the leading edge. Thus, the distance of the CP from the AC decreases with increasing angle of attack up to the stalling angle. This is shown graphically in Figure 4.8

and 11.20. In level flight if the speed is increased the CP moves aft because the angle of attack decreases but the total lift remains constant.

For symmetrical aerofoils at subsonic speeds there is virtually no movement of the CP over the range of working angles of attack because it is independent of the angle of attack.

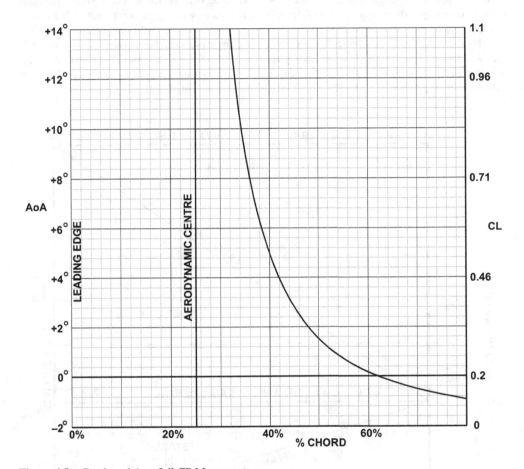

Figure 4.8   Cambered Aerofoil CP Movement.

4.11   Pitching Moments

A moment is a force multiplied by the length of an arm. The pitching moment is the force that causes an aeroplane to move about its lateral axis. The force in this case is the lift and is applied through the CP; if there is no lift then there can be no pitching moment. The arm is the distance of the CP from a specified position. As was shown in Figures 4.7 and 4.8 not only the lift increases with increasing angle of attack up to the stall but also the CP moves with the changing angle of attack. Thus, the moment about a point will also change as the angle of attack changes. The moments about any given point can be determined for a series of angles of attack and plotted on a graph. The slope of the graph line is dependent on the distance of the moment reference point from the leading edge of the aerofoil.

If two points are selected as moment reference positions, one forward at the leading edge (point A) and one aft at the trailing edge (point B) then two sets of moments can be determined for each angle of attack. The following will result:

a. At a low negative angle of attack, for instance $-2°$, the CP is positioned at 80% of the chordline from the leading edge, which although the lift force is weak produces a large moment about A because of the long arm.

b. If the angle of attack is now increased to a low positive angle of attack, for instance to $+4°$, the CP is now at 30% of the chordline from the leading edge and the lift force is increased. Despite the decrease in the length of the arm, the pitching moment about A increases because of the strength of the lift. It will therefore produce an increased nose-down (negative) moment as shown in Figure 4.9.

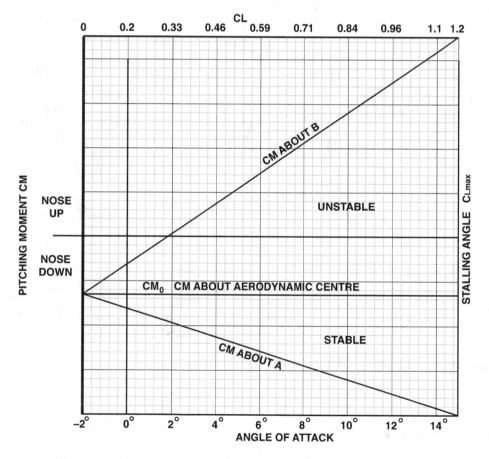

Figure 4.9   Pitching Moments C_M v C_L and Angle of Attack.
(Point A is at the leading edge of the aerofoil and point B is at the trailing edge).

c. Thus, any increase of the angle of attack up to the stalling angle will increase the strength of the lift force, decrease the length of the arm CP from reference point A and increase the nose-down pitching tendency. If trailing-edge flaps are extended the effective angle of attack is increased requiring a decreased IAS; it also produces an increased nose-down pitching moment. However, if the IAS is maintained during flap selection the CP will move aft.

The value of *the pitching moment is directly proportional to the square of the equivalent airspeed* and the pitching moment coefficient (C_M) is equal to the pitching moment divided by the dynamic pressure, the aerofoil plan area and the chord of the aerofoil. It can be deduced from Figure 4.9 that when C_L is zero the pitching moment is negative, i.e. nose-down. When the angle of attack is changed not only does the amount of lift change but also the position of the CP, and consequently the pitching moment C_M also changes.

$$C_M = \frac{\textbf{Pitching Moment}}{\textbf{Dynamic Pressure} \times \textbf{Planform} \times \textbf{Chord Length}}$$

Because of its variable nature, C_M is difficult to use when designing an aerofoil or when determining its longitudinal stability. For such a task it would be beneficial if the C_M remained constant irrespective of the angle of attack. There is such a point called the aerodynamic centre of an aerofoil.

4.12   The Aerodynamic Centre

There is a point along the chordline at which the pitching moment, C_M, remains the same no matter what the angle of attack. In other words, the pitching moment coefficient is constant at the zero lift value, C_{M0}. This is the aerodynamic centre (AC) of the aerofoil. Regardless of camber, aerofoil thickness or angle of attack the AC is located at a point approximately 25% of the chordline from the leading edge of the aerofoil for subsonic flight, rarely is it forward of 23% or aft of 27%. For supersonic flight it is at a point approximately 50% from the leading edge, which increases the static longitudinal stability. It is the point about which all lift changes effectively act.

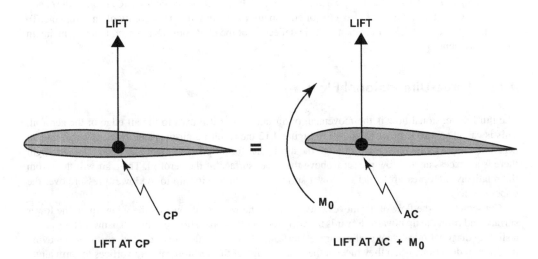

Figure 4.10   The Two Ways of Representing the Lift Force. Lift at the CP = Lift at the AC + M_0.

It should be noted that *a symmetrical aerofoil at zero lift has no pitching moment* about the AC because the upper and lower surface lifts act along the same vertical line. An increase of lift for such an aerofoil does not alter the situation; the CP remains fixed at the AC. Therefore, the total lift may be shown in one of the two ways shown in Figure 4.10.

The constant moment coefficient is the zero lift pitching moment coefficient, C_{M0}. Its value is dependent on the shape of the aerofoil and is:

a. negative or nose-down for a cambered aerofoil.
b. zero for a symmetric aerofoil.
c. positive or nose-up for a reflex curved aerofoil.

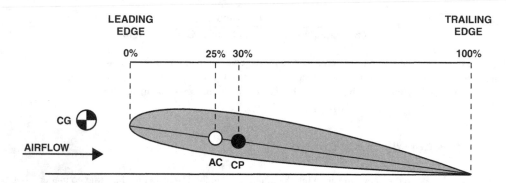

Figure 4.11 A Longitudinally Stable Aerofoil.

On a cambered aerofoil even though the centre of pressure (CP) moves with the changing angle of attack, the lift is assumed to act through the aerodynamic centre. The centre of pressure on a cambered aerofoil is always aft of the AC. The combination of the AC and the pitching moment coefficient are required for the analysis of the longitudinal stability of an aeroplane. See Figure 4.11.

An aeroplane is longitudinally stable at all times that the aerodynamic centre (AC) is aft of the centre of gravity (CG). The tailplane is also a major contributor to the longitudinal stability of an aeroplane. To maintain a particular angle of attack a specific deflection of the elevator is established that will maintain this requirement.

4.13   Three-Dimensional Flow

The third dimensional flow is the movement of air laterally as it moves to the aft edge of the aerofoil. This is called spanwise flow. As shown in Figure 4.12 the resulting air-pressure distribution around an aerofoil, caused by its movement through the air, is that for all positive angles of attack in level flight there is an excess area of low pressure above the upper surface of the aerofoil. Thus, air will flow from the relatively high pressure under the lower surface around the wing tip to the lower pressure over the upper surface.

Consequently, the flow of air moves outward from the wing root towards the wing tip on the lower surface and from the tip inwards towards the wing root on the upper surface. This is spanwise flow and is a slightly diagonal movement of the rearward airflow. Thus, the airflow over the two surfaces is moving in different directions when they meet at the trailing edge of the aerofoil causing vortices to form along its entire length. This causes swept wing aeroplanes to pitch up at low speeds.

4.14   Wing-Tip Vortices

The strength of the vortices at the trailing edge of a wing, where the airflow from the upper and lower surfaces meet, start at zero at the wing root and increase in strength with increased distance from the wing

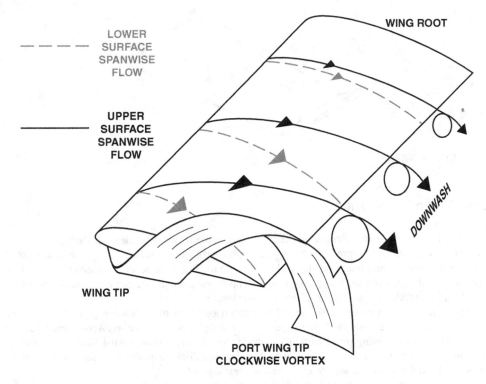

LOWER
SURFACE
SPANWISE
FLOW

UPPER
SURFACE
SPANWISE
FLOW

WING ROOT

DOWNWASH

WING TIP

PORT WING TIP
CLOCKWISE VORTEX

Figure 4.12 Spanwise Flow and Wing-Tip Vortex.

root. Thus, the largest and strongest vortices are at the wing tips. Viewed from behind the port wing-tip vortex rotates clockwise and the starboard vortex rotates anticlockwise. These produce a downward airflow that affects the airflow over the whole wing. Downstream of the wing all of the vortices behind each wing combine to produce very strong wing-tip vortices.

Aerofoils that have a rectangular planform and a low aspect ratio such as those used for small private aeroplanes and basic trainers create large vortices, because the spanwise flow at the wing tip is strong. Wing-tip vortices decrease in size and strength with increased aspect ratio and/or decreased angle of attack. These vortices are induced drag and add to the total drag experienced by the aeroplane.

The efficiency of the wing can be improved if blended winglets (upturned wing tips) are included in the aeroplane design. They reduce the onset and size of wing-tip vortices, decrease the induced drag and consequently reduce the thrust required and the fuel consumption by up to 19%; many modern aeroplanes have them fitted during construction as part of their design. See Figure 4.13.

The induced downwash at the trailing edge of the aerofoil modifies the airflow over the whole wing surface. Its magnitude is determined by the pressure differential between the upper and lower surfaces of the aerofoil; the greater the pressure differential the greater is the spanwise flow, the strength and the size of the wing-tip vortex. It causes:

a. a reduced effective angle of attack that decreases the lift generated;
b. increased induced drag, which increases the thrust required and the fuel flow;
c. a reduced effective angle of incidence of the tailplane that decreases its longitudinal stabilising effect.

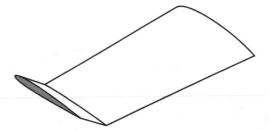

Figure 4.13   A Winglet.

4.15   Wake Turbulence

The strength and size of the vortices produced by an aeroplane are dependent on the amount of lift produced. Strong vortices are produced by large transport aeroplanes that create a significant amount of turbulence behind the aeroplane along the path of the aircraft. The disturbance so caused can extend for a considerable distance and last for several minutes. This is wake turbulence that can seriously affect the controllability of following aeroplanes if they are too close. See Figure 4.14.

The severity of the turbulence so created is categorised to enable following pilots to assess the difficulties that may be encountered. The larger, heavier aeroplanes create the worst turbulence that will cause problems to following pilots particularly when close to the ground during take-off or landing. Pilots are advised by the aviation authorities of the minimum distance and/or time separation from the aeroplane ahead deemed to be safe, to avoid having control problems.

The minimum separation distance and/or time is based on the relative sizes of the aeroplanes concerned. Wake turbulence is affected by the height of the aeroplane and the prevailing wind conditions. The downwash of the airflow behind the aeroplane causes the vortices to descend to one thousand feet below the aeroplane. However, at low height, when the aeroplane is in ground effect, the vortices move outward from the path of the aeroplane over the ground. But any cross wind between 5 kt and 10 kt will prevent this happening to the upwind vortex and will keep it on the aircraft's track. It is particularly important during the take-off phase that the following aircraft complies with the advised separation because of this.

Aircraft mass, speed, aspect ratio and trailing-edge flap setting influence the strength of the vortices. The vortex strength is greatest for heavy aeroplanes at low airspeed and in the clean configuration, which is just after rotation at take-off. Vortices only cease when the nose-wheel touches down during landing ground run. When trailing-edge flaps are extended they produce additional vortices at the flap outboard tips that weaken the wing-tip vortex and accelerate its dissipation. (See page 164).

4.16   Spanwise Lift Distribution

To minimise the amount of induced drag, that is the drag caused by the creation of lift, it is essential to design the wing such that the size and strength of the wing-tip vortices are small. To achieve this the length of the wing chord must be shortest at the wing tip and increases continuously to become the longest at the wing root. In theory a constant downwash along the whole span can be achieved by such a wing planform. Thus, it is the taper ratio that determines the spanwise lift distribution of a wing. See Figure 4.15.

4.16.1   The Effect of Wing Planform

Although the three-dimensional characteristics of a wing are determined by its aspect ratio it is the shape of the wing planform that influences the distribution of the lift. The taper ratio, that is the ratio of the

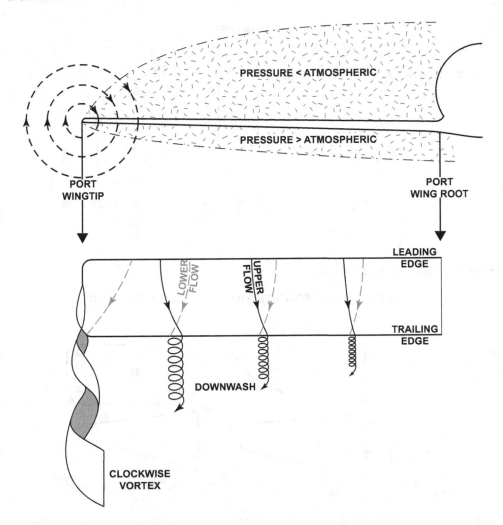

PRESSURE < ATMOSPHERIC

PRESSURE > ATMOSPHERIC

PORT
WINGTIP

PORT
WING ROOT

LEADING
EDGE

LOWER
FLOW

UPPER
FLOW

TRAILING
EDGE

DOWNWASH

CLOCKWISE
VORTEX

Figure 4.14   Wake Turbulence.

tip chord to the root chord of a wing, affects the distribution of lift along the span of a wing. The most convenient method of considering lift distribution is by using the ratio of the local lift coefficient (C_l) to the total lift coefficient for the whole wing (C_L). By plotting this value for each wing planform against the semi-span distance from the wing root the lift distribution can be shown. Figure 4.16 illustrates that:

a. The elliptical wing is the most efficient planform because it produces the same amount of lift along the whole wingspan and incurs the least induced drag.

b. The rectangular wing produces most lift at the wing root and the least at the wing tip. This creates most of its trailing vortices at the wing tip, because it has a constant chord length throughout its span; consequently the greatest downwash is at its wing tips. Of all the wing-shape planforms it produces the largest amount of wing-root lift.

c. A tapered wing on which the chord continuously narrows towards the wing tip generates most of its lift at the approximately 60% of the distance from the wing root to the wing tip, which is also where the trailing vortices are most plentiful for such a wing.

d. The triangular wing generates most of its lift at 70% of the distance from the wing root to the wing tip.

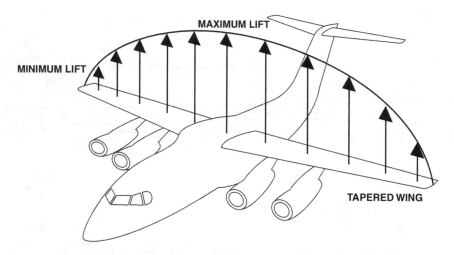

Figure 4.15   The Ideal Elliptical Lift Distribution.

LOCAL LIFT COEFFICIENT (Cl) TO OVERALL LIFT COEFFICIENT (CL)

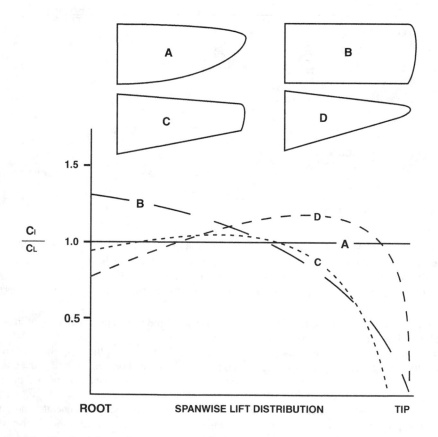

Figure 4.16   Planform Spanwise Lift Distribution.

The most efficient wing for producing an elliptical spanwise lift distribution is an elliptical planform; however, aerofoils of this shape are difficult to manufacture. Nevertheless, a tapered wing planform with a varying aerofoil cross-section has been found to produce an almost perfect elliptical spanwise lift distribution and has no production difficulties. From Figure 4.16 it can be seen that of all the planforms the rectangular planform produces the greatest amount of lift at the wing root. Most large transport aeroplanes have tapered wings because it reduces the structural mass without compromising the strength of the wings.

Self-Assessment Exercise 4

Q4.1 The lift and drag forces of a wing cross-section:
  (a) depend on the pressure distribution around the wing for their magnitude
  (b) are normal to each other at only one angle of attack
  (c) are proportional to each other at all angles of attack
  (d) vary linearly with the angle of attack

Q4.2 When trailing-edge flaps are deployed in level flight the change to the pitching moment is:
  (a) nose-up
  (b) none
  (c) dependent on the CG location
  (d) nose-down

Q4.3 Which of the following statements regarding induced drag and wing-tip vortices is correct?
  (a) The direction of flow on the upper and lower surfaces deviates in the wing tip direction.
  (b) The wing-tip vortices and the induced drag decrease with increasing angles of attack.
  (c) The flow direction on the upper surface has a component towards the wing root and on the lower surface the flow direction has a component towards the wing tip.
  (d) Vortex generators diminish wing-tip vortices.

Q4.4 The induced angle of attack is the result of:
  (a) downwash due to wing-tip vortices
  (b) a large local angle of attack
  (c) downwash due to flow separation
  (d) change in flow direction due to the effective angle of attack

Q4.5 Aft of the transition point between laminar and turbulent flow:
  (a) the mean speed increases and the surface friction decreases
  (b) the boundary layer thickens and the speed decreases
  (c) the mean speed increases and the surface friction increases
  (d) the boundary layer gets thinner and the speed increases

Q4.6 Winglets improve the performance of an aeroplane because they:
  (a) decrease the induced drag
  (b) decrease the static directional stability
  (c) increase the manoeuvrability
  (d) create an elliptical lift distribution

Q4.7 The direction of spanwise flow is from:
  (a) beneath the wing to the top of the wing via the trailing edge
  (b) the top of the wing to beneath the wing via the leading edge
  (c) beneath the wing to the top of the wing via the wing tip
  (d) the top of the wing to beneath the wing via the trailing edge

Q4.8 The CP of a positive cambered wing at an increasing angle of attack will:
  (a) move aft
  (b) move spanwise
  (c) move forward
  (d) not move

Q4.9 The aerodynamic centre of a straight wing when accelerated to supersonic speed will:
  (a) move aft by approximately 10%
  (b) not move
  (c) move slightly forward
  (d) move from 25% to 50% of the aerofoil chord

Q4.10 At a constant IAS and mass in level flight during flap selection:
  (a) the CP moves aft
  (b) both C_L and C_D increase

(c) the stalling speed increases

(d) the whole of the boundary layer becomes laminar

Q4.11 The greatest flow velocity over a cambered wing at a positive angle of attack is:

(a) at the stagnation point

(b) over the upper surface

(c) under the lower surface

(d) in front of the stagnation point

Q4.12 Pitch up at low speed is caused by:

(a) spanwise flow on a swept-back wing

(b) spanwise flow on a swept forward wing

(c) wing-tip vortices

(d) the Mach trim system

Q4.13 For a positively cambered aerofoil when C_L is zero the pitch moment is:

(a) infinite

(b) positive (pitch-up)

(c) negative (pitch-down)

(d) zero

Q4.14 Regarding the boundary layer, which of the following statements is correct?

(a) The turbulent boundary layer has more kinetic energy than the laminar boundary layer.

(b) The turbulent boundary layer is thinner than the laminar boundary layer.

(c) The turbulent boundary layer generates less skin friction than the laminar boundary layer.

(d) The turbulent boundary layer separates more easily than the laminar boundary layer.

Q4.15 Which of the following statements is correct?

(a) The stagnation point moves downward on a wing profile as the angle of attack increases.

(b) The CP is the point on the leading edge of a wing where the airflow splits up.

(c) The stagnation point is another name for the CP.

(d) The stagnation point is always situated above the chordline, the CP is not.

Q4.16 The most important advantage of the turbulent boundary layer over the laminar boundary layer is:

(a) it has less tendency to separate from the surface of the aerofoil

(b) it is thinner

(c) there is less skin friction

(d) there is less energy

Q4.17 Regarding the laminar boundary layer and the turbulent boundary layer, which of the following statements is correct?

(a) friction drag is less in the turbulent layer

(b) the separation point is earlier in the turbulent layer

(c) friction drag is less in the laminar layer

(d) friction drag will be equal in both boundary layers

Q4.18 With increasing angle of attack, the stagnation point moves (i) .... and the point of the lowest pressure moves (ii) ....

(a) (i) down; (ii) forward

(b) (i) up; (ii) aft

(c) (i) down; (ii) aft

(d) (i) up; (ii) forward

Q4.19 The aerodynamic centre of a wing is the point, where:

(a) the pitching moment remains constant irrespective of the angle of attack

(b) the change of lift due to the variation of the angle of attack is constant

(c) aerodynamic forces are constant at the CP

(d) the aeroplane's lateral axis intersects with the CG

Q4.20 If the streamlines of a subsonic airflow converge then at the point of convergence the static pressure will (i) .... and the velocity will (ii) ....
    (a) (i) decrease; (ii) increase
    (b) (i) increase; (ii) increase
    (c) (i) increase; (ii) decrease
    (d) (i) decrease; (ii) decrease

Q4.21 In which type of boundary layer does the largest change of velocity take place close to the surface?
    (a) they are the same
    (b) transition
    (c) turbulent
    (d) laminar

Q4.22 Where streamlines converge in a two-dimensional flow pattern the static pressure will:
    (a) not change
    (b) increase initially and then decrease
    (c) decrease
    (d) increase

Q4.23 On a symmetrical aerofoil, the pitching moment for which C_L is zero is:
    (a) zero
    (b) equal to the moment coefficient for the stabilised angle of attack
    (c) positive (pitch-up)
    (d) negative (pitch-down)

Q4.24 A laminar boundary layer is one in which:
    (a) no velocity components exist, normal to the surface of the aerofoil
    (b) the vortices are weak
    (c) the velocity is constant
    (d) the temperature varies constantly

Q4.25 When the angle of attack increases on an asymmetrical, single curve aerofoil in a subsonic airflow the CP will:
    (a) be unaffected
    (b) move forward
    (c) move aft
    (d) remain at the aerodynamic centre

Q4.26 A symmetrical aerofoil section at a C_L of zero will produce:
    (a) no aerodynamic force
    (b) a positive pitching moment
    (c) zero pitching moment
    (d) a negative pitching moment

Q4.27 Which of the following statements is true?
    (a) A turbulent boundary layer has more kinetic energy than one that is laminar.
    (b) A turbulent boundary layer is thinner than one that is laminar.
    (c) A turbulent boundary layer generates less skin friction than one that is laminar.
    (d) A turbulent boundary layer is more likely to separate.

Q4.28 The airflow that causes wing-tip vortices is most accurately described as:
    (a) flowing from root to tip on the upper surface and from tip to root along the lower surface over the wing tip
    (b) flowing from root to tip on the upper surface and from tip to root along the lower surface over the trailing edge
    (c) flowing from tip to root on the upper surface and from root to tip along the lower surface over the trailing edge
    (d) flowing from tip to root on the upper surface and from root to tip along the lower surface over the wing tip

Q4.29 Wing-tip vortices are caused by:
  (a) air spilling from the upper surface to the lower surface at the wing tip
  (b) air spilling from the lower surface to the upper surface at the wing tip
  (c) air spilling from the lower surface to the upper surface on the port and vice versa on the starboard
  (d) the spanwise flow vector from the wing tip to the root on the lower surface

Q4.30 The unequal pressure distribution around a wing causes wing-tip vortices that result from air flowing from (i) .... surface to (ii) .... surface around the (iii) ....
  (a) (i) lower; (ii) upper; (iii) trailing edge
  (b) (i) upper; (ii) lower; (iii) trailing edge
  (c) (i) lower; (ii) upper; (iii) wing tip
  (d) (i) upper; (ii) lower; (iii) wing tip

Q4.31 Laminar flow has:
  (a) more surface friction than a turbulent flow
  (b) the same surface friction as that of a turbulent flow
  (c) less surface friction than a turbulent flow
  (d) a more forward transition point than a turbulent flow

Q4.32 The advantage of a turbulent boundary layer over a laminar flow boundary layer is that:
  (a) the energy is less
  (b) it is thinner
  (c) it has greater skin friction
  (d) there is less tendency for it to separate

Q4.33 The speed of the airflow at the stagnation point compared to the speed of the airflow at the CP and the speed of the streamline airflow is that:
  (a) The speed at the stagnation point is less than the speed at the CP, which is less than the streamline flow.
  (b) The speed at the stagnation point is zero and the speed at the CP is equal to the streamline flow.
  (c) The speed at the stagnation point is greater than the speed at the CP, which is less than the speed of the streamline flow.
  (d) The speed at the stagnation point is zero and the speed at the CP is greater than the speed of the streamline flow.

Q4.34 Which of the following statements regarding the properties of the laminar boundary layer (i) and turbulent boundary layer (ii) is correct?
  (a) The surface friction for (i) and (ii) are the same.
  (b) The surface friction of (i) is greater than that of (ii).
  (c) The surface friction of (ii) is greater than (i).
  (d) The separation point of (ii) is more forward than that of (i).

Q4.35 Which statement is correct regarding C_L and the angle of attack?
  (a) for an asymmetric aerofoil with positive camber, if the angle of attack is greater than 0, $C_L = 0$
  (b) for symmetric aerofoil, if the angle of attack $= 0$, $C_L = 0$
  (c) for a symmetric aerofoil, if the angle of attack $= 0$, C_L is not equal to 0
  (d) for an asymmetric aerofoil if the angle of attack is $= 0$, then $C_L = 0$

Q4.36 The cause of an induced angle of attack of the tailplane is:
  (a) the downwash from the trailing edge of a wing in the vicinity of the tips
  (b) the change of airflow from the effective angle of attack
  (c) the downwash changing the angle at which the airflow reaches the tailplane
  (d) the upward slope of the airstream at the wing tips

Part 3
Level-flight Aerodynamics

5 Lift Analysis

5.1 The Four Forces

In flight an aeroplane is subject to four types of stress and four forces. The types of stress are compression, tension, shearing and torsional, and these are described in detail in later chapters. The four forces are mass acting vertically downward, drag opposing the forward movement of the aeroplane acting parallel to the relative airflow, thrust acting in a forward direction and lift the upward force acting vertical to the relative airflow that enables the aeroplane to remain airborne. These forces must be in equilibrium during flight. Lift must balance mass and thrust must balance drag. These forces are depicted in Figure 5.1.

5.2 Mass

The downward force exerted on a body by gravity is referred to as its weight. The strength of gravity varies with the position on the earth's surface and with the altitude of the body above the surface of the earth. Hence, the weight of a body varies with its location; this variation is so small that it can be safely ignored. (See Chapter 2). The mass and disposition of individual items (such as the role equipment, the traffic load, the fuel and the crew) carried in or on an aeroplane will affect the balance of the aeroplane. The total of these masses and that of the empty aeroplane is referred to as the gross mass of the aeroplane.

Gravity is assumed to act as a single downward vertical force through a point on the longitudinal axis referred to as the centre of gravity (CG). The CG of an aeroplane is not a fixed point. Its position at any instant is dependent on the location and the mass of individual items within the aeroplane. The CG must always be kept forward of the aerodynamic centre to maintain longitudinal stability and within the safe limits determined by the manufacturers and approved by the licensing authority.

In level flight, lift and mass are directly opposing forces and must balance each other if level flight is to be maintained. If the mass changes, for instance by using fuel, then the lift must be altered to ensure this state of equilibrium is maintained. As fuel is used during flight, the mass of the aeroplane is continually decreasing and if the IAS remains constant then the lift produced will exceed that required to counteract the mass and the aeroplane will slowly climb. This is referred to as a cruise climb. If, however, it is required to maintain level flight then the angle of attack must be reduced and because of this the speed will increase. Should acceleration in level flight not be required then the thrust and IAS must be reduced as the aeroplane mass decreases.

The IAS of an aeroplane in level, unaccelerated flight varies in direct proportion to the square root of the aircraft mass. Because of this, for a given angle of attack the lower the gross mass the lower is

The Principles of Flight for Pilots   P. J. Swatton
© 2011 John Wiley & Sons, Ltd

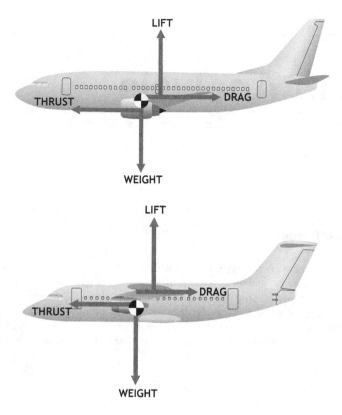

Figure 5.1 The Four Forces in Level Flight.

the IAS. This means the stalling angle, the angle of attack at which the lift generated is insufficient to balance the aeroplane mass and sustain level flight, will be reached at a lower IAS as the mass decreases.

At the optimum angle of attack, a heavy aeroplane mass requires a greater amount of lift, to sustain level flight, than that required by a lighter aeroplane. This has to be generated by increasing the forward speed of the aeroplane; otherwise, the aeroplane will descend. To attain the higher speed the thrust has to be increased, which increases the fuel consumption, resulting in a decreased maximum range and maximum endurance. The increased thrust requirement also decreases the excess thrust available for further acceleration or climbing. Thus, *a high gross mass requires a high IAS, which needs more thrust and causes the fuel flow to increase. Consequently, the maximum range and the maximum endurance decrease.*

5.3 Lift Analysis

Lift is defined as that component of the total aerodynamic force perpendicular to the undisturbed airflow and is generated by the forward movement of the aeroplane through the air. It is mostly produced by the downward acceleration of the air over the wing towards the trailing edge, the downwash, which causes low pressure above the wing (sometimes referred to as the suction over the wing). A small proportion of the total lift is produced in a similar manner by the tailplane. The total lift is defined as the component of the total aerodynamic reaction to the total pressure distribution of an aerofoil that is perpendicular to the relative airflow (flight path) and is assumed to be effectively concentrated through the centre of pressure (CP). It is equal to the amount of air accelerated downward multiplied by the vertical velocity of the air.

The amount of lift generated by an aerofoil is proportional to the magnitude of the negative-pressure area over the upper surface of the aerofoil caused by the downward acceleration of the air over that surface. The greater the angle of attack the larger is the area of negative pressure or suction. *The greatest amount of lift is generated at the stalling angle, at idle thrust, with a forward CG because this increases the size of the negative-pressure area.*

The total reactive force, and therefore the magnitude of the generated lift, is dependent on the following variables:

a. the square of the free airstream speed (V^2);
b. the density of the atmosphere (ρ);
c. wing area (S);
d. coefficient of lift (C_L).

The value of the total lift can be calculated by the formula:

$$\text{Total Lift} = C_L{}^1/_2\rho V^2 S$$

where: C_L = the coefficient of lift that is a mathematical factor derived from wind-tunnel tests and is the ratio of the lift pressure to the dynamic pressure. V = the free airspeed. S = the wing area.

To maintain level, unaccelerated flight this formula must remain in balance. The total lift must balance the mass. The wing area may be assumed to be constant for all conventional aeroplanes in level flight. However, some flap systems increase the wing area when extended, which affects the total lift generated. Total lift is directly proportional to the wing area. If the wing area doubles then the total lift will double.

The part of the formula $^1/_2\rho V^2$ is the dynamic pressure of the air 'q', i.e. pitot pressure minus static pressure. It is, therefore, the IAS and does not vary with changes of altitude. However, because the airspeed indicator measures 'q' and is square speed compensated, if the IAS doubles then 'q' quadruples. Example (1) IAS × 2 = 'q' × 4; (2) IAS × 3 = 'q' × 9.

If, at a constant altitude the total lift remains constant and the speed changes then the C_L must change to maintain the necessary balance. If, for instance, the speed doubles, as in example (1) above, then V^2 is quadrupled (i.e. is multiplied by 4) and the original C_L must be divided by 4 to determine the new C_L. Thus, it is 25% of its original worth. In example (2) if the speed triples, then V^2 is 9 times its original value (i.e. is multiplied by 9) then the original C_L must be divided by 9 to determine the new C_L. Thus, the revised value of C_L is equal to 100 divided by 9, i.e. 11.1% of its original worth.

If the altitude remains constant then '$^1/_2\rho$' does not vary and if the wing area does not change then 'S' also remains constant therefore the total lift must vary in proportion to C_L multiplied by V^2, i.e. Total Lift α $C_L V^2$. So, if the speed doubles the total lift increases by a factor of 4. If the total lift is that at the stalling angle then the coefficient is C_{Lmax} and the speed is the stalling speed and is therefore Vs^2.

$$\text{Maximum Total Lift} = C_{Lmax}{}^1/_2\rho Vs^2 S$$
$$\text{Maximum Total Lift} \ \alpha \ C_{Lmax}Vs^2$$

Example 5.1

If the revised speed of an aeroplane is 1.4 times its original speed determine the revised coefficient of lift, if the total lift remains constant.

Solution 5.1
Revised $V^2 = 1.4 \times 1.4 = 1.96$. Revised $C_L = 100 \div 1.96 = 51\%$ of its original value.

Example 5.2

When an aeroplane is flying at 1.3Vs express the C_L as a percentage of C_{Lmax}.

Solution 5.2.
By transposition of the formula:

$$C_{Lmax} = \frac{\text{Total Lift}}{Vs^2} \quad \text{and} \quad C_L = \frac{\text{Total Lift}}{(1.3Vs)^2} = \frac{\text{Total Lift}}{1.69Vs^2}$$

$$\text{Expressing } C_L \text{ as a percentage of } C_{Lmax} = \frac{\text{Total Lift}}{1.3Vs^2} \times \frac{Vs^2}{\text{Total Lift}} \times 100$$

$$= \frac{1}{1.69} \times 100 = 59.17\%.$$

Therefore, to express C_L as a percentage of C_{Lmax} divide 1 by the square of the multiplicand of Vs and multiply the result by 100.

$$[C_L/C_{Lmax}]\% = [1/\text{Multiplicand}^2] \times 100$$

5.4 The Factors Affecting C_L

The value of the coefficient of lift (C_L) is obtained experimentally at a quoted Reynolds Number (Re) (See Chapter 1) and is dependent on:

a. the angle of attack of the mainplane;
b. the wing shape both cross-section and planform;
c. the aerofoil surface condition;
d. the Reynolds number;
e. the local speed of sound.

5.5 The Effect of Angle of Attack

The angle of attack affects the pressure distribution surrounding an aerofoil. At low angles, such as point A at 8° in Figure 5.2, the airflow is linear and follows the curvature of the surface over which it is flowing almost reaching the trailing edge of the aerofoil before it separates from the surface. This is the separation point (SP). At this point the airflow becomes turbulent over a narrow layer behind the aerofoil.

As the angle of attack increases, C_L increases and the SP advances forward along the upper surface of the aerofoil because the unfavourable pressure gradient increases the depth of the boundary layer. This adversely affects the lifting qualities of the aerofoil shown as point B at 12° in Figure 5.2. At the stalling angle, point C at 16° in Figure 5.2, the SP has moved to its most forward position and the boundary layer is so deep that the airflow is only able to follow the aerofoil upper surface for a very short distance, this is accompanied by a large increase of pressure under the lower surface of the aerofoil, which consequently produces a sudden loss of lift.

The results of tests made using the same Reynolds number and a symmetrical cross-section aerofoil of moderate thickness when plotted against the angle of attack appear as the lift curve shown in Figure 5.2. The subdiagrams illustrate the turbulence for 8° (A), 12° (B) and 16° (C). The angle of attack at which the maximum C_L is attained for any configuration is the critical angle of attack.

5.6 The Effect of the Wing Shape

The wing shape may vary in four different ways and the effect that each of them has on the coefficient of lift is plotted against the appropriate angle of attack and is considered under the following headings:

a. Leading-edge radius;
b. Camber;
c. Aspect Ratio;
d. Wing Planform;
e. Sweepback.

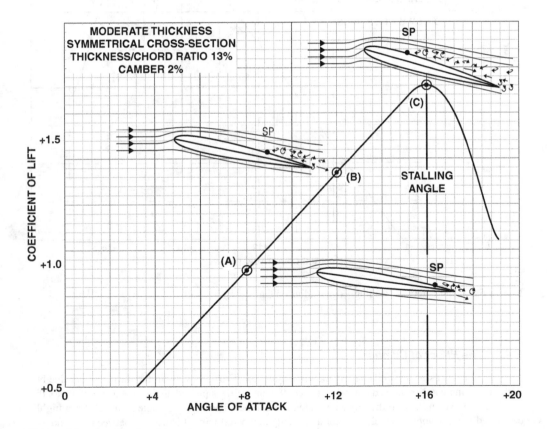

Figure 5.2 C_L v Angle of Attack.

5.6.1 The Effect of Leading-Edge Radius

The lift generated by wings of different leading-edge radii is the same up to an angle of attack of 15°. Beyond that, the radius of the aerofoil's leading edge and the condition of the surface largely determine the stalling characteristics of an aerofoil. The zero-lift condition for all radii symmetrical wings is at 0° angle of attack.

An aerofoil leading edge having a large radius such as the wing of a commercial transport aeroplane will result in the graph curve having a large rounded peak, indicating a slow transition to the stalled condition at approximately 18° with a C_{Lmax} of +1.75. However, an aerofoil having a small-radius leading edge such as a fighter aeroplane will produce an abrupt stall and is shown as a sudden peak to the graph line at a stalling angle of approximately 16° and a C_{Lmax} of +1.73.

This effect is illustrated in Figure 5.3 on which A represents an aerofoil with a large radius leading edge and B a small leading-edge radius. These results may be considerably modified if the surface of the aerofoil is rough, dirty or ice covered and is discussed in more detail later in this chapter.

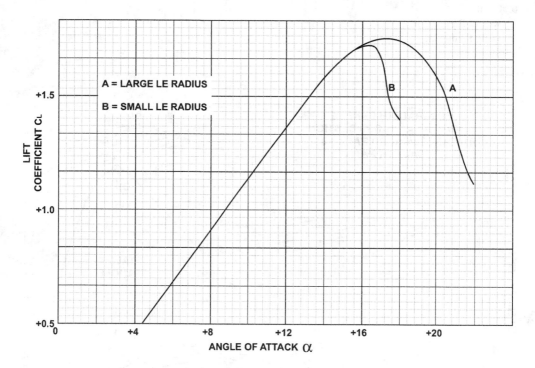

Figure 5.3   The Effect of Leading-Edge Radius on C_L.

5.6.2 The Effect of Camber

A series of trials using aerofoils having the same thickness distribution about the mean camber line but with increasing camber showed that when the coefficient of lift is plotted against the angle of attack the graph line moves up and to the left with increasing camber. The zero-lift angle of attack for a highly cambered wing is approximately –9°, for a normally cambered wing it is usually between –3° and –4° and for a noncambered symmetrical wing it is 0°.

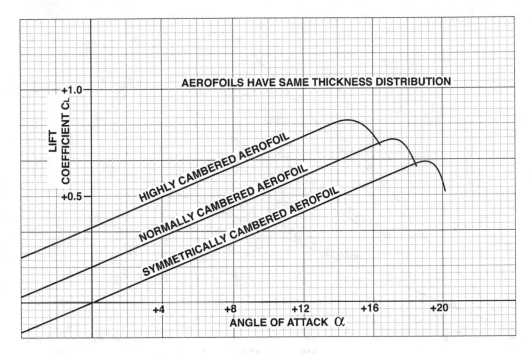

Figure 5.4 The Effect of Camber on C_L.

A cambered wing develops more lift than a symmetrical aerofoil, therefore its C_L curve will be parallel to the symmetrical aerofoil curve but positioned higher up the vertical axis. The maximum lift is developed at the stalling angle, which is:

a. C_L +0.88 at 14.5° for a highly cambered wing;
b. C_L +0.77 at 17° for a normally cambered wing;
c. C_L +0.65 at 19° for a noncambered symmetrical wing. See Figure 5.4.

5.6.3 The Effect of Aspect Ratio

The trailing-edge vortices of an aerofoil cause a downward component or downwash to be induced in the airflow over a wing. This reduces the effective angle of attack. The wing-tip vortices on a low aspect ratio wing are large and strong. They, therefore, cause a large downwash, resulting in a considerably decreased effective angle of attack and a large reduction of lift.

Conversely, a high aspect ratio wing has smaller wing-tip vortices resulting in a relatively small downwash and a smaller reduction in the effective angle of attack that enables it to produce a greater amount of lift than the low aspect ratio wing. When the lift for various aspect ratio wings is plotted against the angle of attack it appears as in Figure 5.5, all example aerofoils have a zero lift angle of attack of −4°.

The maximum lift is developed at the stalling angle as shown in Figure 5.5 is:

a. C_L +1.43 at 13.0° for a high aspect ratio wing;
b. C_L +1.18 at 16.8° for a medium aspect ratio wing;
c. C_L +0.90 at 23.6° for a low aspect ratio wing.

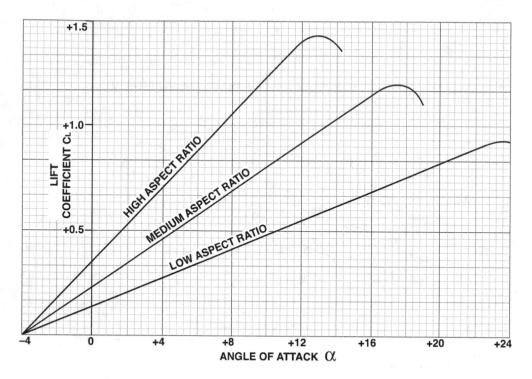

Figure 5.5   The Effect of Aspect Ratio on C$_L$.

5.6.4 The Wing Planform

There are many different wing shapes but those commonly used are the rectangular, the tapered and the elliptical. Of these shapes the one that produces the greatest amount of lift at the wing root is the rectangular. (See page 71).

5.6.4.1 The Effect of Sweepback

The amount of lift generated by a wing is directly proportional to the speed of the airflow passing at right angles over the leading edge of the wing. Because the direction of the airflow over a swept wing is not at right angles to the leading edge the effective camber is reduced and the fineness ratio is increased. The component of the airflow normal to the leading edge of the wing for a swept-wing aeroplane is equal to the speed of the airflow multiplied by the cosine of the sweep angle. Therefore, the lift produced by such a wing is less than that of a straight wing having the same area, aspect ratio, taper and washout at the same angle of attack.

   As depicted in Figure 5.6 the straight wing produces its maximum lift, which is much greater than that of the swept wing, at a much lower stalling angle than that of the swept wing. An amount equal to the cosine of 30° reduces the C$_{Lmax}$ for a sweep angle of 30°, which is approximately 13.4% less than that of the straight-winged aeroplane and 29.3% less for a sweep back of 45°. *The greater the angle of sweepback angle the less lift is produced by the wings.* (See page 364).

   The aspect ratio of a wing is equal to the wingspan squared divided by the wing area. If this calculation is done for a swept-back wing, the aspect ratio so determined is less than the aspect ratio of a straight wing having the same wing area. Thus, the effect on C$_L$ is similar to that of a low aspect ratio wing when

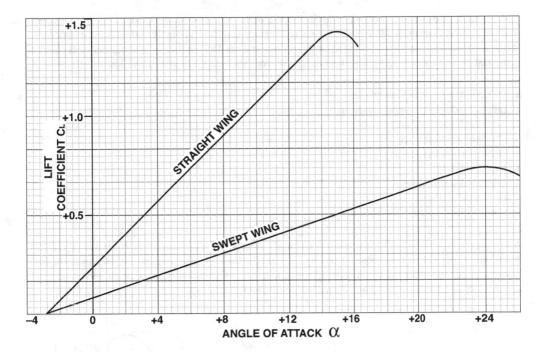

Figure 5.6   The Effect of Sweep Back on C_L.

compared to a high aspect ratio wing; the swept wing being the equivalent of the low aspect ratio wing. This is shown in Figure 5.6.

The air passing over some parts of a swept wing becomes supersonic; this causes a large rise in drag due to compressibility. This disadvantage is offset by the major advantage of a swept wing that it enables an aircraft to fly up to a higher maximum speed before the speed becomes critical.

Both example aerofoils in Figure 5.6 have a zero lift angle of attack of −2.8°. The maximum C_L and angles of attack for each type of wing are:

a. Straight wing +1.42 at 14.4°;
b. Swept wing +0.73 at 24.0°.

5.7   The Effect of Airframe-Surface Condition

The roughness, cleanliness and deposits of ice or moisture all affect the condition of the surface of an aerofoil. However, for this consideration the area of greatest significance with regard to its effect on the C_L of an aerofoil is from the leading edge and up to 20% of the chord aft of the leading edge. Beyond this point the condition of the surface has little effect on the C_L. See Figure 5.7.

The surface condition affects the depth of the boundary layer; the rougher the surface the deeper the boundary layer. This in turn affects the amount of parasite drag experienced by the aerofoil increasing with roughness of the surface.

Both a rough and a smooth surface have the same effect on the value of the C_L up to an angle of attack of approximately 10°. With a rough surface the maximum C_L is attained at an angle of attack of approximately 12.1° and with a smooth surface the maximum C_L is attained at an angle of attack of approximately 14.8°.

Ice and frost deposits always adversely affect the performance of an aeroplane. This is particularly so if the ice forms on the leading edge of the aerofoil at the stagnation point. Its effect is most noticeable at low altitudes and low airspeeds in the landing configuration. During take-off the most critical effect is at the last part of rotation just before becoming airborne. Such deposits have the following effects:

a. increased mass of the aeroplane;
b. increased depth of the boundary layer;
c. increased surface friction;
d. changed aerofoil shape;
e. decreased maximum lift;
f. increased drag;
g. increased stalling speed;
h. decreased stalling angle;
i. possible leading-edge separation.

A frost-covered aerofoil has the same effect on the C_L as a rough surfaced aerofoil. It has a similar effect on aeroplane performance to those of ice formation on the leading edge listed above. However, the loss of lift experienced in a frost-covered aeroplane at take-off is significant because of the increased mass and the increased stalling speed and decreased stalling angle.

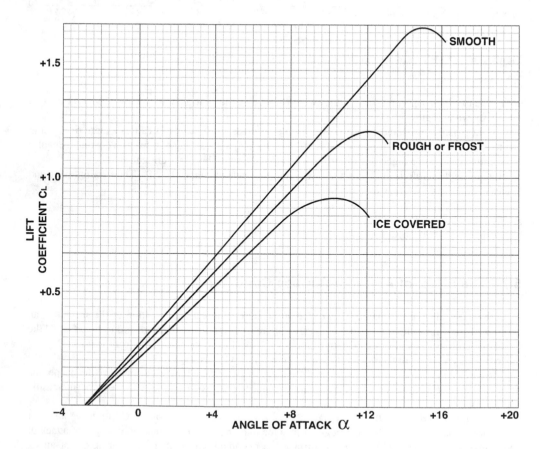

Figure 5.7   The Effect of Surface Condition on C_L.

The zero lift angle of attack for these example aerofoils is −2.8° and the maximum C_L and angle of attack are:

a. smooth Leading edge +1.66 at 15.0°;
b. rough leading edge +1.2 at 12.0°;
c. ice-covered leading edge +0.9 at 10.2°.

5.8 The Effect of Reynolds Number

The ratio between the inertial and viscous (friction) forces is the Reynolds number. It is used to identify and predict different flow regimes such as laminar or turbulent flow. A small Reynolds number is one in which the viscous force is predominant and indicates a steady flow and smooth fluid motion. A large Reynolds number is one in which the inertial force is paramount and indicates random eddies and turbulent flow.

The value of the Reynolds number makes no difference to the coefficient of lift until an angle of attack of 9.75° is attained. The zero-lift angle of attack for all example curves is −2.8°. The maximum C_L and the angle of attack for each example are shown in Figure 5.8 and summarised in Table 5.1 and

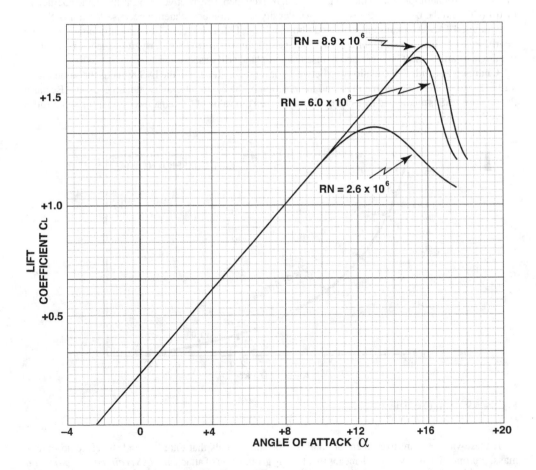

Figure 5.8   The Effect of Reynolds Number on C_L

are:

a. high Reynolds number +1.73 and 16°;
b. medium Reynolds number +1.66 and 15.4°;
c. low Reynolds number +1.36 and 13.2°.

5.9   The Relationship between Speeds, Angles of Attack and C_L

For a given wing area, if the altitude (i.e. air density) and mass remain constant then the remaining factors in the formula, the coefficient of lift (C_L) and the free air velocity (IAS), are variable but have a fixed relationship that is graphically illustrated in Figure 5.9. C_L and IAS must change in such a manner that the total lift formula remains balanced. *For every angle of attack, there is a corresponding IAS, coefficient of lift and coefficient of drag.* At the most efficient angle of attack of 4° the C_L is approximately 0.48 and the C_D is approximately 0.038. *If the IAS is decreased the C_L, the C_D and angle of attack all increase and vice versa.* However, the effect of compressibility reduces the value of the C_L appropriate to any given angle of attack.

Constant AoA. To maintain level flight at a given angle of attack, the IAS is dictated by the aeroplane mass. In the total lift formula, if the angle of attack is fixed then, irrespective of the altitude, the coefficient of lift (C_L) and the IAS (V) are also fixed. Air density is of little consequence because V is an IAS.

Changed IAS. At a constant mass and altitude, if the IAS is increased, the angle of attack must be decreased to maintain level flight otherwise the aeroplane will climb. Similarly, if the speed is reduced, the angle of attack must be increased or else the aeroplane will descend.

Changed Altitude. The relationship between the angle of attack and the IAS remains the same for all altitudes provided the aeroplane mass and the position of the centre of gravity remain constant. Because the value of IAS is unaffected by a changed altitude then the value of C_L remains constant for all altitudes as well.

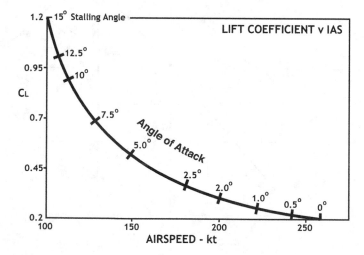

Figure 5.9   Lift Coefficient v IAS.

If these variables are fixed then the total lift formula reveals that total lift, and therefore aeroplane mass, is proportional to V^2, which means that for any given angle of attack, *the IAS is directly proportional to the square root of the mass of the aeroplane.*

Table 5.1   Summary of C_{Lmax} and Angles of Attack.

Variable Factor	Approximate Values		
	Zero Lift Angle of Attack	Stalling Angle of Attack	Maximum C_L
Large LE radius	0°	18.0°	+1.75
Small LE Radius	0°	16.0°	+1.73
Highly Cambered	−9°	14.5°	+0.88
Normal Camber	−4°	17.0°	+0.77
Noncambered	0°	19.0°	+0.65
High Aspect Ratio	−4°	13.0°	+1.43
Medium Aspect Ratio	−4°	16.8°	+1.18
Low Aspect Ratio	−4°	23.6°	+0.9
Straight Wing	−2.8°	14.4°	+1.42
Swept Wing	−2.8°	24.0°	+0.73
Smooth Leading Edge	−2.8°	14.8°	+1.66
Rough Leading Edge	−2.8°	12.1°	+1.2
Ice-Covered Leading Edge	−2.8°	10.0°	+0.9
High Reynolds Number	−2.8°	16.0°	+1.73
Medium Reynolds Number	−2.8°	15.4°	+1.66
Low Reynolds Number	−2.8°	13.2°	+1.36

When flying at a constant Mach number and mass, an increase of altitude will decrease the IAS and will necessitate an increase in the angle of attack to maintain level flight. See Figure 2.5.

5.10   Aerofoil Profiles

The external contour or cross-section of an aerofoil determines its performance. There are three classes of aerofoil as illustrated in Figure 5.10:

a. high-lift;
b. general purpose;
c. high-speed.

5.10.1   High-Lift Aerofoils

The cross-section of a high-lift aerofoil is shown in Figure 5.10(a). These are used mainly for sailplanes and other aircraft where lift is of primary importance and speed is of secondary importance. These aerofoils have a:

a. thick, rounded leading edge;
b. high thickness to chord ratio;
c. pronounced camber;
d. maximum thickness at between 25% and 30% of the chord from the leading edge.

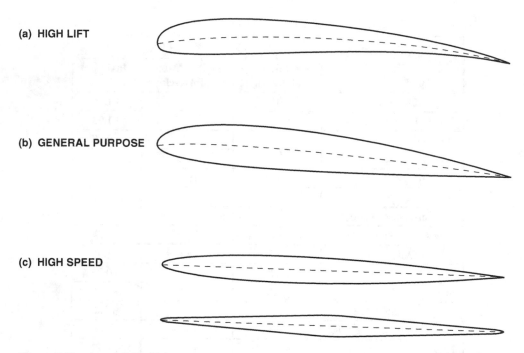

Figure 5.10   Aerofoil Profiles.

The more curved the mean camber line; the greater is the movement of the CP for a given change to the angle of attack. The range of movement of the CP on this type of wing is large, which can be minimised by curving the trailing edge of the wing upward at the expense of a small loss of lift.

5.10.2   General-Purpose Aerofoils

The general-purpose aerofoil is used for aircraft with a maximum TAS of less than 300 kt and is shown in Figure 5.10(b). This aerofoil has less camber, a small leading-edge radius and a lower thickness to chord ratio than the high-lift aerofoil.

The maximum thickness to chord ratio occurs between 25% and 30% of the chordline aft of the leading edge of the aerofoil. Because of lower thickness to chord ratio there is less drag and a lower C_{Lmax} than the high-lift aerofoil.

5.10.3   High-Speed Aerofoils

The high-speed aerofoils have a cross-section that is equally disposed either side of the mean chordline. Some are wedge-shaped, while others are arcs of a circle that are symmetrical about the chordline. Primarily they induce little drag and are in the 5% to 10% thickness to chord ratio band. They are illustrated in Figure 5.10(c).

The high-speed aerofoil cross-section has an extremely low thickness to chord ratio, no camber and a sharp leading edge. The maximum thickness is at 50% of the chordline.

Self-Assessment Exercise 5

Q5.1 On an aerofoil the forces of lift and drag are vertical and parallel respectively to the:
(a) longitudinal axis
(b) horizontal
(c) relative airflow
(d) chordline

Q5.2 Which condition produces the greatest lift in straight and level flight?
(a) Aft CG and idle thrust
(b) Forward CG and take-off thrust
(c) Aft CG and take-off thrust
(d) Forward CG and idle thrust

Q5.3 The lift force acting on an aerofoil:
(a) is mainly caused by increased pressure beneath the aerofoil
(b) is the maximum at an angle of attack of $4°$
(c) is mainly caused by suction on the upperside of the aerofoil
(d) is inversely proportional to the angle of attack up to $40°$

Q5.4 The lift formula is:
(a) $L = nM$
(b) $L = C_L{}^1/_2 \rho V^2 S$
(c) $L = M$
(d) $L = C_L2 \, \rho V^2 S$

Q5.5 Drag is in the direction of and lift is perpendicular to ....
(a) the chordline
(b) the horizontal axis
(c) the longitudinal axis
(d) the relative airflow

Q5.6 When an aeroplane is flying at $1.3V_s$, the C_L as a percentage of C_{Lmax} is:
(a) 59%
(b) 141%
(c) 169%
(d) 85%

Q5.7 If the IAS of an aeroplane in level flight doubles but the total lift is unchanged then C_L is multiplied by:
(a) 0.5
(b) 8.0
(c) 0.25
(d) 2.0

Q5.8 On a C_L v angle of attack graph the curve for a positive cambered aerofoil intersects the vertical axis of the graph at:
(a) the origin
(b) below the origin
(c) no point
(d) above the origin

Q5.9 Which of the following wing planforms produces the greatest amount of lift at the wing root?
(a) Rectangular
(b) Elliptical
(c) Tapered
(d) Positive angle of sweep

Q5.10 Lift is generated when:
    (a) a mass of air is accelerated downward
    (b) the shape of the aerofoil is slightly cambered
    (c) an aerofoil is placed in a high velocity airstream
    (d) a mass of air is retarded

Q5.11 The point at which the aerodynamic lift acts on a wing is:
    (a) the TP
    (b) the CP
    (c) the CG
    (d) the point of maximum wing chord

Q5.12 The terms 'q' and 'S' in the lift formula are:
    (a) static pressure and dynamic pressure
    (b) dynamic pressure and the wing area
    (c) square root of the surface and wing loading
    (d) static pressure and wing surface area

Q5.13 Which of the following creates lift?
    (a) A slightly cambered aerofoil
    (b) the boundary layer
    (c) upwash
    (d) Air accelerated downward

Q5.14 The factor that contributes most to the generation of lift is:
    (a) increased pressure below the wing
    (b) increased airflow velocity below the wing
    (c) suction above the wing
    (d) decreased airflow velocity above the wing

Q5.15 Which of the following statements is correct?
    (a) Lift acts perpendicular and drag horizontally in a rearward direction
    (b) Drag acts parallel to the chordline and in the opposite direction to the movement of the aeroplane and lift acts perpendicular to the chordline
    (c) Lift acts at right angles to the upper surface of the wing and drag acts at right angles to the lift
    (d) Drag acts in the same direction as the relative wind and lift is perpendicular to it

Q5.16 'S' and 'q' in the lift equation represent:
    (a) static pressure and the chord
    (b) wing span and dynamic pressure
    (c) wing area and dynamic pressure
    (d) wing area and static pressure

Q5.17 The position at which it is assumed that the total lift acts is the:
    (a) aerodynamic centre
    (b) CG
    (c) neutral point
    (d) CP

Q5.18 On an angle of attack v C_L graph for a positive-cambered aerofoil, the lift curve intersects the vertical C_L axis at a position:
    (a) above the origin
    (b) below the origin
    (c) at the point of origin
    (d) to the left of the origin

Q5.19 The wing shape that produces most of its lift at the wing root is:
    (a) swept
    (b) rectangular
    (c) elliptical
    (d) tapered

Q5.20 The wing shape that produces the highest local lift coefficient at the wing root is:
  (a) elliptical
  (b) rectangular
  (c) positive sweep angle
  (d) tapered

Q5.21 Mass acts:
  (a) perpendicular to the longitudinal axis
  (b) parallel to the gravitational force
  (c) perpendicular to the chordline
  (d) perpendicular to the relative airflow

Q5.22 C_L can be increased by either the extension of flap or by:
  (a) increasing the angle of attack
  (b) increasing the TAS
  (c) decreasing the 'nose-up' trim
  (d) increasing the CAS

Q5.23 A higher altitude at a constant mass and Mach number requires:
  (a) a lower C_D
  (b) a lower angle of attack
  (c) a higher angle of attack
  (d) a lower C_L

Q5.24 An aircraft increases its speed by 30% in straight and level flight. If the total lift remains constant determine the revised C_L as a percentage of its original value is:
  (a) 77%
  (b) 88%
  (c) 114%
  (d) 59%

6 Lift Augmentation

6.1 Wing Loading

The wing loading of an aeroplane determines the magnitude of unstick speed (V_{US}) during take-off, touchdown speed during landing and the stalling speed. It is defined as the aeroplane mass divided by the wing area and is specified in Newtons per square metre (N/m^2). A high wing loading is the result of a heavy mass or small wing, and is undesirable because it has the following effects:

a. increased take-off and landing speeds;
b. longer take-off ground run and take-off distance;
c. longer landing ground run and landing distance;
d. increased stalling speed;
e. increased glide angle;
f. decreased C_{Lmax};
g. decreased turbulence sensitivity.

The value of the wing loading can be determined from the total lift formula when it has been transposed:

(1) **Total Lift = $C_L.^1/_2\rho.V^2.S$**

(2) **Total Lift = Total Mass**

(3) **Total Mass = $C_L.^1/_2\rho.V^2.S$**

(4) $\dfrac{\textbf{Total Mass}}{\textbf{Wing Area}} = \textbf{C}_\textbf{L}.^1/_2\boldsymbol{\rho}.\textbf{V}^2$

(5) **Wing Loading = Coefficient of Lift × Dynamic Pressure**

From the formula it can be seen that if the coefficient of lift can be increased then the speed does not need to be as high for the same wing loading. This is of great benefit during take-off and landing.

6.2 C_{Lmax} Augmentation

There are four main types of device that may be used to improve the value of C_{Lmax}. They are:

a. slats;
b. built-in slots;

The Principles of Flight for Pilots   P. J. Swatton
© 2011 John Wiley & Sons, Ltd

c. flaps either leading edge or trailing edge;
d. boundary-layer control.

6.3 Slats

A slat is a small highly cambered section attached along a portion of the leading edge of each wing near the wingtip on a swept wing. Its purpose is to increase the critical angle of attack thus enabling a higher angle of attack to be maintained by the aeroplane without stalling. It may operate either automatically or manually. When extended, the slat introduces a small gap between itself and the leading edge of the wing causing a Venturi effect through which the airflow accelerates adding to the kinetic energy and re-energizing the boundary layer. This enables the boundary layer to travel further back along the upper surface of the wing against the adverse pressure gradient before separating from that surface. When used at high angles of attack it smoothes the airflow over the leading edge of the wing, flattens the normal pressure-distribution envelope over the upper surface of the aerofoil and increases the camber of the wing.

Extended slats can increase the C_{Lmax} by as much as 70% whilst simultaneously increasing the stalling angle by up to 10° and decreasing the stalling speed by approximately 20 kt. Thus, when they are used there is a large decrease in stalling speed with little increase in drag. This is shown in Figure 6.1. With slats retracted the stalling angle is 16° and the C_{Lmax} is +1.35 but when they are extended the stalling angle is increased to 26° and the C_{Lmax} is increased to 2.25. Extended slats, unlike trailing-edge flaps,

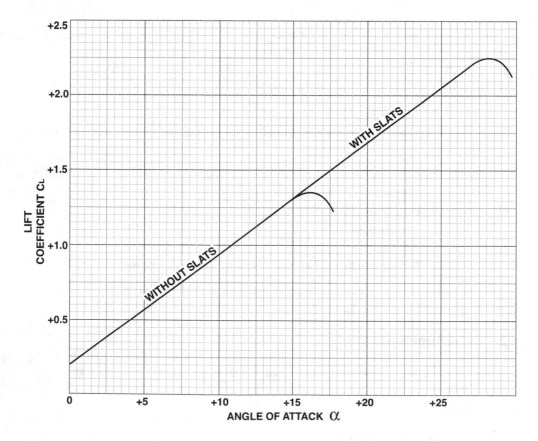

Figure 6.1   The Effect of Slats on C_L.

have no effect on the climb gradient because there is little increase of drag and are therefore retracted after the flaps during the 'clean-up' phase of the climb after take-off.

The magnitude of the increase to C_{Lmax} and the extent of the decrease to the stalling speed are directly proportional to the length of the leading edge of the aerofoil covered by the slat and the chord of the slat itself, i.e. the longer the slat the greater the C_{Lmax} and the lower the stalling speed. The most efficient position for slats to be mounted on a swept-wing aeroplane is outboard of the wing-mounted engines to just inboard of the wing tips.

When used they effectively delay the stall to a greater angle of attack by prolonging the lift curve. At high angles of attack the slat generates a high C_L because of its own camber. This changes the aerofoil negative-pressure-distribution envelope over the upper surface of the wing from a steep marked peak shape to a more rounded shape that spreads the low-pressure area over the whole upper surface of the aerofoil more evenly and moves the peak pressure forward towards the leading edge of the wing. See Figure 6.2 in which the normal stalling angle for this aerofoil is 14° that is increased to 24° when the slats are extended.

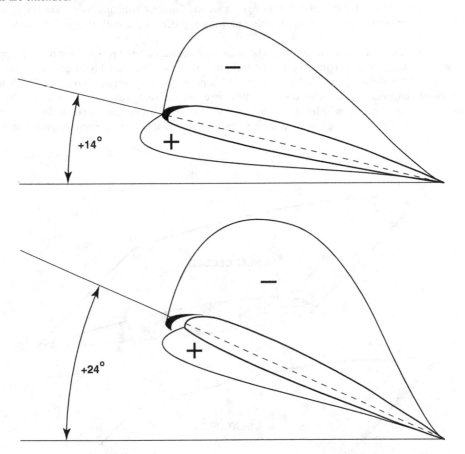

Figure 6.2   The Effect of Slats on Pressure Distribution.

6.3.1 *Automatic Slats*

At normal operating angles of attack a slat is of little use but if it is not flush with the leading edge of the wing it will increase the drag. To minimize this disadvantage the slat is mounted on hinged, movable supporting arms that are able to move between two positions, closed and operational. This type of slat is fully automatic and does not require a separate control. Some slats operate when the trailing-edge flaps setting is altered because it changes the position of the slats simultaneously.

The way in which automatic slats operate is that at normal operating angles of attack the positive air pressure at the stagnation point maintains the slat in the closed position. At high angles of attack, the reduced air pressure above the upper surface of the wing extends over the leading edge of the wing and the migration of the stagnation point under the wing to a point aft of the trailing edge of the slat, enables the spring-loaded slat to open and raise clear of the wing. This produces the required slot between the slat and the leading edge of the wing. The positive pressure beneath the wing accelerates the airflow passing through the slot and that re-energizes the airflow over the upper surface of the wing, in particular the boundary layer.

When extended the slat creates a false leading edge further back along the chordline so that the boundary layer growth starts further back along the upper surface of the wing. This eliminates the unfavourable effect of the initial adverse pressure gradient. Large slats are nearly always the automatic type but some smaller slats that cause negligible drag are fixed and are similar to slots. The slat system is capable of deploying asymmetrically if only one wing approaches the critical angle of attack. See Figure 6.3.

At the high angles of attack the extended slat produces an enlarged low-pressure area over the upper surface of the wing. At approximately 25° the low-pressure area is considerably larger than it would have been had the slat not been used and it generates a proportionate increase to the total lift. When the inevitable happens and the wing does stall this large low-pressure area collapses and the change of attitude of the aeroplane will be abrupt and feel quite violent. If one wing stalls before the other then the stall will be accompanied by a strong tendency to roll that can be very dangerous if it is not counteracted.

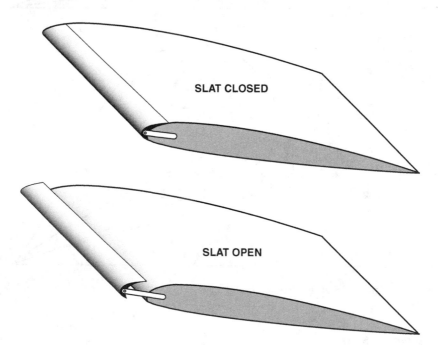

Figure 6.3   The Automatic Slat.

The design of automatic slats for high-performance swept-wing aircraft is modified to account the exaggerated stalling angle that they cause. High angles of attack are deemed unacceptable when landing these aircraft. The slats on these aircraft usually range along the whole length of the leading edge to prevent wing tip stalling, as well as to increase the C_{Lmax}. For these aeroplanes they improve control at low speeds because they operate at speeds much lower than the stalling speed. This feature improves C_L at moderate angles of attack and smoothes the airflow, moving the separation point further aft resulting in a stronger pressure distribution over the whole wing. This characteristic is more evident at high angles of attack.

6.3.2   Manual Slats

The operation of the slats may be under the control of the pilot who can extend them for take-off and landing if so desired. They may be electrically or hydraulically operated and although the pilot has control of this type of slat in normal circumstances, there is an overriding automatic system that will extend the slats if the angle of attack becomes dangerously high approaching the stall. This system is normally included in the design of aircraft that have to operate at low speed and low altitude, such as crop sprayers, because they often have to climb away steeply at very low speeds. Included in the design for such aeroplanes is an in-built safety protection mechanism that prevents asymmetric slat deployment. For an aeroplane fitted with both slats and flaps, after take-off it is customary to retract the flaps first because the slats when extended decrease the stalling speed by a large amount and produce very little drag. Therefore, it is prudent to leave their retraction until last, during the take-off climb.

6.4   Slots

A series of suitably shaped slots built in the wing tip just aft of the leading edge of the wing during manufacture, are referred to as slots. They are an alternative method of smoothing the airflow over the upper surface of the wing to that of the slat system. The mode of operation is similar to that of the slat, effectively creating a second leading edge for the wing and has the same previously stated benefits. During normal flight the slots are closed, however at high angles of attack the slots open to allow air from below the wing to be guided through the slots and to be discharged over the upper surface, re-energizing the boundary layer and increasing the amount of lift developed, thus delaying the stall to a higher angle of attack.

Some specialist low-speed aircraft have slats that are fixed in the extended position, these are also referred to as slots. Others have a long slot just aft of the leading edge that is permanently open. These are usually short-field take-off and landing (STOL) aircraft. See Figure 6.4.

The advantages of using slats or slots are that they delay the stall to a higher angle of attack, increase the C_{Lmax} and decrease the stalling speed. They enable the aeroplane to be operated safely at much lower speeds than would have been possible normally without their use. This feature is of great benefit during take-off and landing. The effect of using slats or slots on the magnitude of C_{Lmax} is shown in Figure 6.5. It can be seen that the stall occurs at a much higher angle of attack and the C_L at this point is greatly increased.

6.5   Leading-Edge Flaps

Leading-edge flaps when extended perform a similar function to that of slats except that they do not create a slot; instead they increase the camber of the wing. Because of this they do not increase the lift at all angles of attack and are not as effective as slats at delaying the stall. Nevertheless, they do, to a lesser extent, increase the critical angle of attack, increase the C_{Lmax} and decrease the nose-down

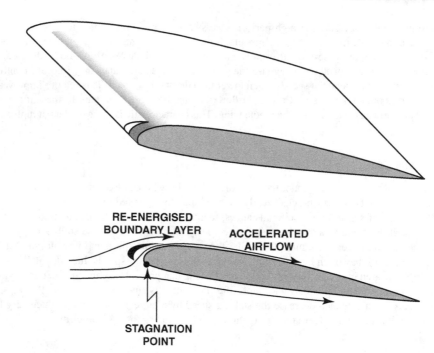

Figure 6.4   The Leading-Edge Slot.

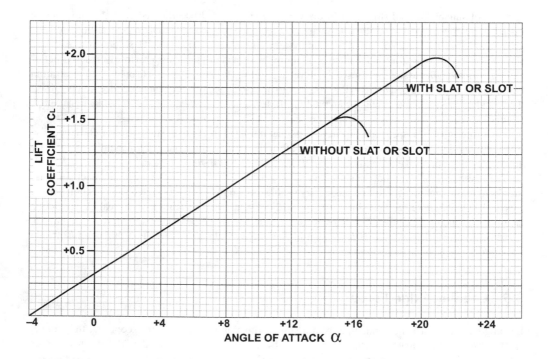

Figure 6.5   The Effect of Slats or Slots on C$_L$.

pitching moment, however, when approaching the stalling angle they have the disadvantage of reducing the stability of the wing. The most efficient position to mount leading-edge flaps on a swept-wing aeroplane is inboard of the wing-mounted engines at the wing roots. There are three types of leading-edge flap, the plain Krueger flap, the drooped leading-edge flap and variable camber Krueger flap. See Figure 6.6.

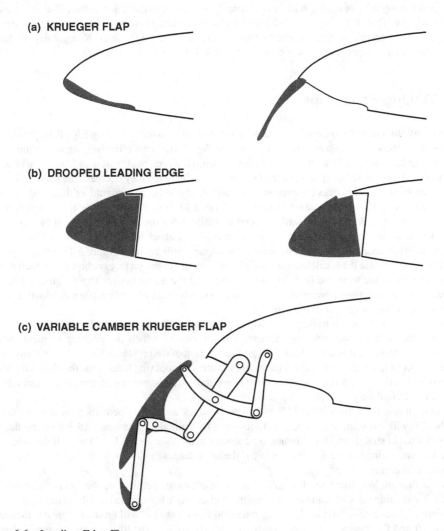

(a) KRUEGER FLAP

(b) DROOPED LEADING EDGE

(c) VARIABLE CAMBER KRUEGER FLAP

Figure 6.6 Leading-Edge Flaps.

6.5.1 The Krueger Flap

The plain Krueger flap is a small flap on the inboard, leading edge, and underside of a wing that is hinged at the leading edge of the wing and extends forward. With the variable camber Krueger flap the amount of downward extension can be selected and controlled. For small angular deflections it reduces the amount of lift produced but when fully extended it can increase the lift by approximately 50% and also increase the stalling angle up to 25°. The pitching moment can vary considerably during extension but when fully extended it will cause a nose-up pitching moment requiring considerable changes of trim.

6.5.2   The Drooped Leading Edge

A drooped leading edge of a wing achieves a similar result to that of the Krueger flap. The whole section of the leading edge of the wing can be depressed in a controlled manner to progressively increase the lift and the stalling angle. It does prevent the trim changes experienced with the Krueger flap but has the disadvantage that it requires extremely complex engineering to enable it to operate.

6.6   Trailing-Edge Flaps

An alternative to increasing the angle of attack to increase lift is to lower trailing-edge flaps provided the speed is at or below the maximum speed for lowering flap (V_{FO}). This effectively increases the camber of the wing, the angle of attack and the coefficient of lift. However, the thrust may have to be increased to overcome the increased drag, despite the fact that when deployed they decrease the magnitude of the wing-tip vortices. Not only does the extension of trailing-edge flaps decrease the critical angle of attack it also increases the C_{Lmax}, increases the total lift generated, increases the total drag, unfavourably affects the lift/drag ratio and decreases the stalling speed no matter what the altitude or mass of the aeroplane.

Unlike slats, trailing-edge flaps increase lift at all angles of attack up to the stall. Thus, if the angle of attack remains constant during flap extension the aeroplane will begin to climb. All trailing-edge flaps when lowered, increase the acceleration of the airflow over the upper surface of the wing, which reduces the pressure above the wing and increases the upwash over the leading edge. Together these influences generate an increased nose-down pitching moment as a result of the altered pressure distribution around the flaps and the aft movement of the wing CP.

However, when deployed trailing-edge flaps also increase the downwash over the tailplane, which causes an opposing nose-up pitching moment. The amount by which the pitching moment changes because of this phenomenon depends on the size and position of the tailplane. The resultant change to the pitching moment is determined by the relative sizes of the two opposing influences, the changed pressure distribution and the downwash. The dominant feature will establish what trim change is required when flaps are lowered usually it results in a pitch-down moment.

In straight and level flight if the IAS and angle of attack are maintained when the flap is extended then the CP will move aft and the C_L will increase. To maintain a constant IAS whilst the flaps are being retracted in straight and level flight it is necessary to increase the angle of attack. If the same angle of attack is maintained as the flaps are retracted the aeroplane will sink or when they are extended the aeroplane will climb.

As the flap angle is increased the critical angle of attack decreases and the C_{Lmax} increases. See Figure 6.7 and Table 6.1. Consequently, the minimum glide angle is increased and the resulting maximum glide distance is decreased. Typically, a flap extension from 0° to 20° will produce a greater increase to the total lift and C_L, than an increased extension from 20° to 30°. The lift/drag ratio is most adversely affected by the change from a 30° flap extension to a 40° flap extension because the increase to the total lift is far less than the increase to the total drag.

There are several different types of trailing flaps currently employed on modern aeroplanes. They include:

a. plain;
b. split;
c. slotted and double slotted;
d. Fowler;
e. combinations.

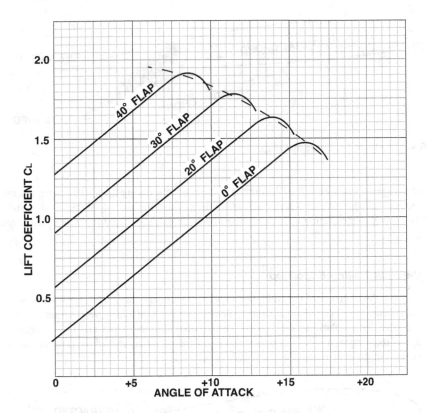

Figure 6.7   The Effect of Typical Trailing-Edge-Flap Angle on C_L.

Table 6.1   The Effect of a Typical Trailing-Edge-Flap Extension. **(As shown in Figure 6.7)**

Flap Angle	C_{Lmax}	Critical AoA
0°	+1.48	16°
20°	+1.63	14°
30°	+1.78	11.5°
40°	+1.92	8.5°

6.6.1   The Plain Trailing-Edge Flap

The plain flap is the simplest of all flap systems. It is a large part of the trailing edge of a wing that rotates on a hinge and can be extended in a controlled manner by the pilot. The use of this type of flap does not alter the wing area but it does increase the effective wing camber when it is lowered. The centre of pressure moves aft with increasing degrees of flap selection that also produces an increase of lift the amount of which is dependent on the flap angle selected; it also causes a nose-down pitching moment. The increase of C_{Lmax} a plain flap can produce is approximately 21% at a given IAS. See Figure 6.8.

The most significant advantage of this type of flap is that when it is fully extended it decreases the stalling angle to approximately 14° but C_{Lmax} is limited because of boundary-layer separation. Thus, for any given value of C_L less nose-up pitch is required at a lower IAS, which gives the pilot a better visual aspect of the runway during a slower approach to land.

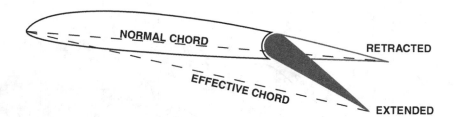

Figure 6.8   The Plain Trailing-Edge Flap.

The advantage experienced during landing is of little benefit during take-off because the large flap setting used for landing produces a large increase of drag. Thus, the amount of flap used for take-off is considerably less than for landing but still has the advantage of a significant increase of lift at relatively low airspeeds.

6.6.2   The Split Trailing-Edge Flap

The split flap is a development of the plain flap; they are no longer common but are fitted to some small twin-engined aeroplanes. Their main advantage is that they are structurally very strong and as a result they can be extended at relatively high speed. During operation the upper and lower surfaces of the trailing edge of the wing separate. The upper surface remains stationary in its fixed position but the lower surface operates in the same manner as the simple flap.

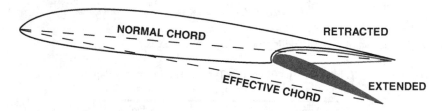

Figure 6.9   The Split Trailing-Edge Flap.

As with the plain flap the use of this type of flap does not alter the wing area but does increase the effective angle of attack when it is lowered. However, the loss of energy in the boundary layer leads to separation of the airflow at an early point over the upper surface of the aerofoil.

When operated the centre of pressure moves aft with increasing degrees of flap selection and causes a nose-down pitching moment. Its use also produces an increase of lift, the amount of which is dependent on the flap angle selected. The increase of C_{Lmax} a split flap can produce is approximately 32% at the cost of a large increase of drag. The stalling angle is only marginally decreased by the use of these flaps to approximately 15°. See Figure 6.9.

6.6.3   The Slotted Trailing-Edge Flap

The separation of the streamline flow over the upper surface of the aerofoil when either the plain or split flap is used occurs at an early point. If the boundary-layer energy is restored, in a similar manner to that of the use of slats, the airflow separation can be delayed until a point further aft along the aerofoil upper surface with flap extended. If such is the case the lift developed can be maintained until a much greater angle of attack because the stalling angle will be increased. These desirable qualities can be achieved by introducing the benefits found to exist with the slot produced by a slat but by producing the slot between the trailing edge of the aerofoil and the trailing-edge flap. See Figure 6.10.

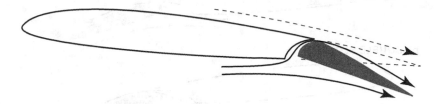

Figure 6.10   The Slotted Trailing-Edge Flap.

The slot so created allows the air from the relatively high-pressure area beneath the wing to flow to the low-pressure area above the wing. The slot acts as a Venturi and accelerates the air passing through it adding kinetic energy, which re-energizes the boundary layer and causes the separation point to migrate aft. Thus, the use of these flaps increase C_{Lmax} by approximately 65% and marginally decreases the stalling angle to 14°. Because the drag produced by the employment of these flaps is less than the plain or split flap they are more suitable for both take-off and landing.

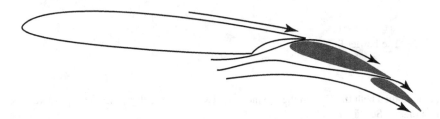

Figure 6.11   The Double-Slotted Trailing-Edge Flap.

The advantages of the slotted flap may be further enhanced if the flap is also constructed to produce a slot within its own area when extended. Thus, there are two slots when the flaps are used the first between the flap and the wing and the second within the flap. This double-slotted flap enables the aircraft to make a relatively steep approach at an attitude that is shallower but provides the pilot with a good visual aspect of the runway. See Figure 6.11.

When the flaps are fully extended they increase the camber of the wing and there is a marked nose-down pitching moment because of the rearward migration of the CP. This feature is normally counteracted by the use of leading-edge flaps or slats that are simultaneously and automatically deployed when the flaps are selected. The double-slotted flap can increase C_{Lmax} by 70% and increase the stalling angle to approximately 18°. The advantages of the double-slotted flap system can be further increased if the flap is triple slotted.

6.6.4   The Fowler Flap

The Fowler flap is commonly fitted to light aeroplanes and is a development of the slotted flap; it slides rearwards before hinging downward. In this manner it increases the wing area as well as increasing the camber and chord making the wing more suitable for operating at low airspeeds. It is the most efficient flap system and when extended not only increases the effective camber but also increases the nose-down pitching moment. The increase to the C_{Lmax} is approximately 90%; and the stalling angle is decreased to approximately 12.5°. However, if the same angle of attack is maintained after the flaps have

(a) SINGLE SLOTTED

(b) DOUBLE SLOTTED

(c) TRIPLE SLOTTED

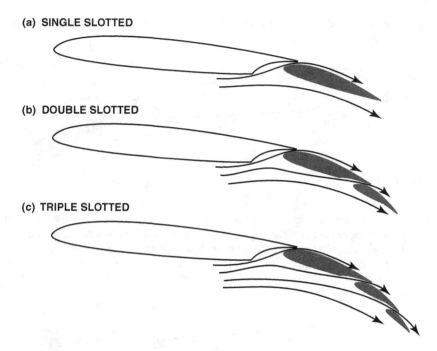

Figure 6.12   The Fowler Flap.

been extended then both the C_L and the C_D increase. Extending the flap produces a positive nose-down pitching moment. See Figure 6.12.

There are several different versions of the Fowler flap currently in use with larger aeroplanes they include the single-slotted, the double-slotted and the triple-slotted types. The greater the number of slots the higher is the value of C_{Lmax}. A similar system that moves bodily downward first before moving aft and rotating is the Fairey–Youngman flap. When deployed they also increase the wing area and the camber.

6.6.4.1   *The Effect of Trailing-Edge Flaps*

A summary of the effect of trailing flaps on the maximum coefficient of lift and the stalling angle of attack is shown in Figure 6.13 for all of the types of flap so far described.

The effect of extending the various lift augmentation devices is summarized in Table 6.2. It can be seen that Fowler flaps produce the highest C_{Lmax} and the lowest stalling angle.

6.6.5   *Leading- and Trailing-Edge Combinations*

High-lift systems for modern aeroplanes are often quite complex involving linked combinations of trailing-edge and leading-edge devices. A leading-edge slat is linked with a trailing-edge double or triple-slotted Fowler flap so that when flap is selected both are deployed simultaneously. The amount by which both are extended is dependent on the degree of flap selected. The maximum amount of deployment is that used for landing as shown at (c) in Figure 6.14. The lift/drag ratio is most adversely affected by the largest angled deployment of the flap system.

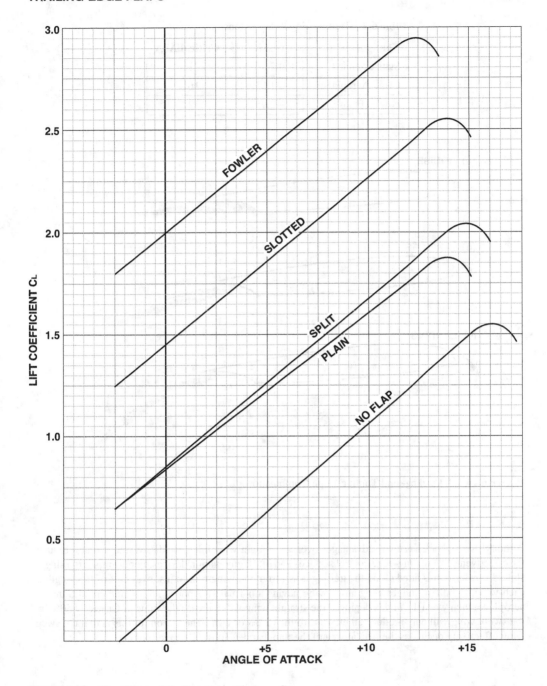

Figure 6.13 The Effect of Trailing-Edge Flaps on C_L.

Table 6.2 Trailing-Edge-Flap Extension.

Configuration	C_{Lmax}	Critical AoA
Slats extended	+2.00	21°
Clean Aeroplane	+1.55	16°
Fowler Flaps	+2.95	12.5°

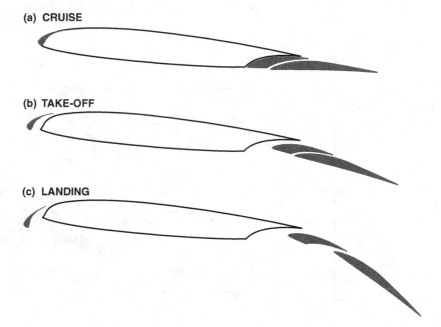

(a) CRUISE

(b) TAKE-OFF

(c) LANDING

Figure 6.14 A Leading- and Trailing-Edge Combination.

6.6.5.1 The Effect of Sweepback on Flap

Although sweepback has many advantages over the straight wing it has the disadvantage that it reduces the effectiveness of trailing-edge control surfaces and lift-augmentation devices. For example, if a single-slotted Fowler flap is fitted along the trailing edge, from the wing root to just over half of the length of each wing, to both a straight wing and a wing having a normal sweepback for a commercial transport aeroplane; then when the flap is deployed the straight wing will increase the value of C_{Lmax} by 50% but the swept wing will only produce an increase to the C_{Lmax} of 20%. The reason for this is that the swept wing effectively decreases the frontal area of the flap that develops the increase of lift.

Self-Assessment Exercise 6

Q6.1 Trailing-edge-flap extension will (i) ....the critical angle of attack and (ii) ....the value of C_{Lmax}.
  (a) (i) decrease; (ii) decrease
  (b) (i) increase: (ii) decrease
  (c) (i) decrease; (ii) increase
  (d) (i) increase; (ii) increase

Q6.2 In straight and level flight at a constant IAS if flap is selected, it will increase the:
  (a) lift and the drag
  (b) C_{Lmax} and the drag
  (c) lift coefficient and the drag
  (d) stalling speed

Q6.3 Which of the following statements regarding the difference between Krueger flaps and slats is correct?
  (a) A slot is created by deploying Krueger flaps but not by deploying a slat.
  (b) The critical angle of attack increases by deploying a slat but not by deploying a Krueger flap.
  (c) The critical angle of attack increases by deploying a Krueger flap but not by deploying a slat.
  (d) A slot is created by deploying a slat but not by deploying a Krueger flap.

Q6.4 After take-off in an aeroplane fitted with both slats and flaps the reason that it is customary to retract the flaps first is:
  (a) Extended flaps decrease the stalling speed by a large amount with relatively little drag.
  (b) Extended slats decrease the stalling speed by a large amount with relatively little drag and have no effect on the climb gradient.
  (c) Extended slats give the pilot a better view from the cockpit than does extended flaps.
  (d) Extended slats enable a more favourable V_{MCA} to be used.

Q6.5 When extended, the trailing-edge flaps:
  (a) significantly increase the angle of attack for the maximum lift
  (b) significantly decrease the total drag
  (c) worsen the best glide angle
  (d) increase the zero-lift angle of attack

Q6.6 Extension of the Fowler-type trailing-edge flaps will produce:
  (a) a force that reduces drag
  (b) a nose-down pitching moment
  (c) no pitching moment
  (d) a nose-up pitching moment

Q6.7 A plain flap will increase C_{Lmax} by:
  (a) increasing the camber of the aerofoil
  (b) increasing the angle of attack
  (c) boundary-layer control
  (d) movement of the CP

Q6.8 At a constant IAS in straight and level flight, if flaps are deployed the magnitude of the tip vortices will eventually:
  (a) increase
  (b) remain the same
  (c) increase or decrease depending on the initial angle of attack
  (d) decrease

Q6.9 The function of the slot between an extended slat and the leading edge of the wing is to:
  (a) decrease the wing loading
  (b) slow the airflow in the slot to create more pressure under the wing
  (c) cause a Venturi effect that energizes the boundary layer
  (d) allow space for slat vibration

Q6.10 The most efficient flap system is the:
  (a) plain flap
  (b) single-slotted flap
  (c) Fowler flap
  (d) split flap

Q6.11 A slat will:
  (a) increase the boundary-layer energy and prolong the stall to a higher angle of attack
  (b) increase the camber of the aerofoil and divert the flow around a sharp leading edge
  (c) increase the lift by increasing the wing area and the camber
  (d) provide a boundary-layer suction on the upperside of the wing

Q6.12 The use of a slot at the leading edge of a wing enables the aeroplane to fly at a slower speed because:
  (a) it changes the camber of the wing
  (b) it delays the stall to a higher angle of attack
  (c) the laminar part of the boundary layer gets thicker
  (d) it decelerates the upper surface boundary layer

Q6.13 The purpose of an automatic slat system is to spontaneously:
  (a) operate at a predetermined high angle of attack
  (b) retract the slats after take-off
  (c) extend the slats when 'ground' is selected on the ground/flight system
  (d) assist the ailerons when rolling

Q6.14 Deploying a Fowler flap will cause it to:
  (a) turn down and then move aft
  (b) move aft
  (c) turn down
  (d) move aft and then turn down

Q6.15 A deployed slat will:
  (a) decrease the boundary-layer energy and decrease the maximum negative pressure on the slat, so that C_{Lmax} is attained at a lower angle of attack
  (b) increase the camber of the aerofoil and increase the effective angle of attack, so that C_{Lmax} is attained at a higher angle of attack
  (c) increase the boundary-layer energy and move the maximum negative pressure from the wing to the slat, so that C_{Lmax} is attained at a higher angle of attack
  (d) increase the boundary-layer energy and increase the maximum negative pressure on the wing, so that C_{Lmax} is attained at a higher angle of attack

Q6.16 The maximum angle of attack for the flaps-extended configuration when compared with the flaps-retracted configuration is:
  (a) smaller or larger, dependent on the angle of flap deflection
  (b) smaller
  (c) larger
  (d) unchanged

Q6.17 Full extension of a Fowler flap will:
  (a) increase the wing area only
  (b) increase the wing area and increase the camber
  (c) increase the camber only
  (d) not affect the C_D

Q6.18 Which of the following statements is correct?
  (a) Extension of the flaps will increase the maximum L/D ratio causing the minimum rate of descent to decrease.
  (b) Extension of the flaps does not affect the minimum rate of descent because only the TAS has to be accounted.
  (c) Spoiler extension decreases the stall speed and the minimum rate of descent but increases the minimum descent angle.
  (d) Extension of the flaps reduces the stalling speed and decreases the maximum glide distance.

Q6.19 To maintain a constant airspeed in straight and level flight whilst the flaps are being retracted, the angle of attack must:
  (a) remain constant
  (b) increase or decrease dependent on the type of flap
  (c) increase
  (d) decrease

Q6.20 At a constant angle of attack, during flap retraction the aeroplane begins to:
  (a) climb
  (b) yaw
  (c) sink
  (d) bank

Q6.21 In straight and level flight, when flaps are extended the lift coefficient will:
  (a) remain constant
  (b) increase
  (c) decrease
  (d) increase at first and then decrease

Q6.22 Deflection of the leading-edge flaps will:
  (a) decrease the drag
  (b) not affect the critical angle of attack
  (c) increase the critical angle of attack
  (d) decrease C_{Lmax}

Q6.23 At a constant angle of attack, during flap extension the aeroplane begins to:
  (a) climb
  (b) bank
  (c) sink
  (d) yaw

Q6.24 Slat extension will:
  (a) increase the critical angle of attack
  (b) reduce wing-tip vortices
  (c) create gaps between the leading edge and engine nacelles
  (d) decrease the energy in the boundary layer on the upper surface of the wing

Q6.25 An efficient arrangement of the slats and/or leading-edge flaps on a swept-wing aeroplane is wing roots (i) . . . . wing tips (ii) . . . . :
  (a) (i) L.E. Flaps; (ii) no devices
  (b) (i) L.E. Flaps; (ii) slats
  (c) (i) slats; (ii) L.E. flaps
  (d) (i)slats; (ii) no devices

Q6.26 A slotted flap will increase the C_{Lmax} by:
  (a) decreasing the skin friction
  (b) increasing the camber of the aerofoil only
  (c) increasing the critical angle of attack
  (d) increasing the camber of the aerofoil and improving the boundary layer

Q6.27 At a constant angle of attack, when trailing-edge flaps are extended C_L will:
  (a) increase
  (b) decrease
  (c) remain constant
  (d) vary as the square of the IAS

Q6.28 After the leading-edge slats have been deployed the boundary layer is (i) . . . . and the peak pressure moves (ii) . . . .
  (a) (i) re-energised; (ii) moves forward onto the slat
  (b) (i) de-energised; (ii) moves forward onto the slat
  (c) (i) re-energised; (ii) moves forward towards the wing leading edge
  (d) (i) de-energised; (ii) moves forward towards the wing leading edge

Q6.29 Where are Krueger flaps fitted?
  (a) outboard on the wing leading edge
  (b) outboard on the wing trailing edge
  (c) inboard on the wing trailing edge
  (d) inboard on the wing leading edge

Q6.30 When a Fowler flap moves back:
  (a) wing area decreases and camber decreases
  (b) wing area is unaffected and camber is unaffected
  (c) wing area increases and camber increases.
  (d) wing area increases and camber decreases

Q6.31 The most effective flap system is:
  (a) split
  (b) slotted
  (c) plain
  (d) Fowler

Q6.32 In straight and level flight the effect that deploying flaps has on the pitching moment is:
  (a) pitch-up
  (b) pitch-down
  (c) dependent on the CG position
  (d) dependent on speed

Q6.33 When trailing-edge flaps are extended:
  (a) the critical angle of attack decreases and the C_{Lmax} increases
  (b) C_{Lmax} increases and the critical angle of attack increases
  (c) C_{Lmax} increases and the critical angle of attack remains the same
  (d) Vs increases and the critical angle of attack remains constant

Q6.34 An aeroplane is fitted with slats and flaps that have the following settings 0°, 15°, 30° and 45°. The greatest negative influence on the C_L/C_D ratio is caused by:
  (a) 0° to 15° flap setting
  (b) 30° to 45° flap setting
  (c) 15° to 30° flap setting
  (d) deploying slats

Q6.35 In which location on a swept-wing aeroplane would the following devices be fitted leading-edge flaps (i) . . . . leading-edge slats (ii) . . . .
  (a) (i) outboard of the inboard engines: (ii) inboard of the inboard engines
  (b) (i) inboard of the inboard engines: (ii) inboard of the inboard engines
  (c) (i) inboard of the inboard engines; (ii) outboard of the inboard engines
  (d) (i) outboard of the inboard engines: (ii) outboard of the inboard engines

Q6.36 The purpose of a leading-edge slat is to:
  (a) increase the camber of the wing
  (b) decelerate the air over the upper surface of the wing
  (c) thicken the laminar boundary layer over the upper surface of the wing
  (d) permit a greater angle of attack

Q6.37 The effect of deploying leading-edge flaps is to:
  (a) decrease the critical angle of attack
  (b) decrease the C_{Lmax}
  (c) increase the critical angle of attack
  (d) not affect the critical angle of attack

Q6.38 If all other factors remain constant the effect of extending trailing-edge flaps whilst maintaining a constant angle of attack is that the aeroplane will:
  (a) roll
  (b) climb
  (c) yaw
  (d) suddenly sink

Q6.39 To maintain straight and level flight when trailing-edge flaps are retracted the angle of attack must:
  (a) be increased
  (b) be decreased
  (c) remain constant
  (d) be increased or decreased dependent on the type of flap

Q6.40 When Fowler flaps are deployed on an aeroplane with a 'T' tailplane the pitching moments generated will be:
  (a) nose-down
  (b) nose-up
  (c) will depend on the relative positions of the CP and the CG
  (d) unchanged

Q6.41 The effect of leading-edge slats is to:
  (a) increase the boundary-layer energy, move the maximum negative pressure onto the slat and to increase the C_{Lmax} angle of attack
  (b) increase the boundary-layer energy, increase the maximum negative pressure on the main wing section and to move the C_{Lmax} to a higher angle of attack
  (c) increase the camber, increase the maximum negative pressure on the main wing section, increase the effective angle of attack and to move the C_{Lmax} to a higher angle of attack
  (d) decrease the boundary-layer energy, move the maximum negative pressure onto the slat and to move the C_{Lmax} to a lower angle of attack

Q6.42 When trailing-edge flaps are deployed the following occurs:
  (a) C_{Lmax} increases and the critical angle of attack increases.
  (b) The critical angle of attack remains constant but C_{Lmax} increases.
  (c) The critical angle of attack decreases and C_{Lmax} increases.
  (d) The critical angle of attack remains constant and the stalling speed increases.

Q6.43 Which of the following lists of configurations gives an increasing angle of attack with the sequence?
  (a) clean wing, flaps extended, slats extended
  (b) slats extended, clean wing, slats extended
  (c) clean wing, slats extended, flaps extended
  (d) flaps extended, clean wing, slats extended

Q6.44 Which of the following statements is correct regarding the deployment of slats/Krueger flaps?
  (a) Slats increase the critical angle of attack, Krueger flaps do not.
  (b) Slats form a slot, Krueger flaps do not.
  (c) Krueger flaps form a slot, slats do not.
  (d) Krueger flaps increase the critical angle of attack, slats do not.

Q6.45 A slat on an aerofoil:
   (a) increases the energy of the boundary layer and decreases the critical angle of attack
   (b) increases the leading-edge radius by rotating forward and downward from the retracted position on the underside of the leading edge
   (c) increases the energy of the boundary layer and increases the maximum angle of attack
   (d) extends automatically because of the increased pressure at the stagnation point at high angles of attack and low IAS

Q6.46 The effect of deploying trailing-edge flaps is:
   (a) increased minimum glide angle
   (b) decreased minimum glide angle
   (c) increased gliding endurance
   (d) decreased sink rate

Q6.47 In straight and level flight the effect that extending trailing-edge flaps has on the magnitude of the wing-tip vortices is that they:
   (a) decrease in magnitude
   (b) increase in magnitude
   (c) are unaffected
   (d) increase or decrease dependent on the angle of attack

Q6.48 In straight and level flight the effect that extending trailing-edge flaps has on the maximum coefficient of lift is that it:
   (a) decreases
   (b) increases
   (c) increases and eventually decreases
   (d) remains unchanged

Q6.49 The movement of Fowler flaps when they are extended is that they first move (i). . . . . . . and then move (ii) . . . .
   (a) (i) downwards; (ii) backwards
   (b) (i) backwards; (ii) downwards
   (c) (i) forwards; (ii) downwards
   (d) (i) downwards; (ii) forwards

Q6.50 The purpose of a leading-edge slat is to:
   (a) permit a greater angle of attack
   (b) decelerate the airflow over the upper surface of the wing
   (c) deepen the laminar boundary layer over the upper surface of the wing
   (d) increase the camber of the wing

Q6.51 An aeroplane fitted with slats and flaps that have four settings, 0°, 15°, 30° and 45°. Which of them when selected will have the greatest negative influence on the L/D ratio?
   (a) flaps from 15° to 30°
   (b) flaps from 30° to 45°
   (c) the slats
   (d) flaps from 0° to 15°

7 Drag

The force directly opposing the forward movement of the aeroplane through the air is referred to as drag. It is the resistance to forward motion and is assumed to act along a line parallel to the longitudinal axis. The total drag may act through a line above or below the thrust line depending on the position of the engines. There are two major components that make up the total drag acting on an aeroplane. They are parasite drag, sometimes called profile drag or zero-lift drag, and induced drag. Parasite drag is that which is produced even when lift is not produced and induced drag is that which is caused when the aeroplane's aerofoils are producing lift. The lift and drag forces acting on an aerofoil depend on the pressure distribution around the aerofoil. At a normal angle of attack C_L is greater than C_D. The sum of parasite drag and induced drag is the total drag, which can be calculated from the formula:

$$\text{Total Drag} = C_D{}^1\!/_2\rho V^2 S$$

where: C_D = the coefficient of drag (a mathematical factor) which is the ratio of the drag pressure to the dynamic pressure. V = the free airspeed. S = the wing area.

ρ = the density of the atmosphere. $^1\!/_2\rho V^2$ = the dynamic pressure 'q'.

From this formula it can be determined that if the stream velocity (V) is increased by a factor of 4 then the aerodynamic drag is increased by a factor of (V^2) 16. If the air density (ρ) is halved then the total drag is halved. If all other factors remain constant increased dynamic pressure increases total drag.

7.1 Parasite (Profile) Drag

The element of the total drag not directly attributed to the procurement of lift is parasite drag, sometimes called profile or zero lift drag. Its value varies as the square of the EAS and comprises three components:

a. Surface-friction drag.
b. Form drag.
c. Interference drag.

7.2   Surface-Friction Drag

Surface-friction drag is that portion of parasite drag caused by a boundary layer of almost static air that adheres to the skin of the aeroplane. The boundary layer is defined as that part of the airflow in which the airspeed is less than 99% of the speed of the freestream airflow and is either laminar or turbulent.

Air directly in contact with the surface of the aeroplane will be dragged along at a slightly lower speed than that of the aeroplane because of the inherent cohesion of the air. The air further away from the surface of the aeroplane will be pulled along but at a lower speed than the surface air. This dragging effect decreases with increasing depth of the layer of air until at a particular depth the movement of the aeroplane does not affect the air at all. The depth of the layer of air from the aeroplane surface to the depth at which the air is unaffected is the boundary layer.

The disposition of the boundary layer determines the amount of surface friction drag, C_{Lmax}, the stalling characteristics of the aerofoil, the value of the form drag and to a certain extent the high-speed peculiarities of the aeroplane. The drag force within the boundary layer is unaffected by pressure and/or density variations to the normal of the surface of the aeroplane.

The value of surface-friction drag varies and is determined by:

a. the total surface area of the aeroplane;
b. the coefficient of the air viscosity;
c. the rate of change of flow speed over the depth of the boundary layer.

7.2.0.1   Surface Area

There is a boundary layer of air covering the entire surface of the aeroplane. The larger the area the greater is the value of the surface-friction drag.

7.2.0.2   Coefficient of Viscosity

A measure of the adhesive qualities of the air is the coefficient of viscosity. The greater the viscosity of the air the greater is the surface-friction drag. The viscosity of air is affected by its temperature and increases with a rise of temperature.

7.2.0.3   Rate of Change of Airspeed

The airflow ahead of the aeroplane is laminar and becomes turbulent at a point along the aeroplane surface referred to as the transition point. The rate at which the speed of the airflow changes is greatest at the surface of the aeroplane in the turbulent flow not in the laminar flow. The magnitude of the amount of surface friction is highest where the *rate of change* of speed is greatest. This is the transition point to turbulent flow, but even when the flow is turbulent there is an extremely shallow layer of the boundary layer next to the surface that remains laminar called the laminar sublayer. This is important when considering the reduction of the total amount of parasite drag that can be achieved by smoothing the surface.

7.2.1   Flow Transition

The nearer the transition point is to the leading edge of an aerofoil the greater is the amount of surface friction. The location of the transition point from laminar flow to turbulent flow is dependent on four factors:

a. surface condition;
b. speed of flow;
c. size of the object;
d. adverse pressure gradient.

7.2.1.1   Surface Condition

The depth of the boundary layer, both laminar and turbulent, increase with distance from the leading edge of an aerofoil. Generally the depth of the turbulent boundary layer is approximately ten times that of the laminar boundary layer. The precise depth varies from surface to surface. A rough, dirty, wet, or ice-covered surface will increase the depth of this layer causing greater interference with the airflow over the aerofoil. Polishing the aerofoil surface minimizes the effect of this type of drag. In winter it is essential to ensure the aeroplane is properly and thoroughly deiced before take-off to reduce the effect of surface friction.

7.2.1.2   Speed and Size

According to Reynolds the transition from laminar flow to turbulent flow occurs when the speed attains a value that is inversely proportional to the thickness of the body placed in the airflow. In other words, the thicker the body the lower the transition speed. When applied to an aerofoil of a given depth, the faster the airflow, the nearer the transition point moves towards the leading edge. This increases the amount of surface friction drag but the turbulent airflow has greater kinetic energy, which will move the separation point rearward and increase the value of C_{Lmax}.

7.2.1.3   Adverse Pressure Gradient

The transition point on the curved upper surface of an aerofoil is where the lowest surface pressure is located, which is usually at the point of maximum aerofoil thickness. Therefore, the pressure aft of this point increases with increasing distance from the transition point and is an adverse pressure gradient in the direction of the flow. If such is the case it is impossible to maintain a laminar flow without mechanical assistance to reduce the surface friction drag.

7.3   Form (Pressure) Drag

The component of the parasite drag caused by the shape of the aeroplane and its aerofoils is form drag, which is also referred to as pressure drag by EASA; use of the ailerons affects this type of drag the most. The magnitude of this type of drag is directly proportional to the size of the frontal area and is dependent on the separation point of the airflow over the surface of the wing and streamlining. Extending the undercarriage in flight increases the form drag and increases the nose-down pitching moment as a consequence of the drag on the undercarriage acting below the CG. Surface friction reduces the speed of the air passing over the upper surface of an aerofoil and the kinetic energy of the air within the boundary layer. The kinetic energy of the boundary layer is further reduced if the surface is curved by the adverse pressure gradient.

As a result, of these factors, the boundary layer creeps slowly aft until towards the trailing edge of the aerofoil the boundary layer stops moving altogether. This causes wake turbulence because the airflow ceases to be laminar and separates from the surface over which it was passing. The point at which the flow change from laminar to turbulent flow is total is the separation point. Aft of this point the faster-moving air above the boundary layer mixes with the turbulent boundary layer and therefore has greater kinetic energy than the laminar flow above. See Figure 7.1.

The decrease of pressure experienced between the stagnation point on the leading edge of the aerofoil and the separation point of the boundary layer causes drag; this is 'pressure drag.' The whole of the aeroplane is subject to this type of drag and is a large proportion of the total drag on the aeroplane. To delay the separation as long as possible and reduce the amount of form drag it is essential to streamline the aerofoil. This will increase the fineness ratio; reduce the curvature of the surfaces and the adverse pressure gradient.

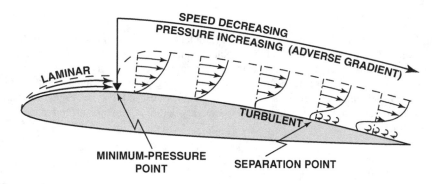

Figure 7.1   Boundary-Layer Separation.

7.3.1   Interference Drag

The total drag experienced by an aircraft is greater than the sum of the drag caused by individual parts of the aeroplane. This additional drag is caused by flow interference at the junction of the various parts of the aeroplane such as the wing/fuselage and wing/nacelle junctions. Interference drag, as it is known, is caused by the mixing of the boundary layers at the junction of adjoining aerofoil surfaces. It creates a thicker and more turbulent boundary layer in which there is a greater loss of energy than is normal for a boundary layer. Adding fairings at the junctions, e.g. trailing-edge wing roots and blended winglets, can reduce its effect.

Parasite drag is therefore caused by:

a. the shape of the aeroplane and its aerofoils;
b. boundary-layer surface friction;
c. poor and/or inadequate streamlining;
d. contaminated aerofoil surface condition.

Parasite drag can be minimized by:

a. design improvements during manufacture;
b. polishing the surfaces of the aeroplane before flight; and
c. ensuring that in low ambient temperatures all surfaces of the aeroplane are ice free.

The magnitude of parasite drag is directly proportional to the square of the EAS. Therefore, as speed increases parasite drag has an increasingly detrimental effect. This is often referred to as the 'speed-squared law' because if the speed is doubled the profile drag is quadrupled. *The value of parasite drag is not directly affected by mass.*

7.4   Induced Drag

The drag caused by the movement of the aeroplane through the air comprises parasite drag and induced drag. Lift-dependent drag (i.e. that which is caused in the production of lift) although it comprises mainly induced drag it also contains elements of parasite drag. It is the result of the pressure differences over and under a lifting aerofoil. Both lift and drag forces are conditioned by the pressure distribution around the aerofoil. The total lift, the reactive force, is assumed to act upward at 90° to the airflow as it passes over the upper surface of the aerofoil.

Induced drag, sometimes referred to as vortex drag, is the component of the aerodynamic reaction acting on an aerofoil in a direction parallel and opposite to the direction of flight and is approximately equal to the total lift multiplied by the sine of the angle of attack. See Figure 7.2. The magnitude of induced drag is dependent on three variable factors and two fixed factors. The variable factors are the lift coefficient (C_L), the speed of the aeroplane and the mass of the aeroplane. The fixed factors on conventional aeroplanes are the planform and the aspect ratio. Wing twist is often used in the aeroplane wing design to reduce the induced drag and improve the stall characteristics. (See page 48)

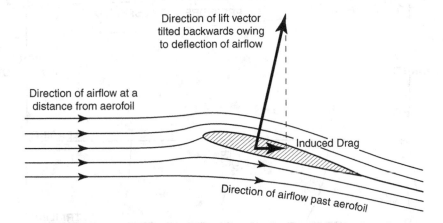

Figure 7.2   Induced Drag.

As an aerofoil moves through the air, the pressure differential between the upper and lower surfaces of the wing causes air to flow from the underside high-pressure area, to the upperside low-pressure area, by means of spillage at the wing tips. This causes the air to have a spanwise flow from the wing root towards the wing tip on the lower surface of the wing and from the wing tip towards the wing root on the upper surface of the wing. This spanwise flow is strongest at the wing tip and decreases in strength towards the wing root. It comprises the main portion of the induced drag. See Figure 7.3.

As a result, the airflow at any point along the trailing edge of the wing leaves the upper surface travelling in a different direction to the air leaving the lower surface of the wing. This creates vortices the magnitude of which increases with distance from the wing root; the larger the vortex, the greater is the induced drag. The largest vortices occur at the wing tip and viewed from behind rotate clockwise on the port wing and anticlockwise on the starboard wing. Thus, the wing tip has the greatest influence on the magnitude of induced drag.

The efficiency of a wing can be improved if blended winglets (upturned wing tips) are included in the aeroplane design. They reduce the onset and size of wing-tip vortices, and consequently reduce the thrust required and the fuel consumption by up to 7%. See Figure 4.13.

7.4.1   The Effect of Speed

In level flight at a given mass the lift generated is the same at all forward speeds above the stalling speed. However, the pressure distribution around the wing is different as shown in Figure 4.7. This shows that the pressure difference between the upper and lower surfaces of the wing increases with increasing angle of attack. As a result, at any point along the trailing edge of the wing the induced downwash velocity is constant. However, because the angle of attack must decrease with increasing speed the same downwash speed has less effect at a high forward speed than it does at a low forward speed because it has a smaller downwash vector. Thus, the downwash angle is less at higher forward speeds than it is at lower forward

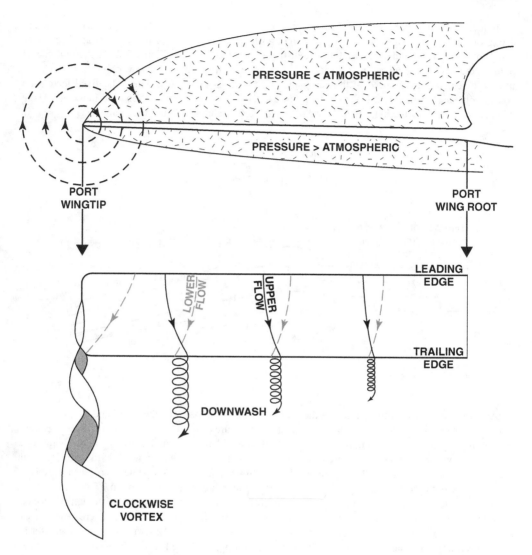

Figure 7.3   Vortex Formation Viewed from Behind.

speeds. Therefore, the induced drag is decreased at high forward speeds. As a result, of the downwash the effective angle of attack of the tailplane is reduced. This decreases its effectiveness. See Figure 7.4.

Large vortices are created at high angles of attack and vice versa. When flying at low speeds the angle of attack is high, therefore, the induced drag is high. *Maximum induced drag occurs at the highest possible angle of attack and at the lowest possible speed, which is the stalling speed in the landing configuration (Vso).* The stalling EAS increases with increased mass and, if compressibility is ignored, remains virtually constant for the same mass at all altitudes.

Mathematically, *induced drag is inversely proportional to the square of the EAS and the coefficient of induced drag* C_{DI} *is directly proportional to* C_L^2, i.e. if the EAS is doubled the induced drag is a quarter of the original value and the C_{DI} will be one sixteenth of its original value. If the EAS is halved the induced drag is increased fourfold and coefficient of induced drag, C_{DI}, would increase by factor of 16.

Induced drag decreases with increased speed. At a constant aeroplane mass in level flight, if the Mach number is increased, the TAS and the EAS will increase so the angle of attack must be reduced and consequently the induced drag will decrease. The wing tip affects the value of induced drag the most and the strength of the wing-tip vortex is directly proportional to the angle of attack. At high angles of attack induced drag accounts for 75% of the total drag.

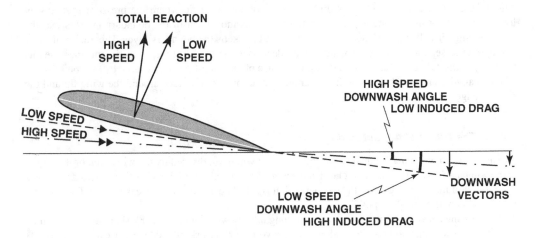

Figure 7.4   Effect of Speed on Induced Drag.

7.4.2   The Effect of Mass

Induced drag is directly proportional to C_L^2 *and therefore to the square of the aeroplane gross mass.* But for any given mass, as speed increases, C_L reduces and the induced drag diminishes. With a constant load factor induced drag increases with increased aeroplane mass. At high aeroplane masses, the induced drag for a given speed will be greater than that experienced at lower aeroplane masses. In other words, if the EAS remains constant then induced drag is affected by aeroplane mass.

7.4.3   The Effect of Planform

The greatest amount of induced drag occurs where the vortices are largest, at the wing tips. An elliptical wing produces the smallest wing-tip vortices because it has the smallest wing tip and therefore generates the least amount of induced drag. To decrease the induced drag it is necessary to have an elliptical spanwise pressure distribution because it will produce a steady downwash at a constant angle. See Figure 4.15.

This ideal is impossible to achieve because of the interference of the fuselage, the engines and the propellers with the airflow together with the influence of other factors. Nevertheless, a combination of taper and washout can be used to obtain elliptical loading at a specified angle of attack. Winglets are often used to decrease the induced drag they also decrease the value of V$_{IMD}$ but have the penalty of increased parasite drag; fitting wing tip fuel tanks has the same benefits and penalties. In straight and level flight at a constant IAS wing-tip vortices are also diminished when flap is deployed, however, the total induced drag remains unchanged. See also Figure 8.7.

7.4.4   The Effect of Sweepback

Although the spanwise airflow around a wing is from root to tip on the underside of the wing and from tip to root on the upper surface of the wing so that the air flows from relatively high pressure below the wing to relatively low pressure above the wing this is not so for the boundary layer. On both the upper and lower surface of a swept wing the boundary layer moves, albeit very slowly, initially in the direction of the free-flowing airstream back over the wing towards the lowest pressure, which for a cruising angle of attack is positioned at approximately 60% of the chord measured from the leading edge as shown in Figure 4.7.

Once past this point the boundary layer, because of the adverse pressure gradient towards the trailing edge of the wing, gradually changes direction towards the wing tip where the pressure is relatively lower than that of the trailing edge. The outcome of this movement of the boundary layer is that the slow-moving air collects in a pool at the wing tip and thickens the boundary layer. This makes it more susceptible to separation from the wing, which decreases the C_{Lmax} at the wing tip and increases its tendency to tip stall. It also adds to the strength and size of the wing-tip vortex caused by the combination of the spanwise airflow of air from the under surface to the upper surface around the wing tip and the leading edge separated laminar flow.

7.4.5   *The Effect of Aspect Ratio*

The aspect ratio of a wing is equal to the square of the wingspan divided by the wing area or it is equal to the wingspan divided by the mean chord of the wing. The magnitude of the wing-tip vortices determines the amount of induced drag that will be generated. It is clearly desirable to reduce the size of the wing-tip vortex to become as small as possible.

If two wings have the same plan area, one a long narrow wing and the other a short wide wing, then the long narrow wing will produce smaller wing-tip vortices than a short wide wing because there is less width at the wing tip for the air to spill over and produce the vortex. The strength of the wing-tip vortex decreases with increased aspect ratio. The best shape to diminish induced drag is an elliptical wing.

The spillage of air around the tip of a high aspect ratio wing is a small proportion of the overall airflow over the wing and causes less downwash. Consequently, the induced drag is decreased to less than that for a low aspect ratio wing having the same wing area. The implication of this is that the low aspect ratio wing, the short stubby one, has the greater amount of induced drag and a higher stalling angle than the long thin wing with the high aspect ratio. Therefore it is the wing-tip that affects the magnitude of the induced drag the most.

Generally, the slower the cruising speed of an aeroplane the higher the aspect ratio that can be usefully employed in its design. The higher the aspect ratio the smaller is the induced drag incurred, consequently, the stalling speed increases but the stalling angle decreases. *Induced drag is, therefore, inversely proportional to the aspect ratio.* If the aspect ratio is doubled then the induced drag is halved. Thus, V_{IMD} is decreased and speed stability is increased (described later in this chapter).

7.4.6   *The Effect of Flap*

When flap is lowered a vortex forms on the outboard corner of the flap and rotates in the same direction as the wing-tip vortex. Between the vortices, the rotations directly oppose one another. This reduces the size and strength of the wing-tip vortex. However, the lowering of flap has virtually no effect on the overall magnitude of the induced drag because the formation of the flap vortex effectively nullifies the benefit of the reduced strength of the wing-tip vortex. There are now two smaller vortices instead of one large one. See Figure 8.7.

7.4.7   *The Effect of the CG Position*

The position of the centre of gravity directly affects the amount of induced drag generated because it determines the amount of elevator trim that must be used to counteract any nose-down tendency of the aeroplane in level flight. If the CG is in an aft position then less elevator trim is required, thus both the induced drag and the stalling speed decrease because the download on the tailplane is reduced. This reduces the thrust required and, consequently the fuel flow; which increases the maximum range and increases the maximum endurance.

7.4.8 Effects Summary

In summary, induced drag is:

a. Proportional to the square of the coefficient of lift. (C_L^2).
b. Proportional to the square of the mass (M^2). (i.e. increases with increasing mass).
c. Inversely proportional to the square of the EAS ($1/V^2$). (i.e. decreases with increasing speed).
d. Inversely proportional to the aspect ratio (1/Aspect Ratio).
e. Decreased by elliptical lift distribution. (See Chapter 4).

7.5 Ground Effect

When an aeroplane flies at a height equal to or less than the length of its wingspan the efficiency of its wings is considerably improved. A low-wing aeroplane, just before touchdown, experiences a 50% reduction of its total drag. This phenomenon is ground effect, which is only experienced for conventional aeroplanes during the initial climb phase after take-off and the final descent phase during landing. *CS 25.111(d)*.

At low airspeeds it is considered that the air is noncompressible and, because of the close proximity of the earth's surface, the upwash of air ahead of the wing is reduced. Thus, because of the diminished negative lift, the downwash over the upper surface of the wing does not have to be as large as it would be normally. This enables the aeroplane to maintain level flight at a lower than usual angle of attack, which decreases the induced drag and improves the efficiency of the wing. See Figure 7.5.

For ground effect to have a noticeable effect on the performance of an aeroplane, it must be operating at a height much lower than half the wingspan, although CS 25 considers an aeroplane to be in ground effect when it is at or less than a height equal to its wingspan. The consequences of operating **in** ground effect at a constant IAS are:

a. Decreased upwash and downwash angles produce smaller wing-tip vortices.
b. Increased C_L.
c. Significantly decreased induced drag coefficient; by approximately 8% at a height equal to half wingspan and reduced by considerably more at any lower height.
d. Decreased total drag (no effect on parasite drag).
e. Decreased angle of attack for a given lift coefficient.
f. Increased angle of attack for a constant IAS.
g. Decreased stalling angle of attack.
h. Decreased induced angle of attack.
i. Increased lift, if the pitch attitude remains constant.
j. Increased speed for a constant thrust setting.
k. Increased float distance on landing, if the angle of attack is constant, increasing the total landing distance.
l. The lift vector is inclined forward, which decreases the thrust required.
m. Less thrust and power are required to fly at the same speed.

The effects on the performance of an aeroplane of **leaving** ground effect are:

a. Decreased lift during take-off, if the angle of attack remains constant.
b. The effective angle of attack decreases.

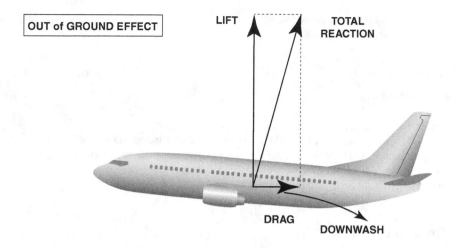

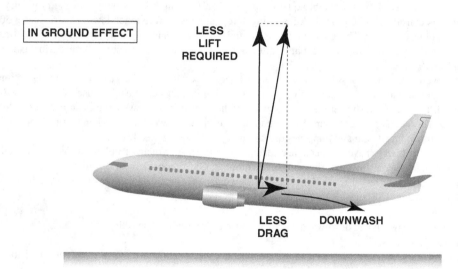

Figure 7.5 Ground Effect.

c. The induced angle of attack increases.
d. Induced drag increases.
e. The downwash increases.

7.6 Wing-Tip Design

There are many different methods used by aircraft designers in an attempt to reduce the drag caused by wing-tip vortices. The methods currently in use include:

a. On short stubby wings the wing tip is turned downward to encourage the separation of the airflow at the tip reducing the size of the tip vortex.

b. Some small aeroplanes have endplates fitted to the wing tip, which modify the wing-tip vortices in a beneficial manner.
c. Some small passenger jet aeroplanes are fitted with wing tip long-range fuel tanks that have the same effect as endplates and the added advantage of providing a wing relieving load, reducing the wing bending stress in flight.
d. Larger passenger jet aeroplanes are fitted with wing tip winglets. These may be mounted vertically or at an angle to the mainplane. The vortex now forms at the tip of the winglet. Because this is above the level of the mainplane the downwash from the mainplane is considerably reduced, which decreases induced drag by up to 19%.

7.7 Wingspan Loading

The lift distribution of a tapered wing is dependent on the taper ratio as shown in Figure 4.16 and produces the greatest amount of lift at a point between 60% and 70% of the span distance measured from the wing root. Therefore, the downwash produced is principally between this point and the wing root and smallest at the wing tip. Such characteristics are undesirable because:

a. the wing tip will stall before the wing root;
b. the wing loading is greatest at the wing tip;
c. it causes twisting wing bending stresses.

These unpleasant peculiarities can be minimized on a tapered wing at the design and construction stage by causing the angle of incidence to progressively decrease towards the wing tip so that the effective angle of attack remains constant along the whole wingspan. This is known as 'washout'. See Figure 2.1(b).

7.8 The Coefficient of Induced Drag (C_DI)

As the speed decreases below V_{IMD} the induced drag increases, as does the total drag. The coefficient of induced drag for an elliptical spanwise pressure distribution is given by the formula:

$$C_{DI} = \frac{C_L{}^2}{\pi A}$$

where C_{DI} = Coefficient of Induced Drag: C_L = Coefficient of Lift; A = Aspect Ratio

This formula shows that, excluding constants, the coefficient of induced drag is the ratio $C_L{}^2$ to the aspect ratio and as such that it is inversely proportional to the aspect ratio and directly proportional to $C_L{}^2$. If the speed doubles then the induced drag is a quarter of the value at the original speed. Therefore, as stated before, from the above formula the C_{DI} is equal to $1/4^2 = 1/16$ of its original value. It also shows that:

a. there is no induced drag on any wing at zero lift;
b. there is less induced drag for a wing with a high aspect ratio;
c. there is no induced drag if only two-dimensional flow is considered.

For any other planform a correction factor **k** is introduced, the value of which is dependent on the exact planform. The formula then becomes:

$$C_{DI} = \frac{k C_L{}^2}{\pi A}$$

7.9   Total Drag

The total drag affecting an aeroplane at any specific EAS is equal to the sum of the induced drag and the parasite drag. Thus, the formula for the total drag is:

$$C_D = C_{DP} + C_{DI}$$

Therefore:

$$C_D = C_{DP} + \frac{kC_L^2}{\pi A}$$

For any given wing then the part of the formula $\dfrac{k}{\pi A}$ can be replaced by the letter **K**

The total drag formula then becomes: $C_D = C_{DP} + KC_L^2$

Total drag is also equal to $C_D{}^1/_2\rho V^2 S$.

At low angles of attack the major component of total drag is parasite drag, whereas; at high angles of attack the major component of total drag is induced drag. At the minimum drag point where the speed is VIMD the total drag comprises induced drag and parasite drag in equal amounts. The relationship of induced, parasite and total drag in level flight can be illustrated by a graph of 'Drag v EAS'. Such a graph applies only to one mass and one configuration but the same graph applies to all altitudes. See Figure 7.6. *Total drag is directly proportional to the gross mass of the aeroplane*. Thus, the EAS at the lowest point of the total-drag curve, *VIMD, varies in direct proportion to the mass*. Figure 7.8(a) shows the effect that an increased gross mass has on the graph.

7.10   Analysis of the Total-Drag Curve

For an aeroplane at constant mass, acceleration from Vs to VNE causes the total drag to decrease until VIMD is attained and then increase as shown in Figure 7.6. It can be seen that the value of parasite, profile or pressure drag varies in proportion to the square of the IAS, whereas the value of the induced drag is inversely proportional to the square of the speed. The total-drag curve has three points that are of great importance. They are the speed at the maximum lift/drag ratio, the speed at the maximum EAS/drag ratio and the minimum power speed.

7.11   The Velocity of Minimum Drag (VIMD)

The lowest point on the total drag (or thrust required) curve is the point at which the thrust required is least and total drag is minimum. It is also the point at which the maximum lift/drag ratio is attained and parasite drag is equal to induced drag. The angle of attack at which the minimum drag and the maximum lift/drag ratio occur is approximately 4° in level flight and is the same for all masses and at all altitudes. The speed (EAS) at which this occurs is referred to as the minimum drag speed or VIMD.

This EAS has a constant value for all altitudes, provided the gross mass and configuration do not change and compressibility is ignored. It is shown as point A on Figure 7.6. If this point is depicted on the power-required curve it occurs at the speed where the tangent from the origin touches the curve and is referred to as VMD, which is a TAS, and for a constant mass increases in value with increased altitude. See Figure 9.7(b).

In the clean configuration, when related to the stalling speed (Vs), which is the speed at which the lift generated in level flight is insufficient to sustain the mass, VIMD for a piston/propeller-driven aeroplane is approximately 1.3 times the stalling speed for the mass being considered. Whereas, for a jet-engined aircraft it is approximately 1.6 times the stalling speed for the mass being considered. From this it can be

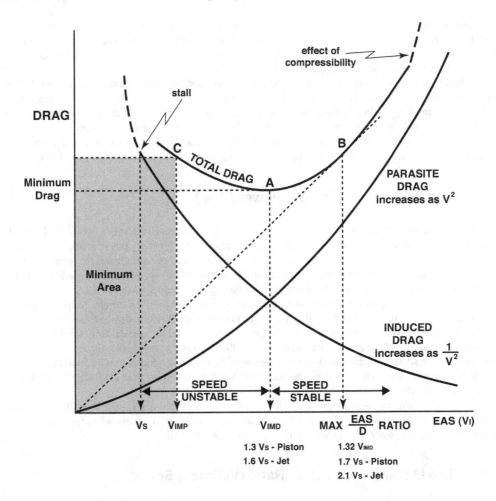

Figure 7.6 The Drag v EAS Curves.

deduced that to maintain speed stability in level, unaccelerated flight a jet aeroplane has to be flown at a relatively high speed compared to the propeller-driven aeroplane. V_IMD is *always* a higher speed than the minimum power required speed, V_IMP (described below).

The value of V_IMD in level flight at a given angle of attack increases with increased mass and decreases with decreased mass but not linearly because parasite drag is not affected to the same degree as induced drag by mass changes. V_IMD is directly proportional to the aeroplane mass. If the speed is decreased below V_IMD the total drag increases because induced drag increases. See Figure 7.8(a). If the speed is greater than V_IMD, then the induced drag will decrease and the parasite drag will increase from their appropriate values at V_IMD.

For any flight condition, other than level flight, *the best lift/drag ratio is not coincident with V_IMD.* This is because the lift required in a climb or descent is less than it is in level flight. The speed attained at the best lift/drag ratio does produce the maximum glide distance from a specific altitude.

When the thrust graphs are drawn for a specific mass and pressure altitude for a piston/propeller aeroplane, the V_IMD obtained is used as the basis to determine the maximum range for propeller-driven aircraft in those conditions. The greatest range is obtained at the altitude at which the engines are most efficient. For aerodynamic reasons the maximum range for a jet aeroplane is attained at the higher speed of that which is obtained where the EAS/Drag ratio is maximum.

For a jet-engined aeroplane, V_{IMD} at the highest practical altitude is used to obtain the maximum endurance, thus it is used to fly for as long as possible using all of the available fuel. This speed is used because the thrust required is least and results in the lowest fuel flow but it increases the flight time. V_{IMD} is also the holding speed for jet aeroplanes.

In the event of one power unit of a multiengined aeroplane becoming inoperative, the total drag increases because of the increased rudder trim drag necessary to counteract the yaw effect caused by the dead windmilling engine. The aeroplane, as a result of the increased total drag and reduced thrust available may not be able to maintain altitude. If such is the case and there is high terrain to be cleared, then the greatest vertical clearance will be obtained by flying at the one-engine-inoperative speed to obtain the best lift/drag ratio, which will be close to V_{IMD} for this configuration. (See page 193).

7.12   The Velocity of Minimum Power (V_{IMP})

Power is a measure of the work done per unit of time and is equal to force × velocity. In level, unaccelerated flight, the force is thrust, which is equal to total drag. Therefore, power required is equal to total drag multiplied by TAS. The lowest point on the power-required curve indicates the speed at which the product of the speed and total drag (thrust required) are least. The speed attained at the lowest point on the power-required curve is referred to as the minimum power speed V_{MP} and is a TAS.

Minimum power required = minimum value of total drag × TAS in level flight. Minimum power speed can be shown on the total drag (thrust required) curve as the EAS that results from a horizontal drawn from the left vertical axis to intercept the total-drag curve and the vertical from this intersection to the horizontal axis of the graph that contains the smallest area on the graph, that is the thrust required times EAS. The speed at which this occurs is referred to as V_{IMP} when located on the total drag (thrust required) curve. From Figure 7.6 it can be seen that V_{IMP} occurs at a lower EAS than V_{IMD} and is labelled C on the diagram.

Because the total drag is the same at all altitudes at the same gross mass, then V_{IMP} for a constant mass does not vary with altitude. However, V_{MP} for a constant mass will increase with increased altitude due to the effect of density on the speed conversion from EAS to TAS.

7.13   The Maximum EAS/Drag Ratio (V_I/Dmax) Speed

The maximum range, that is the greatest distance that can be travelled using all of the fuel available, for a *jet aeroplane* is realised by flying at the speed at which the maximum EAS/Drag ratio occurs, often referred to as V_I/Dmax. This is the highest EAS that can be obtained in relationship to the amount of drag incurred and will produce the highest TAS and enables the greatest distance to be travelled for a given amount of fuel.

V_I/Dmax can be determined by drawing the straight-line tangent from the origin of the graph to the total drag (thrust required) curve. A vertical from this intersection to the carpet of the graph shows the EAS at which this ratio is greatest. It is approximately equal to 1.32 V_{IMD} for both types of aeroplane.

When related to the stalling speed (V_s), in the clean configuration, V_I/Dmax for a propeller-driven aeroplane equates to approximately $1.7V_s$ and for jet-engined aeroplanes to approximately $2.1V_s$. This is point B in Figure 7.6.

The total drag for a constant mass is the same at all altitudes, and the EAS at which this occurs remains the same. When this EAS is converted to a TAS it produces a much higher value at high altitudes than at low altitudes becoming almost double its mean sea-level value at 40 000 feet.

Because the fuel consumption for a jet aeroplane is approximately proportional to the thrust available, the fuel efficiency is improved at high altitude because of the decreased thrust produced. Furthermore, because of the high TAS attained at high altitude, the distance travelled is considerably increased for a given quantity of fuel, as is the maximum range attained. Hence, it is essential to fly jet aeroplanes at the highest practicable altitude, at a speed of V_I/Dmax, to obtain the maximum range.

7.14   Speed Stability and Instability

An aeroplane is considered to be flying at a stable speed if a disturbance causes an undemanded speed change and it naturally returns to its original speed when the disturbance ceases. If the speed continues to deviate from the original speed after the disturbance ceases, the speed is considered to be unstable. It is considered to be neutrally stable if, after the disturbance has ceased, the speed remains at the new value, neither increasing nor decreasing.

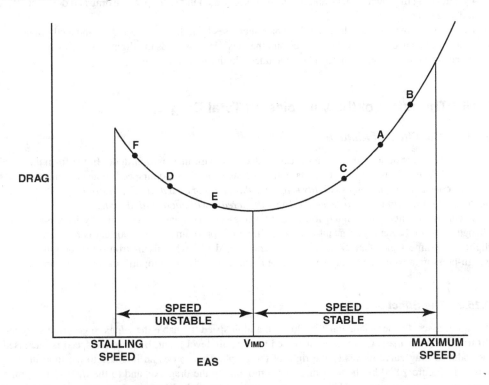

Figure 7.7   Speed Stability and Instability.

The speeds in the range between VIMD and the maximum speed are deemed stable because if a disturbance causes the speed to:

a. Increase, for example from A to B in Figure 7.7, and the thrust remains unchanged then the drag will exceed the thrust and the speed will automatically decrease to its original value.
b. Decrease, for example from A to C in Figure 7.7, and the thrust remains constant then the thrust will exceed the drag and the aeroplane will automatically accelerate to its original speed.

There are two speed ranges considered to be unstable; they are between M0.80 and M0.98 (See Chapter 15) and the speeds in the range between VIMD and the stalling speed because if a disturbance causes the speed in the low speed range to:

a. Increase, for example from D to E in Figure 7.7, the drag will decrease and if the thrust remains constant it will exceed the drag and the aeroplane will accelerate.
b. Decrease, for example from D to F in Figure 7.7, the drag will increase and if the thrust is unchanged then the drag will exceed the thrust and the aeroplane will decelerate further.

Any acceleration increases the speed stability of an aeroplane and any deceleration decreases the speed stability. The area of the graph to the left of V$_{IMD}$ is referred to as the area of 'reversed command.' This means that if the airspeed is decreased from this point more thrust is required. It is sometimes called the region of speed instability or colloquially referred to as 'the backside of the drag curve.'

To increase the range of stable speeds and decrease the range of unstable speeds, V$_{IMD}$ must be decreased in value. This can be achieved:

a. by extending the undercarriage and/or speed brakes, which increases parasite drag and does not affect induced drag.
b. by extending the flaps to a large angle, which increases both the parasite and the induced drag.
c. as a consequence of the fuel being used and the mass decreasing as the flight progresses.
d. by ensuring the design of an aeroplane includes a high aspect ratio wing.

7.15   The Effect of the Variables on Total Drag

7.15.1   The Effect of Altitude

The density of the atmosphere and air pressure decrease with increased altitude. In the formula: Total Drag $= C_D {}^1/_2 \rho V^2 S$. If V in the formula is the IAS as shown on the airspeed indicator, and the mass remains constant, then *if compressibility is ignored the values of total drag, C$_D$, V$_{IMD}$ and V$_S$ (IAS) remain unchanged for all altitudes; however, the TAS will increase with increased altitude.*

If compressibility is accounted for, then the values of V$_{IMD}$ and V$_S$ will slightly increase but this change does not become significant until an altitude of approximately 30 000 feet is reached. In level flight at the same angle of attack, ambient temperature and TAS when flying from a low-pressure area to a high-pressure area the total drag will increase because it is directly proportional to the air pressure.

7.15.2   The Effect of Mass

At high masses, the aeroplane must be flown at a high speed to enable the wings to generate sufficient lift to support the mass. C$_L$ must be increased to maintain level flight. Because the speed is increased, the induced drag curve moves to the right of the graph. This causes an increase to the total drag, as shown in Figure 7.8(a) by the total-drag curve moving up the drag axis and to the right of the graph. Thus, the upward movement of the total-drag curve to the right increases the value of the minimum drag and increases V$_{IMD}$ proportional to the mass increase. This movement of the total-drag curve also causes an increase in the stalling speed (V$_S$). *The value of the stalling speed also varies in proportion to the mass.*

7.15.3   The Effect of Flap

Extending flap has a twofold effect. It increases both the lift and the profile drag but not proportionately. The total drag increases at a greater rate than the total lift. The use of flap effectively increases the angle of attack. A large flap angle will cause a large increase in the effective angle of attack, so the aeroplane must be flown at a lower speed. The maximum value of parasite drag and total drag are therefore experienced at lower speeds. Thus, the total-drag curve moves to the left.

If the angle of attack is decreased to maintain constant lift the induced drag will not increase but the parasite drag curve moves to the left, as shown in Figure 7.8(b). This causes the total-drag curve to move upward and to the left. Thus, by extending flap it decreases the values of V$_{IMD}$ and the stalling speed and increases the range of stable speeds. However, flap cannot be lowered at speeds in excess of V$_{FO}$, the design speed for lowering flap. The aeroplane must not be flown with flap extended at speeds in excess of V$_{FE}$, the maximum speed with flaps extended.

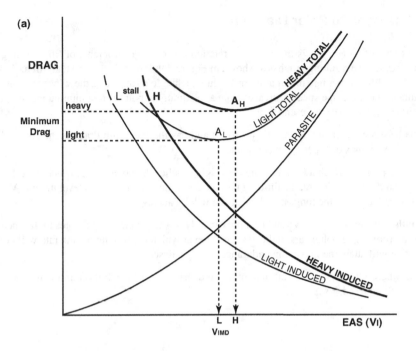

Figure 7.8(a) The Effect of Increased Mass on Drag.

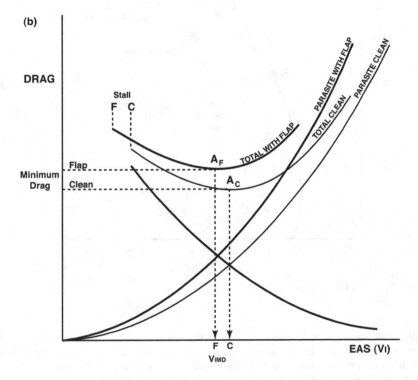

Figure 7.8(b) The Effect of Increased Flap on Drag.

7.16   The C∟ v C◠ Polar Diagram

The polar curve of an aerofoil is the graphical relationship of the coefficients of lift and drag plotted against each other, the resulting graph is as shown in Figure 7.9. It reveals that the total lift and total drag increase from point A to a maximum at point D, thus the angle of attack of the wings increases in the same manner. Therefore, the speed decreases from a maximum at point A to a minimum at point D. The points marked on the curve represent:

Point A is between VIMD and the maximum speed and has the minimum drag coefficient. It therefore approximates VI/Dmax or VY for a jet aeroplane.

Point B is the point at which a tangent from the origin touches the polar curve, which is at the point at which the total drag is minimum, the lift/drag ratio is maximum and the IAS in level flight is VIMD. This is also the speed at which the longest glide distance will be achieved.

Point C falls between VIMD and Vs and is approximately VIMP, which approximates to the speed of VY for a piston/propeller aeroplane and indicates the value at which the minimum sink rate will occur. It is the speed that will attain the maximum endurance during a descent.

Point D is where CL and CD are both maximum and the point at which the minimum speed in level flight, the stalling speed, Vs is attained.

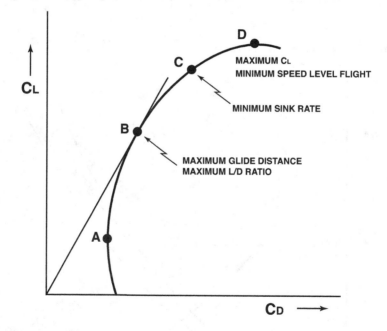

Figure 7.9   The CL v CD Diagram.

The effect that the aeroplane configuration has on this curve is shown in Figure 7.10. It can be seen that the clean aeroplane, i.e. the aeroplane with undercarriage and flaps retracted, has the least drag. Lowering flap increases the amount of drag experienced but significantly increases the lift generated. The second curve is moved to the right if the undercarriage is lowered, with the flap extended, but there is no increase of lift. However, if the undercarriage is lowered without flap then the curve produced is

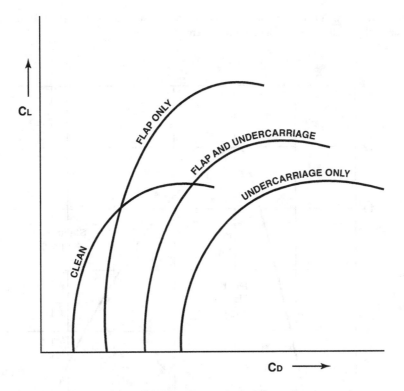

Figure 7.10 The Effect of Configuration on the C$_L$ v C$_D$ Diagram.

the same as the first curve but significantly moved to the right. This shows a considerable increase of drag but no increase of lift. *Identification of the curves in Figures 7.9 and 7.10 has been required in past examination questions.*

7.17 Analysis of the Lift/Drag Ratio

For any given mass and altitude, level flight can be maintained at any speed, between the maximum and minimum limits, by adjusting the angle of attack to keep the lift formula in balance, to ensure the lift developed is equal to the mass. However, as the angle of attack increases so does the induced drag, this increases the total drag. To counteract the increased total drag a greater amount of thrust is required. This will increase the fuel consumption and is clearly an inefficient way to maintain level flight.

The most efficient manner in which to fly an aeroplane is with an angle of attack that creates the greatest amount of lift whilst causing the least amount of drag. Therefore, the most efficient angle of attack is that which produces the highest lift/drag ratio. This angle of attack is the same for all masses and altitudes and is usually 4°. The EAS at which this occurs is the most economical cruise EAS and varies in direct proportion to the gross mass of the aeroplane. In a glide descent from a specific altitude in still air the speed at the maximum lift/drag ratio will enable the aeroplane to attain the maximum glide distance, which is close to but not V$_{IMD}$.

The lift/drag ratio can be calculated for a range of angles of attack and plotted as depicted in Figure 7.11. As shown the highest lift/drag ratio of 12:1 occurs at an angle of attack of 4°. At the stalling angle of 15°, where the lift developed is insufficient to support the mass of the aeroplane, the lift/drag ratio is approximately 4.8:1 which is the minimum lift/drag ratio. The specific ratios vary for different aircraft types but the angles of attack are constant for most aeroplane types.

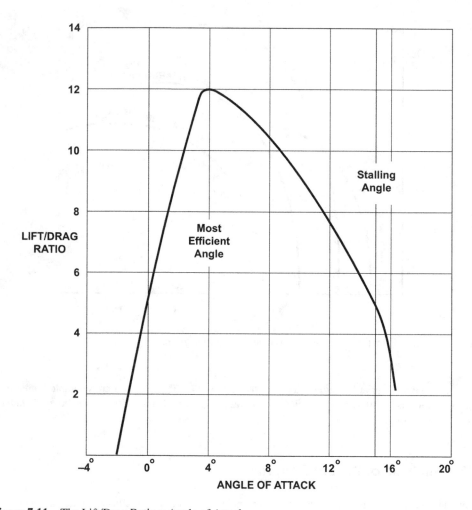

Figure 7.11   The Lift/Drag Ratio v Angle of Attack.

7.17.1   The Effect of Flap

Although lift increases with greater flap angles so also does drag, but to a much greater extent, particularly at angles of attack above that which produces the best lift/drag ratio. Thus, *lowering flap usually decreases the lift/drag ratio* and, although it is possible to fly at a lower speed, the thrust required increases. Selecting the largest flap angle causes the greatest negative influence on the lift/drag ratio. Leading-edge devices when used incur a relatively small drag penalty for the increase of lift attained; consequently the lift/drag ratio is improved at high angles of attack over that achieved without using this device.

7.17.2   The Effect of Aspect Ratio

The maximum value of the Lift/Drag ratio is determined by the aspect ratio. High aspect ratio wings produce a high maximum Lift/Drag ratio value and vice versa. Because of this, aeroplanes with short wings must be flown at low angles of attack to be efficient; these are attained at relatively high airspeeds. Aeroplanes having a high aspect ratio are flown relatively slowly at high angles of attack to remain efficient. Hence, sailplanes fly at a low airspeed and a high angle of attack to fly for maximum endurance or the airspeed at the maximum lift/drag ratio for maximum range.

7.17.3   The Effect of Mass

An increased aeroplane mass requires the speed to be increased to produce sufficient lift to sustain the new mass. The increased speed increases the induced drag, thus the total drag is increased, which means that the Lift/Drag ratio is unaffected by mass changes.

7.18   Drag Augmentation

It is sometimes necessary to deliberately increase the total drag to decrease V_{IMD}. Using one or more of the following high-drag devices achieve this aim by increasing the parasite drag:

a. airbrakes;
b. spoilers;
c. barn-door flaps;
d. drag parachutes.

7.19   Airbrakes

Airbrakes are a very basic mechanical method of dramatically increasing the profile drag and decreasing V_{IMD} without destroying the lift during landing and are predominantly used on military jet aircraft. They all are hydraulically operated but vary considerably in design. Some consist of a number of finger-like metal bars that extend approximately three inches from the upper surface of the wing in a line along the wing at the point of maximum thickness, whilst others are operated by splitting the tail cone.

On some American fast jets a special kind of aileron, known as a deceleron, that functions normally in flight but on landing, when selected, it can split with the upper surface hinging upward from its leading edge and simultaneously the lower surface hinging downward from its leading edge together they produce a considerable amount of drag. They are not used in their pure form on civilian aircraft but their characteristics are combined with those of spoilers to the maximum benefit of the aircraft on landing.

7.20   Spoilers

The method most commonly used on large passenger-carrying aircraft to destroy the lift and increase the drag, at the same angle of attack, is the aerodynamic spoiler sometimes referred to as a lift dumper. Spoilers are flat metal plates mounted on the upper surface of the wing that can be extended upward into the airflow to diminish the lift generated, i.e. C_L is decreased, and to increase the drag, i.e. C_D is increased. They are extended symmetrically to decrease the speed and/or increase the rate of descent. There are three types of spoiler in current use:

a. flight spoilers;
b. ground spoilers;
c. roll spoilers.

7.20.1 Flight Spoilers

In flight the spoilers are raised symmetrically to slow the aeroplane in level flight or to increase the rate of descent without increasing the airspeed. However, their use for this purpose on airliners is often limited because the noise and vibration may give the passengers cause for concern. The spoilers can be deployed to enable an aircraft to descend in a controlled manner at a speed above the maximum speed for flap operation (VFO). When extended the spoilers destroy some of the total lift generated and create a moderate increase to the total drag, which together diminish the lift/drag ratio. If the speed and the load factor remain constant after the extension of the flight spoilers then CD increases but CL is not affected.

7.20.2 Ground Spoilers

During the landing ground run after touchdown the aeroplane is still travelling at a relatively high speed that could cause the aeroplane to become airborne again should there be a sudden gust of wind. To prevent such an occurrence ground spoilers may be used after touchdown that totally destroy the remaining lift. They also cause the support of the mass to be transferred from the wings to the undercarriage. This increases the effectiveness of the brakes and decreases the likelihood of the aeroplane skidding when the brakes are applied. These spoilers can only be deployed if the weight-on ground sensing switch on the undercarriage oleo is made first. Together with reverse thrust, the form drag, caused by the spoilers, directly assists in halting the aeroplane.

7.20.3 Roll Spoilers

Roll spoilers are flat panels mounted in the upper surface of the wings that deploy asymmetrically to assist the downward wing aileron effect a turn by decreasing the lift on that wing. Thus, if a roll right is demanded then the right aileron will move up and the right spoiler will reduce the lift generated by the right wing. Usually, on transport aeroplanes, the roll spoilers are connected to one of the two control columns and the ailerons to the other control column. The two control yokes are mechanically linked through a clutch so that they operate in harmony with the ailerons. In this manner the manufacturers ensure that if either the ailerons or the roll spoilers become jammed the clutch will disconnect the jammed control surfaces from the other control surfaces to enable one of the pilots to retain a reasonable measure of roll control.

The roll-control spoilers on many aeroplanes operate as a function of the airspeed. Two sets of spoilers are installed on each wing one set operates below a specific airspeed and the other above that airspeed. The system used by high-speed swept-wing aircraft for roll control is the spoiler-mixer control. At high speed the use of aileron for roll control induces wing twisting that is clearly undesirable. Instead of using solely the ailerons to control roll; a combination of ailerons and spoilers is used. The spoilers are deployed asymmetrically by an amount appropriate to the demand made by the pilot through the aileron control. When extended they decrease the lift of that wing, which generates a rolling moment and increases the drag that suppresses the adverse yaw.

The complex control system is able to select spoilers only for high-speed flight and ailerons for flight at an intermediate speed. During the landing and take-off phase of flight to ensure an adequate rate of roll is attained a combination of aileron and spoiler has to be used. In normal cruise flight the inboard ailerons are active, the outboard ailerons are inactive and the roll-control spoilers are active. See Figures 7.12 and 7.13.

Some aircraft use spoilers in combination with or instead of ailerons for roll control, primarily to decrease adverse yaw when high speed limits the rudder input. This type of spoiler is known as a spoileron and is used as a control surface asymmetrically. On the wing on which it is raised it decreases the lift and speed and causes the aeroplane to roll and yaw.

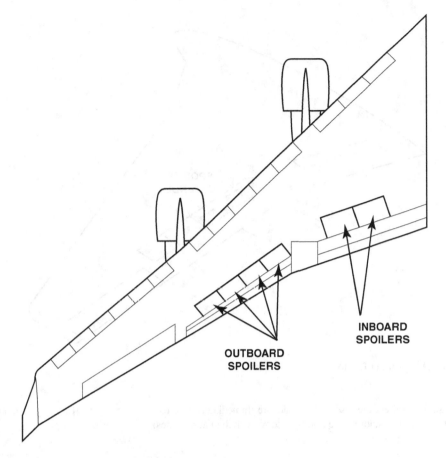

Figure 7.12   Trailing Edge-Spoilers.

7.21   Barn-Door Flaps

The term used to describe all flap systems that can be adjusted to such an angle that they produce very little increase of lift but a massive increase of drag is 'barn-door flaps.' When deployed to their maximum extension these flaps act as a large barn door impeding the airflow. Included in this group are Fowler flaps, Fairey–Youngman flaps and other double-slotted and triple-slotted systems.

The advantage of using this system is that on the approach, apart from slowing the aeroplane and decreasing the stalling speed, they allow a much steeper approach which gives the pilot a better view of the runway.

7.22   Drag Parachutes

Braking drag parachutes are only used on military aeroplanes that have extremely high touchdown speeds. They can only be used once and have to be jettisoned after use. They are therefore an effective but severe method of stopping an aeroplane during the landing ground run. A ground support crew must

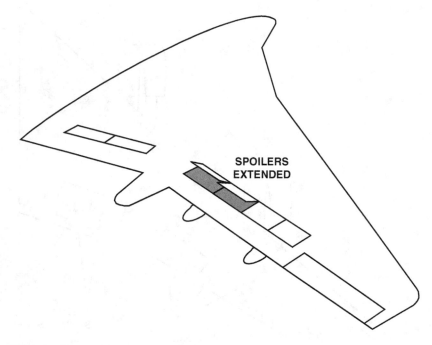

Figure 7.13 Spoiler Deployed.

be present to collect the jettisoned parachute immediately to clear the runway for further use. Clearly then it is a most unsuitable method for use with civilian aeroplanes.

Self-Assessment Exercise 7

Q7.1 The polar curve of an aerofoil is the graphical relationship between:
  (a) TAS and the stalling speed
  (b) the angle of attack and C_L
  (c) C_D and the angle of attack
  (d) C_L and C_D

Q7.2 The frontal area of a body, placed in an airstream is three times that of a body having precisely the same shape. Its aerodynamic drag will be . . . . times that of the smaller body.
  (a) 1.5
  (b) 3.0
  (c) 9.0
  (d) 6.0

Q7.3 If an aeroplane flies in ground effect the result will be:
  (a) the lift increases and the drag decreases
  (b) the effective angle of attack is decreased
  (c) the induced angle of attack is increased
  (d) drag and lift decrease

Q7.4 When spoilers are used as speed brakes:
  (a) the C_{Lmax} of the polar curve is unaffected
  (b) they do not affect wheel braking during landing
  (c) at the same angle of attack, C_L remains constant
  (d) at the same angle of attack, C_D is increased and C_L is decreased

Q7.5 Which point marks the value for the minimum sink rate?

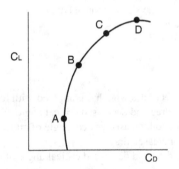

  (a) point B
  (b) point A
  (c) point D
  (d) point C

Q7.6 The induced drag coefficient, C_{DI} is directly proportional with:
  (a) C_{Lmax}
  (b) C_L^2
  (c) C_L
  (d) $\sqrt{C_L}$

Q7.7 In straight and level flight at a constant mass, parasite drag is directly proportional to:
  (a) the square of the speed
  (b) speed
  (c) angle of attack
  (d) the square of the angle of attack

Q7.8 One of the factors on which total drag is dependent is:
   (a) airstream speed
   (b) specific body mass
   (c) body mass
   (d) location of the CG

Q7.9 Comparing C_L with C_D at a normal angle of attack:
   (a) C_L is much lower than C_D
   (b) C_L is much higher than C_D
   (c) C_L is approximately equal to C_D
   (d) C_L is lower than C_D

Q7.10 The result that 'ground effect' has on landing distance is that it:
   (a) decreases
   (b) has no effect
   (c) increases only if flaps are fully extended
   (d) increases

Q7.11 When flying at a speed greater than V_{IMD} the effect that increased speed has on induced drag is
   (i)........and on parasite drag is (ii)...........
   (a) (i) decreases; (ii) decreases
   (b) (i) increases; (ii) decreases
   (c) (i) decreases; (ii) increases
   (d) (i) increases; (ii) increases

Q7.12 The lowest total drag in level flight is when:
   (a) induced drag is zero
   (b) induced drag is lowest
   (c) parasite drag is equal to induced drag
   (d) parasite drag is twice the size of induced drag

Q7.13 The total drag formula is:
   (a) $C_D.2\rho V^2 S$
   (b) $C_D.^1/_2\rho VS$
   (c) $C_D.^1/_2.1/\rho V^2 S$
   (d) $C_D.^1/_2\rho V^2 S$

Q7.14 The effect that high aspect ratio wing has compared with low aspect ratio wing is:
   (a) decreased induced drag and decreased critical angle of attack
   (b) increased total lift and increased critical angle of attack
   (c) increased lift and increased drag
   (d) increased induced drag and decreased critical angle of attack

Q7.15 Induced drag:
   (a) increases as the aspect ratio increases
   (b) is not related to the lift coefficient
   (c) increases in direct proportion to the size of the wing-tip vortices
   (d) increases as the lift coefficient increases

Q7.16 The effect that a high aspect ratio wing has on the induced drag is that it is:
   (a) decreased because the effect of the wing-tip vortices is reduced
   (b) increased because of the greater wing frontal area
   (c) unaffected because there is no relationship between them
   (d) increased because of the greater downwash

Q7.17 The part of an aeroplane that affects the induced drag the most is:
   (a) wing tip
   (b) engine cowling
   (c) wing root
   (d) landing gear

Q7.18 The ratio of induced drag to profile drag at a speed of V_{IMD} is:
  (a) 1:1
  (b) It varies according to aeroplane type
  (c) 2:1
  (d) 1:2

Q7.19 The polar diagram of an aerofoil is a graph that shows:
  (a) wing thickness as a function of the chord
  (b) the relationship between C_L and C_D
  (c) the relationship between C_L and the angle of attack
  (d) the relationship between the horizontal and vertical speeds

Q7.20 During an approach to land floating due to ground effect will occur:
  (a) when a higher than normal angle of attack is used
  (b) at a speed approaching the stall
  (c) when the height above the ground is equal to less than half the wing span
  (d) when the height above the ground is equal to less than twice the wing span

Q7.21 The information that can be obtained from a polar diagram is:
  (a) minimum L/D ratio and the minimum drag
  (b) the maximum L/D ratio and the C_{Lmax}
  (c) the minimum drag and the maximum lift
  (d) the minimum C_D and the maximum lift

Q7.22 The maximum L/D ratio is used to obtain:
  (a) the distance taken to climb to a specified altitude
  (b) the maximum distance taken to glide from a specified altitude in still air
  (c) the glide distance from a specified altitude
  (d) the minimum distance for the cruise

Q7.23 In straight and level flight at a constant mass the value of induced drag varies linearly with:
  (a) $1/V$
  (b) $1/V^2$
  (c) V^2
  (d) V

Q7.24 At which point on the polar diagram does it show the C_L for the minimum horizontal flight speed?

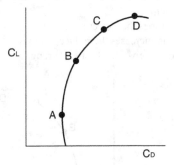

  (a) point D
  (b) point A
  (c) point B
  (d) point C

Q7.25 At a constant angle of attack, OAT and TAS the effect that increased air pressure will have on the total drag is:
    (a) it depends on the temperature
    (b) it increases
    (c) none
    (d) it decreases

Q7.26 Spoiler deflection causes:
    (a) increased lift only
    (b) decreased lift and decreased drag
    (c) increased drag and decreased lift
    (d) increased lift and increased drag

Q7.27 An aeroplane accelerates from 80 kt to 160 kt. The induced drag coefficient (i) and the induced drag (ii) change by the following factors:
    (a) (i) $^1/_2$. (ii) 1/16
    (b) (i) 4. (ii) $^1/_2$
    (c) (i) 1/16. (ii) $^1/_4$
    (d) (i) $^1/_4$ (ii) 2

Q7.28 A high aspect ratio wing produces:
    (a) an increased stalling speed
    (b) increased induced drag
    (c) decreased induced drag
    (d) less sensitivity to gust effects

Q7.29 Induced drag can be reduced by:
    (a) a thinner wing tip
    (b) increased aspect ratio
    (c) increased wing taper ratio
    (d) decreased aspect ratio

Q7.30 Excluding constants, the coefficient of induced drag is the ratio of:
    (a) C_L^2 to AR (aspect ratio)
    (b) C_L to C_D
    (c) C_L to b (wingspan)
    (d) C_L^2 to S (wing area)

Q7.31 The relationship between the induced drag and the aspect ratio is:
    (a) a decrease in aspect ratio increases the induced drag
    (b) there is no relationship
    (c) induced drag = 1.3 times the aspect ratio
    (d) increased aspect ratio increases the induced drag

Q7.32 Which point of the diagram gives the best glide range?

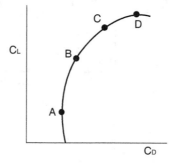

(a) point C
(b) point D
(c) point B
(d) point A

Q7.33 At a constant IAS induced drag is affected by:
(a) engine thrust
(b) aeroplane mass
(c) aeroplane wing location
(d) the angle between the wing chord and the fuselage centreline

Q7.34 In an acceleration from Vs to V_{NE} at a constant mass, the total drag:
(a) increases
(b) increases and then decreases
(c) decreases and then increases
(d) decreases

Q7.35 Induced drag is created by the:
(a) downwash across the wing
(b) spanwise flow pattern resulting in wing-tip vortices
(c) interference of the airstream between the wing and the fuselage
(d) separation of the boundary layer

Q7.36 The effect on induced drag of changes of mass is (i).......... and speed is (ii).......
(a) (i) increases with increasing mass; (ii) increases with decreasing speed
(b) (i) decreases with decreasing mass; (ii) decreases with increasing speed
(c) (i) decreases with increasing mass; (ii) decreases with decreasing speed
(d) (i) increases with decreasing mass; (ii) increases with increasing speed

Q7.37 Interference drag is caused by:
(a) the addition of induced and parasite drag
(b) the interaction of the airflow between adjoining aeroplane parts
(c) downwash behind the wing
(d) separation of the induced vortex

Q7.38 Which of the following statements regarding the L/D ratio is correct?
(a) The L/D ratio always increases as the lift decreases.
(b) The greatest value of the L/D ratio is when lift is equal to the aeroplane mass.
(c) At the greatest value of the L/D ratio the total drag is lowest.
(d) The greatest value of the L/D ratio is when the lift is zero.

Q7.39 If all other factors remain constant the effect that increased dynamic pressure has on total drag is:
(a) none
(b) that it decreases
(c) that it is only affected by the groundspeed
(d) that it increases

Q7.40 If the air density decreases to half of its original value the total drag will decrease by a factor of:
(a) 8
(b) 1.4
(c) 2
(d) 4

Q7.41 The effect that the extension of a spoiler has on a wing is:
(a) C_D is increased and C_L is decreased
(b) C_D is unaffected and C_L is decreased
(c) C_D is increased and C_L is increased
(d) C_D is increased and C_L is unaffected

Q7.42 Which of the following reduces induced drag?
  (a) flap extension
  (b) elliptical lift distribution
  (c) low aspect ratio
  (d) high angles of attack

Q7.43 Which of the following will happen in ground effect?
  (a) A significant increase of thrust is required.
  (b) The induced angle of attack and the induced drag decreases.
  (c) The wing downwash on the tail surfaces increases.
  (d) An increase in the strength of the wing-tip vortices.

Q7.44 The point on the diagram that gives the lowest speed in horizontal flight is:

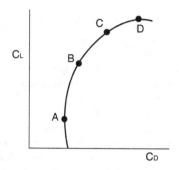

  (a) point B
  (b) point C
  (c) point D
  (d) point A

Q7.45 A high aspect ratio wing:
  (a) increases induced drag
  (b) decreases induced drag
  (c) has greater structural strength than a low aspect ratio wing
  (d) stalls at a higher angle than a low aspect ratio wing

Q7.46 When accelerating from C_{Lmax} to the maximum speed the total drag:
  (a) increases
  (b) increases and then decreases
  (c) decreases
  (d) decreases and then increases

Q7.47 If the IAS is increased by a factor of 4, the total drag would increase by a factor of:
  (a) 4
  (b) 8
  (c) 12
  (d) 16

Q7.48 If the IAS is halved, the C_{DI} would (i).........by a factor of.........
  (a) (i) increase; (ii) 4
  (b) (i) decrease; (ii) 4
  (c) (i) increase; (ii) 16
  (d) (i) decrease; (ii) 16

Q7.49 Which of the following phenomena causes induced drag?
   (a) wing-tip vortices
   (b) wing-tip tanks
   (c) the high pressure at the leading edge
   (d) the spanwise flow, inward under the wing and outward above the wing around the tip

Q7.50 If the aeroplane's undercarriage is lowered in flight, the (i) . . . . drag will (ii) . . . . and the nose-down pitching moment will (iii) . . . .
   (a) (i) form; (ii) increase; (iii) be unchanged
   (b) (i) induced; (ii) increase; (iii) increase
   (c) (i) form; (ii) increase; (iii) increase
   (d) (i) induced; (ii) decrease; (iii) increase

Q7.51 An aeroplane accelerates from 80 kt to 160 kt and the TAS also doubles. The induced drag coefficient C_{DI} (i) . . . . and the induced drag D_I (ii) . . . . change by the following factors:
   (a) (i) 2. (ii) 2
   (b) (i) 4. (ii) 2
   (c) (i) $^1/_4$. (ii) 4
   (d) (i) 1/16 (ii) $^1/_4$

Q7.52 When the airspeed is doubled the aerodynamic drag increases by a factor of . . . .
   (a) 4
   (b) 16
   (c) 1
   (d) 2

Q7.53 If the OAT and TAS are constant, and the air pressure increases then total drag will:
   (a) increase
   (b) decrease
   (c) remain constant
   (d) drag is not affected by pressure

Q7.54 In straight and level flight, if air density halves then aerodynamic drag will:
   (a) increase by a factor of two
   (b) decrease by a factor of two
   (c) increase by a factor of four
   (d) decrease by a factor of four

Q7.55 In straight and level flight at a constant IAS, when flaps are lowered the induced drag:
   (a) is unchanged
   (b) decreases
   (c) increases
   (d) increases or decreases depending on aeroplane type

Q7.56 At a normal angle of attack, the relationship of C_L and C_D is:
   (a) C_L is less than C_D
   (b) C_D is greater than C_L
   (c) C_L is approximately equal to C_D
   (d) C_L is much greater than C_D

Q7.57 Regarding induced drag and wing-tip vortices, which of the following statements is true?
   (a) There is a flow component towards the wing root on the upper surface of the wing and a flow component towards the wing tip under the lower surface of the wing.
   (b) Vortex generators diminish wing-tip vortices.
   (c) Both decrease at high angles of attack.
   (d) Induced drag increases but wing-tip vortices decrease at high angles of attack.

Q7.58 If all other factors remain constant, the effect an increase of dynamic pressure has on total drag is that it:
    (a) increases
    (b) decreases
    (c) remains the same
    (d) has no effect at speed above V_{MD}

Q7.59 The effect that winglets have is that they:
    (a) decrease the stalling speed
    (b) increase static longitudinal stability
    (c) reduce induced drag
    (d) increase the value of V_{MD}

Q7.60 The effect that winglets have is that they:
    (a) decrease the value of V_{IMD}
    (b) decrease profile drag
    (c) decrease the stalling speed
    (d) increase static directional stability

Q7.61 Induced drag is decreased by:
    (a) wing fences
    (b) anhedral
    (c) winglets
    (d) low aspect ratio

Q7.62 In straight and level flight, which of the following statements regarding the lift/drag ratio is correct?
    (a) L/D ratio is maximized when lift and mass are equal
    (b) L/D ratio decreases with increasing lift
    (c) L/D ratio is maximized when the lift is zero
    (d) L/D ratio is maximized at the speed for the minimum total drag

Q7.63 When an aeroplane enters ground effect:
    (a) lift decreases and drag decreases
    (b) lift increases and drag decreases
    (c) the induced angle of attack increases
    (d) the cushion of air partially supports the aeroplane mass

Q7.64 Ground effect could begin to be felt during the approach to land:
    (a) at a height equal to twice the wingspan above the ground
    (b) at a height equal to one wingspan above the ground
    (c) when the angle of attack is increased
    (d) at a height equal to half the wingspan above the ground

Q7.65 Ground effect has a significant effect on performance:
    (a) at a height equal to twice the wingspan above the ground
    (b) at a height equal to one wingspan above the ground
    (c) when the angle of attack is increased
    (d) at a height equal to half the wingspan above the ground

Q7.66 The lift and drag forces of an aerofoil:
    (a) vary linearly with the angle of attack
    (b) depend on the pressure distribution around the aerofoil
    (c) are equal in level flight
    (d) act at right angles to one another at only one angle of attack

Q7.67 When an aeroplane enters ground effect the lift vector is inclined (i)........., which (ii)........... the thrust required.
    (a) (i) forward (ii) reduces
    (b) (i) forward (ii) increases
    (c) (i) rearward (ii) reduces
    (d) (i) rearward (ii) increases

Q7.68 On entering ground effect induced drag..........
    (a) increases
    (b) remains constant
    (c) decreases
    (d) increases and profile drag decreases

Q7.69 On entering ground effect:
    (a) more power is required
    (b) the power required is not affected
    (c) less power is required
    (d) lift decreases

Q7.70 When the speed brakes are used as spoilers:
    (a) for the same angle of attack C_L is unaffected
    (b) C_{Lmax} is not affected
    (c) on landing the braking action is unaffected
    (d) for the same angle of attack C_D is increased and C_L is decreased

Q7.71 In normal cruise flight, which of the following devices will be active for an aeroplane fitted with (i) inboard ailerons, (ii) outboard ailerons and (iii) roll-control spoilers.
    (a) (i) inactive (ii) inactive (iii) active
    (b) (i) inactive (ii) active (iii) may be active
    (c) (i) active (ii) active (iii) may be active
    (d) (i) active (ii) inactive (iii) may be active

Q7.72 Deploying spoiler surfaces results in drag (i)..........and lift (ii)..............:
    (a) (i) increasing; (ii) increasing
    (b) (i) increasing; (ii) decreasing
    (c) (i) decreasing; (ii) increasing
    (d) (i) decreasing; (ii) decreasing

Q7.73 For a turbojet aeroplane the speed (the TAS) for the minimum power is:
    (a) less than V_{MD}
    (b) equal to V_{MD}
    (c) greater than V_{MD}
    (d) slower in a climb

Q7.74 Which of the following statements is correct regarding V_{MD}?
    (a) Parasite drag is greater than induced drag.
    (b) The ratio of lift to drag is maximum.
    (c) The best glide range is achieved at this speed.
    (d) It is the best endurance speed for a piston-engined aeroplane.

Q7.75 V_{IMD} for a jet aeroplane is approximately equal to:
    (a) $1.3V_s$
    (b) $1.7V_s$
    (c) $1.6V_s$
    (d) $2.1V_s$

Q7.76 The maximum EAS/Drag ratio is approximately:
    (a) $1.3V_{IMD}$
    (b) $1.32V_{IMD}$
    (c) $1.6V_{IMD}$
    (d) $1.8V_{IMD}$

Q7.77 The maximum EAS/Drag ratio for a piston-engined aeroplane is:
    (a) $1.3V_s$
    (b) $1.32V_s$
    (c) $1.6V_s$
    (d) $1.7V_s$

Q7.78 The angle of attack for a turbojet flying at the maximum range speed:
  (a) is equal to that for the maximum L/D ratio
  (b) is less than that for the maximum L/D ratio
  (c) is more than that for the maximum L/D ratio
  (d) is equal to that for C_{Lmax}

Q7.79 What point on the polar diagram is that at which the maximum gliding range is achieved?

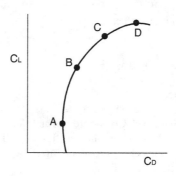

  (a) point C
  (b) point D
  (c) point A
  (d) point B

8 Stalling

8.0.1 The Stall

When the lift developed by an aeroplane wing is no longer sufficient to sustain the mass of the aeroplane in flight, in the configuration being considered, the wing is stalled. It is caused by the total drag restricting the speed of the aeroplane to less than that required to produce the necessary lift. The stalling characteristics of a wing are determined by the nature of the boundary layer of air on the surface of the wing and the type of entry to the stall. There are five types of entry to a stall; they are the low-speed stall or the high incidence angle stall, the power-on stall, the accelerated stall, the deep or superstall and the high-speed or shock stall. The low-speed stall is fully described below and the remainder is described later in this chapter but all of them have one common feature, that is that the boundary layer becomes detached from the upper surface of the wing.

8.1 The Boundary Layer

Although the boundary layer was dealt with in considerable detail in Chapter 4 it is essential to repeat the salient facts in this chapter. There is a layer of air that clings to the surface of a wing and has a resistance to flow. This resistance is the viscosity of the air, which although being very small, is sufficient for the molecules of the air in contact with the surface to stick to that surface. The air on the wing surface is therefore almost static, which causes the laminar streamline flow adjacent to this layer to slow down; consequently this produces turbulence. The speed of the airflow changes from being zero at the surface of the wing to the full speed of the freeflow airstream a few millimetres away from the surface. This thin layer of air attached to the surface of the wing, in which the speed change takes place, from being almost stationary to the full laminar flow over the wing, is the boundary layer.

In aerodynamics the boundary layer is important when considering wing stall. The depth of the boundary layer is dependent on:

a. The Reynold's number, which accounts for the speed, density, viscosity and compressibility of the free airflow. (See Chapter 1).
b. The cleanliness and the condition of the aerofoil surface.

The Principles of Flight for Pilots   P. J. Swatton
© 2011 John Wiley & Sons, Ltd

8.2 Boundary-Layer Separation

Boundary layers are classified as laminar (layered) or turbulent (disordered). A low Reynold's number indicates that the boundary layer is laminar and that the speed of the airflow changes gradually with increasing distance from the surface of the aerofoil. Higher Reynold's numbers indicate a turbulent and swirling flow within the boundary layer. If such is the case the boundary layer may lift off or 'separate' from the surface of the aerofoil and create an effective shape different from the physical shape of the aerofoil surface. Despite this, of the two, the laminar boundary layer is most likely to separate first because it lacks the kinetic energy necessary to remain attached to the surface of the aerofoil. The effect that the boundary layer has on lift is accounted for in the coefficient of lift and its effect on drag is accounted for in the coefficient of drag.

Boundary-layer separation develops because its slow movement aft is opposed by the increasing pressure towards the trailing edge of the wing that creates an adverse pressure gradient above the surface of the wing. The kinetic energy of the air at the surface of the wing is insufficient to enable the boundary layer to flow in the opposite direction to the pressure gradient. Instead, it flows in the reverse direction to the free-flowing airstream. The point at which this occurs is the separation point. Aft of the separation point little or no lift is generated.

The development of the boundary-layer separation is shown in Figure 8.1. It is the profile of the speed of the air above a wing that is depicted and is such that at point:

a. The air has a normal typical profile.
b. The adverse pressure gradient has modified the velocity profile.
c. The surface flow has ceased and this is the separation point.
d. The flow has reversed direction and the air is eddying and turbulent and has separated from the surface of the wing with a mean velocity opposite to the free-stream flow.

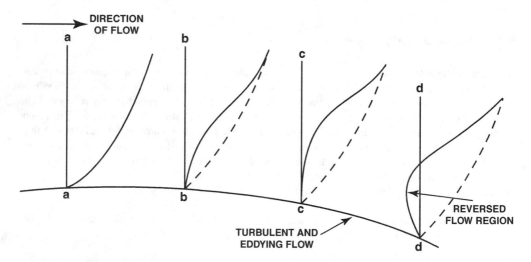

Figure 8.1 Boundary-Layer Speed Profile.

8.2.1 *Trailing-Edge Separation*

At low angles of attack, for a normal cambered subsonic wing, no flow separation occurs before the trailing edge of the wing, instead there is an attached turbulent boundary layer in this area. As the angle of attack increases the adverse pressure gradient intensifies and the boundary layer begins to separate from the surface of the wing before it reaches the trailing edge as shown in Figure 8.2. Speed does not affect the point at which separation occurs. As speed increases the kinetic energy of the airflow increases, but so also does the adverse pressure gradient because both are a function of dynamic pressure.

The separation point and the CP move forward along the upper surface of the wing towards the leading edge as the angle of attack of the wing increases. This movement of the separation point continues until at a specific angle of attack, approximately 15° or 16°, the airflow over the upper surface of the wing becomes completely separated from the wing at a point close to the leading edge of the wing. At this angle of attack the wing is stalled, the airflow is completely broken down and turbulent, furthermore now the wing is generating only a very small amount of lift, which is insufficient to maintain level flight. This is the stalling angle; at angles of attack beyond this the CP moves abruptly aft.

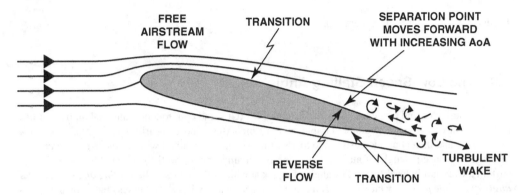

Figure 8.2   Trailing-Edge Separation.

8.2.2 Leading-Edge Separation

There is a second type of flow separation that occurs on wings having a thin cross-section and a sharp pointed leading edge. This is laminar-flow separation, which is the airflow detaching itself from the upper surface of the wing at point close to the leading edge before it has become turbulent. It forms as a vortex in which the air is turbulent and extends along the complete span of the wing. These vortices form under the separated laminar layer above and behind the leading edge of the wing. Because of their stability and predictability they can be controlled and utilised to produce lift. Swept-wing and delta-planform aeroplanes, particularly at high angles of attack, are most likely to encounter leading-edge separation. See Figure 4.5.

The stationary vortex rotates clockwise on the port wing viewed from the wing tip and vice versa on the starboard wing. This type of vortex is often referred to as a 'bubble.' It can vary in shape and size depending on the profile of the aerofoil but its magnitude is smallest at the wing root and largest at the wing tip. Although they are likely to extend along the whole wingspan, their width and depth may only reach aft a short distance, albeit in particular circumstances they can stretch rearward along the whole length of the aerofoil chordline. See Figure 8.3.

The aft spread of the bubble width can affect the pressure distribution over the wing. Short bubbles have little effect, however, long bubbles significantly affect the pressure distribution, even at low angles of attack. Long bubbles cause the stall of such a wing to be a more gradual event but if it bursts it causes an abrupt stall.

The airflow aft of this stationary vortex may reattach itself to the upper surface of the wing as a turbulent boundary layer behind the stationary vortex at a distance that is short near the wing root and at an increasingly greater distance towards the wing tip. At the wing tip the airflow does not reattach itself to the upper surface of the wing but the stationary vortex links with the wing-tip vortex adding to its strength.

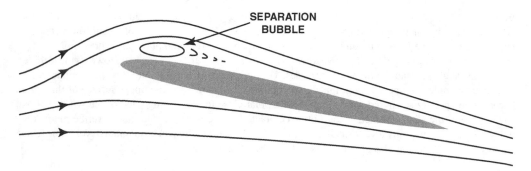

Figure 8.3 Leading-Edge Separation.

8.3 The Low-Speed Stalling Angle

At a constant mass and air density, the coefficient of lift (C_L), and consequently the total amount of lift, is dependent on the angle of attack. Graphically, lift increases almost linearly from an angle of attack of 0° to 4°, and then at an increasing rate to 15°. Lift cannot be obtained without incurring the penalty of drag. At a specific angle of attack, depending on circumstances, the lift generated is insufficient to sustain the aircraft in level flight. The angle of attack at which this occurs is the *stalling angle or critical angle and is the point at which C_{Lmax} is attained.* An aeroplane will not stall unless the critical angle is exceeded.

The magnitude of the critical angle is independent of the mass of the aeroplane, it is therefore the same for all aeroplane masses; *mass only affects the value of the airspeed at which the stall will occur.* The critical angle can be anywhere between 8° and 20° to the relative airflow; its exact value is dependent on the type of stall entry, the wing profile, the planform and the aspect ratio. Except at high Reynolds numbers, *an aeroplane, in the same configuration, in subsonic flight will always stall at the same critical angle of attack* (unless shock-induced separation occurs – see Chapter 15). The commencement and spread of the stall is dependent on the wing design, this is described later in this chapter.

At angles of attack greater than the critical angle, the total drag and C_D continue to increase but the total lift decreases dramatically, as shown in Figure 8.4. A further consequence of increasing the angle of attack beyond the critical angle is that the stagnation point will move aft along the lower surface of the wing. The wing will continue to produce some lift, albeit very little, up to an angle of attack of 90°.

The low-speed stall occurs at a particular angle of attack of the wing to the relative airflow, **not** at a specific airspeed. As the angle of attack and C_L increase, the IAS must be reduced to maintain the lift formula in balance. The speed attained at the stalling angle is the stalling speed, which is directly proportional to the aeroplane mass.

8.4 Factors Affecting the Low-Speed Stalling Angle

8.4.1 Slat/Flap Setting

The use of leading-edge slats, be they automatic or manual, cause negligible drag but do delay the stall to a higher angle of attack than would be possible without their use. They re-energise the boundary layer, enlarge the low-pressure area over the upper surface of the wing, thus producing a larger C_{Lmax}, and

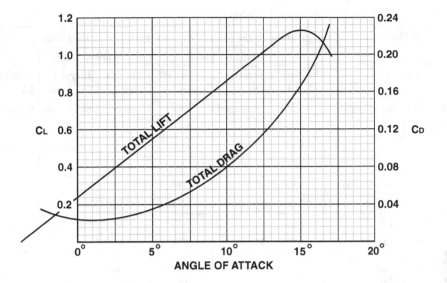

Figure 8.4   The Lift and Drag Curves.

cause the separation point to move further back from the leading edge of the wing. Leading-edge flaps have a similar effect and can increase the maximum lift generated by 50%.

Trailing-edge flaps when deployed will decrease the critical angle of attack to between 8.5° and 14.0° and increase C_{Lmax} by between 20% and 80%, depending on flap type but the increase may be limited by the boundary-layer separation. A large extension of trailing-edge flaps usually causes a considerable increase of drag and produces a low stalling angle but contrary to this, extension of the triple slotted trailing-edge flap, such as the Fowler flap, increases the critical angle to 22.0° and significantly increases the C_{Lmax} by approximately 110%. The greater the number of slots the higher is the value of C_{Lmax}. See Figure 6.13. A comprehensive list of the effect that each type of flap, slat or slot has on the stall is shown in Table 8.1.

8.4.2 Ice Accretion

The formation of ice on the surface of an aeroplane has a dramatically adverse effect on its performance; frost has a similar effect but not to the same extent, as shown in Table 8.1. The most dangerous type of ice accretion is that which occurs when flying through cumulo-nimbus cloud or through supercooled water droplets ahead of a warm front because of its rapid accumulation. The effect that ice accretion has on an aeroplane's performance includes:

a. increased mass;
b. decreased total lift;
c. decreased C_{Lmax} that is the most serious effect;
d. increased total drag;
e. up to 25% increased stalling speed;
f. reduced stalling angle;
g. changed aerofoil shape;
h. abnormal stalling characteristics.

Table 8.1   The Effect of Configuration on the Stall.

The Effect of Configuration on the Stall Compared with a Standard Wing						
		Effect on C$_{Lmax}$			Stalling Angle	
Factor		**Approx. Max. C$_L$**	**Increase/ Decrease**	**Approx% Change**	**Approx. Value**	**Increase/ Decrease**	**Stalling IAS**
Standard Clean Wing		**+ 1.55**	**Standard**	**0%**	**16.0°**	**Standard**	**Standard**
Trailing	20°	+ 1.6		+ 5%	14.0°		
Edge Plain	30°	+ 1.8		+ 15%	11.5°	Decrease	Decrease
Flaps	40°	+ 1.9		+ 25%	8.5°		
	Single Fowler	+ 2.95		+ 80%	15.0°		
	Slotted	+ 2.55		+ 65%	15.0°		
Flap Type	Split	+ 2.05		+ 30%	14.0°	Decrease	Decrease
	Plain	+ 1.85	Increase	+ 20%	12.0°		
Leading	Krueger Flap	+ 2.33		+ 50%	25.0°		
Edge	Slat	+ 2.50		+ 60%	22.0°		
	Slotted Wing	+ 2.15		+ 40%	20.0°	Increase	Decrease
Triple Slotted Fowler		+ 3.25		+ 110%	22.0°		
Slat and Slotted Flap		+ 1.80		+ 75%	25.0°		
Slat and Triple Slotted Fowler		+ 3.57		+ 130%	28.0°		
Wing	Rough	+ 1.2		− 25%	12.1°		
Surface	Frost	+ 1.2	Decrease	− 25%	12.1°	Decrease	Increase
	Ice Covered	+ 0.9		− 40%	10.0°		

8.4.3   Effect on Take-off and Landing

For an aeroplane in the take-off or landing configuration, ice accretion is *extremely dangerous* because of the close proximity of the normal operating speed to the stalling speed. Consequently, there is a very high risk of the aeroplane stalling when it is close to the ground.

8.4.3.1   Take-Off

The total effect of the aforementioned factors is to produce a much lower level of performance for any aeroplane experiencing icing conditions. Before take-off the aeroplane must be thoroughly deiced and the take-off must be completed before the effectiveness time of the deicing fluid has passed. All of the aeroplane's anti-icing and/or deicing systems should be selected 'on' before commencing the take-off.

8.4.3.2   Landing

All of the aeroplane's engine and airframe anti-icing and/or deicing systems should be used throughout the descent when icing conditions prevail. This is particularly important when the engines are throttled back during the descent and the intake air temperature is in the icing range, this can occur even in summer. The throttles should be opened up at periodic intervals during the descent to exercise the engines to ensure that full power is still available should the speed become too low.

8.4.3.3   *Reduced Stalling Angle*

The increased mass and increased total drag caused by the ice accretion can decrease the stalling angle of attack by up to 25%. Thus, the aeroplane could stall at an angle of attack as low as 8°. This means that any automated stall-warning devices will not be activated before the aeroplane has actually stalled. Such devices may also be inoperative because they are frozen solid.

The recovery from a stall in these conditions will result in a greater height loss when accelerating to a higher than normal speed. Consequently, operating the aeroplane at low speed close to the ground, as for take-off and landing, in these meteorological conditions is extremely dangerous and should be avoided if possible.

8.4.3.4   *Abnormal Stalling Characteristics*

If ice has formed on the aeroplane wings it will have also formed on the tailplane, modifying its aerodynamic effectiveness and increasing the possibility of the stabiliser stalling. As a result, the changes caused by the ice accretion will adversely affect the behaviour of the aeroplane during the stall, which may well be abnormal and unexpected.

8.4.4   Heavy Rain

Heavy rain has the following effects on aeroplane performance:

a. The impingement and rapid accumulation of water sufficiently distorts the shape of the upper wing surface and diminishes the total lift by up to 30% of its normal value.
b. The impact of the rain on an aeroplane in the landing configuration decreases the aircraft's forward speed.
c. When operating in such conditions a jet engine is slow to respond to rapid demands for thrust.
d. The mass of the rain can increase the aircraft mass momentarily.

The total effect of these factors for any aeroplane experiencing such weather conditions is to produce a much lower level of performance. The mass increase raises the stalling speed, whilst the reduced lift and increased drag together reduce the forward speed achieved. For an aeroplane in the take-off or landing configuration, this situation is *extremely dangerous* because of the close proximity of the normal operating speed to the stalling speed. There is, therefore, a very high risk of the aeroplane stalling when it is close to the ground.

If there is a possibility of encountering *heavy rain* during take-off, it is advisable to *delay* the *departure*. If the danger were during the landing phase, it would be prudent to divert to an alternate aerodrome or to hold off until the rain has cleared the area. It could prevent a disaster, because not only is there a likelihood of stalling but there is also the possibility that aquaplaning (hydroplaning) could be an additional hazard and cause the aeroplane to overrun the end of the runway.

8.5   The Effect of Wing Design on the Low-Speed Stall

As the angle of attack increases, because of the movement of the CP the pitching moment of the wing changes and affects the angle of the downwash impinging on the tailplane thereby changing its pitching moment. Because the greatest amount of lift is generated at the wing roots, on entering the stall the *CP moves aft on a straight-winged aeroplane and, because the wing tips stall first, forward on a swept-winged aeroplane.*

The overall effect varies with aeroplane type; however, most aeroplanes are designed to have a nose-down pitching moment at the stalling angle. This occurs because the CP moves rapidly aft at the stall, increasing the wing restoring moment and reducing the angle of attack. This induces the wing to produce lift once more and is therefore a self-correcting characteristic.

The shape of the wing determines the point of stall commencement and also the spread of the stall across the wing surface. Those wing shapes in common use are shown in Figure 8.5, on which the spread of the stall across each of them is depicted.

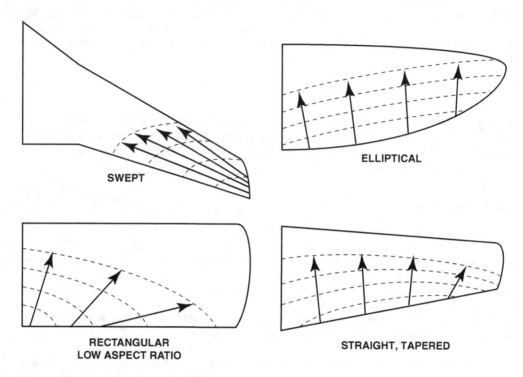

Figure 8.5   The Spanwise Spread of the Stall.

8.5.1 Swept Wings

Of all wing shapes, swept wings have the greatest tendency to tip stall first because of the spanwise flow of the boundary layer (See Chapter 7). The three elements that combine to form the wing-tip vortex on swept-wing aeroplanes are leading-edge separation, flow around the wing tips and spanwise boundary layer flow. The greater the sweepback the greater is the boundary-layer spanwise flow and the magnitude of the vortex drag. Increased angle of attack also increases the spanwise flow of the boundary layer.

Swept wings produce a nose-up pitch tendency due to the separation of the thickened boundary layer from the upper surface of the wing tip causing it to stall first, which induces the stall to spread across the wing from the tip to the root on the upper surface; consequently the CP moves forward and in towards the wing root. This results in a decreased wing-restoring moment and the maximum downwash being concentrated inboard, which increases its effect on the tailplane. Although the effectiveness of the elevators is diminished by the turbulent airflow over the tailplane, the increased downwash on the tailplane causes the tail moment to exceed the wing moment and causes the nose to pitch-up.

Some swept wings also have a high aspect ratio, which have the advantages that it decreases the critical angle of attack, decreases the size of the wing-tip vortex and decreases the induced drag but has the disadvantage that it increases the stalling speed and is more likely to tip stall. The C_L for a swept wing is diminished in direct proportion to the cosine of the sweep angle. This is because the effective fineness ratio of the wing to the airflow is increased, which decreases the acceleration of the airflow over the upper surface of the wing. (See Table 15.4).

8.5.2   Elliptical Wings

The commencement of the stall of an elliptical wing is evenly spread along the trailing edge and progresses forward uniformly towards the leading edge. The reason for this is that the wing has a constant lift coefficient from root to tip and all sections of the wing reach the stalling angle at the same time. It is the most efficient wing design but is difficult to manufacture.

8.5.3   Rectangular Wings

The stall of an unswept rectangular wing does not occur over the whole wing simultaneously, it commences at one particular point usually at the wing root near the trailing edge and spreads outward and forward to the rest of the wing. This pattern of progression is caused by the lift coefficient being much greater at the wing root than at the wing tip.

An untapered wing has less downwash moving inboard from tip to root and therefore stalls at the root first, which is a desirable feature. At or near the stall the wing-tip vortices are large enough to affect the airflow over the tailplane decreasing its effective angle of attack. This causes a tendency for the aeroplane to pitch nose-up.

The separation of the boundary layer from the upper surface of the wing for a rectangular, low aspect ratio, wing usually commences at the wing root at a point close to the trailing edge of the wing when the angle of attack is approximately 8°. It then gradually spreads outward and moves forward with increasing angle of attack until at approximately 15° angle of attack it is located at between 15% and 20% of the length of the mean aerodynamic chord from the leading edge of the wing.

8.5.4   Straight Tapered Wings

Because the lift distribution of a straight tapered wing is similar to that of an elliptical wing, the commencement of the stall is evenly spread along the trailing edge and progresses forward towards the leading edge but at a greater rate at the wing root than at the wing tip. This wing shape is easy to manufacture and is very efficient.

8.6   Spanwise-Flow Attenuation Devices

The retardation of the free airflow over the upper surface of the wing is caused by the boundary layer. The thicker the boundary layer the greater is the retardation and the more likely is the wing tip to stall, particularly on swept-wing aeroplanes. To lessen the likelihood of such an occurrence and to improve the low-speed handling characteristics there are four design features that may be incorporated in the construction of an aeroplane. Their common purpose is to prevent or diminish the spanwise flow of the boundary layer. They are known as spanwise-flow attenuation devices and are:

a. the wing fence;
b. the sawtooth leading edge;
c. the notched leading edge;
d. vortex generators.

8.6.1 The Wing Fence

The wing fence is a thin metal plate having a height of two or three centimetres mounted edge-on to the airflow and positioned across the wing width at a distance of one third of the semi-span of the wing from the wing tip and extending from just beneath the leading edge of the wing along the upper surface of the wing for approximately two thirds of the chord length. It acts as a barrier and obstructs the spanwise flow of the boundary layer along the wing surface and prevents the entire wing from stalling at once. It also stops the spanwise growth of the leading-edge separation bubble. Often, wing fences are used as an addition to or instead of slats to improve the low-speed characteristics of an aeroplane. (See Figure 8.6(a)).

8.6.2 The Sawtooth Leading Edge

At approximately three quarters of the length of the wing, measured from the wing root, the chord length, which decreases in length with increasing distance from the wing root because of the taper, is significantly increased in length. In plan view the wing leading edge appears as a sawtooth or that it has a leading-edge extension and markedly decreases the thickness/chord ratio in the wing tip area. This has the advantage of increasing the critical Mach number, which is defined on page 352.

The shape of the wing produces a large vortex behind the sawtooth leading edge that obstructs the spanwise flow and diminishes the size of the wing-tip vortex. Because of its shape, the mean centre of pressure of this type of wing is further forward than is normal for a swept wing. The likelihood of a wing tip stall is reduced and if the wing does stall it will be less pronounced than it is for a normal swept wing, thus diminishing the pitch-up tendency. See Figure 8.6(b).

8.6.3 The Notched Leading Edge

At a position of two thirds of the wing length from the wing root a notch is designed into the leading edge of the wing that extends aft from the leading edge three or four centimetres. The shape of the wing produces a large vortex behind the notched leading edge that obstructs the spanwise flow of the boundary layer. It diminishes the size of the wing-tip vortex and produces the same benefits as the sawtooth leading edge.

On some aircraft the sawtooth and notched leading edges are used together to magnify the inboard vortex and completely inhibit the spanwise flow of the boundary layer from the wing root. This is a potent combination for preventing the wing tip from stalling. See Figure 8.6(c).

8.6.4 Vortex Generators

Vortex generators are a series of small metal plates mounted end on to the airflow, in a straight line along the upper surface of the wing at a distance of one third of the chord length from the leading edge. The plates extend from one third of the wing length from the root to three quarters of the wing length from the wing root. Each plate induces a small vortex that transfers energy from the free airflow to re-energise the boundary layer. Thus, a series of small vortices are produced along the upper surface of the wing. They do not decrease the size of the wing-tip vortex but they do decrease the intensity of the shockwave induced separation and reduce the spanwise flow of the boundary layer. See Figure 8.6(d).

The prime purpose of the vortex generators is to reinvigorate a sluggish boundary layer and prevent or delay its detachment from the upper surface of the wing. The turbulent airflow they cause introduces kinetic energy to re-energize the boundary layer by drawing down the fast-moving air from above the boundary layer and mixing the two, this enables it to counter the adverse pressure gradient more easily. This weakens any shockwaves, reduces shock drag and markedly reduces shock-induced boundary-layer separation. Often, they are used at airframe junctions to prevent boundary-layer separation.

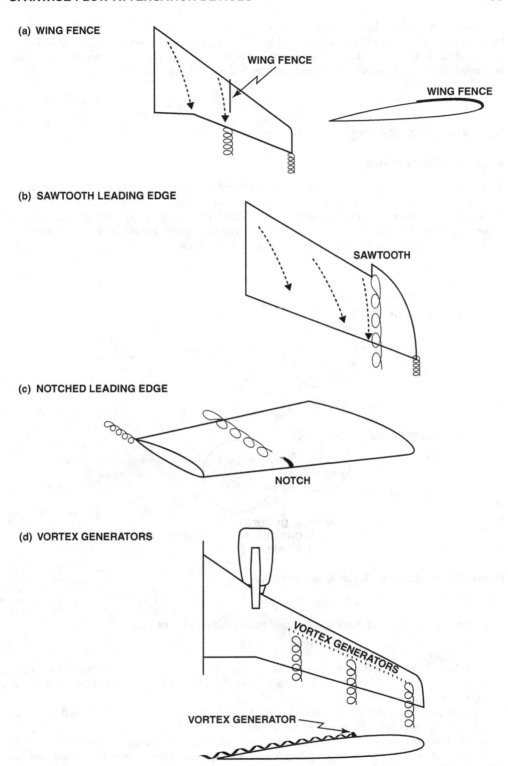

Figure 8.6 Spanwise Flow-Attenuation Devices.

Although this method incurs the penalty of a small increase to the total drag, it more than offsets this disadvantage by the considerable amount of boundary layer drag that is saved. Wings having a large thickness/chord ratio benefit the most from vortex generators. Some vortex generators are air-jets that are turned off when not required thus eliminating the drag penalty. (See page 60).

8.7 Wing-Tip Stalling

8.7.1 The Effect of Flap

Flap extension increases profile drag but has no significant effect on induced drag. At high angles of attack a swept-wing aeroplane is more likely to experience tip stalling than a straight-winged aeroplane. When flap is lowered the downwash over the flaps produces a balancing upwash over the outer portion of the wing that can be sufficient to increase the effective angle of attack to the stalling angle at the wing tip, thus making it more likely to stall, which is an undesirable feature.

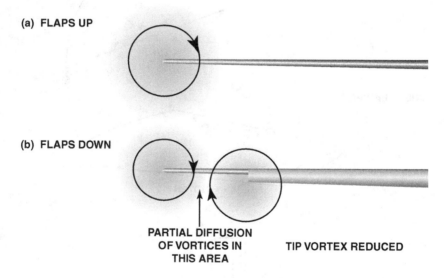

Figure 8.7   The Effect of Flap on Wing-Tip Vortices.

However, the lowered flap more than compensates for this effect in two ways:

a. It produces increased suction at the wing roots, thus restricting the spanwise outward flow of the boundary layer of the wing. This is a beneficial effect because it weakens the initial strength of the wing-tip vortex.
b. It generates a vortex at the outboard end of the lowered flap that rotates in the same direction as the wing-tip vortex but the interaction between the vortices decreases the magnitude of the wing-tip vortex. Thus, the already debilitated wing-tip vortices decrease in size and are further weakened. However, the total amount of induced drag is unaffected because there are now two vortices instead of one on each wing. See Figure 8.7.

8.7.2 The Prevention of Wing-Tip Stalling

Wing-Tip stalling is an undesirable feature because it can lead to control difficulties. It can be prevented or its likelihood reduced by the inclusion of any one or more of the following design features in the construction of the wing:

8.7.2.1 a. Washout.

The wing is constructed so that the angle of incidence decreases from root to tip; in this manner the wing is slightly twisted. This ensures that the wing root reaches the critical stalling angle of attack before the wing tip for unswept wings.

8.7.2.2 b. Root Spoiler.

The leading edge of the wing root is constructed so that it is much sharper than the rest of the wing leading edge. Consequently, it is more difficult for the airflow to follow the leading edge and separates before the rest of the wing. A stall inducer, stall strip or root spoiler can be fitted to the inboard section of the leading edge of the wing so that at high angles of attack the airflow over the wing is spoilt and this induces the wing root to stall first.

8.7.2.3 c. Changing Camber.

The camber of the wing is constructed in such a manner that it gradually increases from the root to the tip. Thus, the tip produces proportionately more lift than does the root, which delays the stall of the tip behind that of the root.

8.7.2.4 d. Slats and Slots.

The inclusion of a slat or a slot in the design of the outer portion of the wings effectively increases the stalling angle of the wing tips to greater than that of the wing root. The use of slats significantly increases the stalling angle and C_{Lmax} of the whole wing.

8.7.2.5 e. Aspect Ratio.

High aspect ratio wings reduce the strength and size of the wing-tip vortices and cause an increased C_{Lmax}, it also causes the stalling speed to increase and the stalling angle decrease to be less than that of a low aspect ratio wing. Induced drag is inversely proportional to the square of the aspect ratio. If the aspect ratio is doubled then the induced drag is halved.

8.8 Stalling Characteristics

8.8.1 Ideal Stalling Characteristics

As previously stated, it is desirable that the wing stalls first at the wing root and then spreads spanwise towards the wing tip. Such a wing has the following advantages:

a. Full aileron control is available up to the point of the stall but use of the ailerons at very high angles of attack may initiate a tip stall because they are located at the trailing-edge extremities of the wing.
b. Stall warning is given by the flow separation at the wing root buffeting the tailplane.
c. There is a decreased probability of experiencing a wing drop at the stall.

d. There is less likelihood of the stall inducing a spin because the rolling moment, that may arise if one wing stalled before the other, is minimized.

8.8.2 Swept-Wing Stalling Characteristics

Notwithstanding the fact that some or all of the spanwise-flow prevention devices are fitted to aeroplanes with a significant amount of sweepback, tip stalling will still occur before the rest of the wing. A swept wing when subject to additional loading not only produces an upward bend but a twist as well. This is because, by design, most of the wing structure is aft of the wing root and results in a reduced angle of attack at the wing tip. This reduces the amount of lift produced by the wing tip added to which the pooling of the spanwise flow of the boundary layer and its separation from the wing upper surface cause the wing tip to stall first. This causes the CP to move inward toward the wing root.

As the stall spreads towards the wing root the total lift generated by the wing gradually decreases and the CP slowly moves forward. As a result, of the movement of the CP the aircraft nose will pitch-up when it stalls and the downwash over the horizontal tail causes a lack of elevator response which may lead to a deep stall condition, which is extremely undesirable.

8.9 Summary of Factors Affecting the Stalling Angle

The stalling angle is affected by the following factors:

Stalling (Critical) Angle is	
Increased by:	**Decreased by:**
a. Leading-edge slats.	a. Trailing-edge flaps.
b. A small amount of thrust.	b. Ice accretion.
c. Low aspect ratio.	c. High aspect ratio.
d. Swept wing.	d. Straight wing.
e. Large leading-edge radius.	e. Heavy rain.
f. High Reynolds number.	f. Low Reynolds number.
g. Symmetrical aerofoil.	g. Highly cambered aerofoil.

8.10 Aerodynamic Stall Warning

To ensure the pilot is able to maintain positive control of the aeroplane at all times, it is essential that the speed at which this is no longer possible, be known for each configuration of the aircraft. This speed is the stalling speed. The aeroplane may be considered stalled when the behaviour of the aeroplane gives a clear distinctive indication of an unacceptable nature to the pilot that the aeroplane is stalled.

Acceptable indications of a stall, occurring either individually or in combination are:

a. A nose-down pitch that cannot be readily arrested.
b. Buffeting, of a magnitude and severity that is a strong and effective deterrent to further speed reduction. or
c. The pitch control reaches the aft stop and no further increase in pitch attitude occurs when the control is held full aft for a short time before recovery is initiated. *CS 25.201(d)*.

The effectiveness of the controls, in particular the ailerons, will decrease as separation occurs. The most reliable aerodynamic indication of the approaching stall is the vibration felt on the control column and/or the rudder pedals caused by the elevators and/or the rudder being buffeted by the separated turbulent airflow over those control surfaces. Within a few degrees of the stalling angle the buffeting will become noticeable and will give adequate warning of the onset of the stall.

The severity of the prestall buffet is dependent on the position of the tail surfaces with respect to the turbulent wake. If trailing-edge flap is used it may decrease the severity of the buffet and therefore diminish the amount of warning.

8.11 Mechanical Stall Warning

To provide automatic stall warning to the pilot, the device installed has to be activated when the angle of attack attained is just less than the stalling angle; this occurs when the stagnation point moves rearward along the lower surface of the wing. Either of two systems may be employed to produce such a warning, they are the flapper switch and the angle of attack sensor. They must activate for each normal configuration at a speed of Vsw (velocity of stall warning) + 5 kt or Vsw + 5%, whichever is the greater or when the speed is decreasing at a rate of less than 1 kt/s Vsw + 3 kt or Vsw + 3% whichever is the greater. Beware that in icing conditions the system fitted to the aeroplane is likely to be frozen and may not operate. *CS 25.107(c) and (d)*.

The automatic procedure adopted for recovery action from the stall warning is:

a. Reduce the angle of attack below the warning-activation value by pushing the nose-down and preventing the stall developing.
b. Simultaneously, apply maximum thrust/power to minimize height loss.

8.11.1 The Flapper Switch

The flapper switch is a vane fitted to the leading edge of the wing and positioned just below the airflow stagnation point in level flight. The vane is fitted with an electrical contact, which when made activates the stall warning in the cockpit. For angles of attack up to a preset angle of attack, which is just less than the stalling angle, the vane is closed and remains flush with the surface of the wing leading edge. The switch remains open and the electrical circuit incomplete.

As the angle of attack increases the stagnation point moves aft along the lower surface of the wing downward from the leading edge of the wing. The vane is operated by the direction of the airflow over it, which is downward until the stagnation point on the leading edge of the wing is below the flapper vane. When this occurs the airflow reverses its direction and flows upward, which opens the vane and closes the electrical switch that completes the electrical circuit to the stall-warning device in the cockpit.

8.11.2 The Angle of Attack Sensor

If the system requires a stick shaker and/or a stick pusher to be activated as well as the stall-warning device then it is necessary to employ a more sophisticated accurate sensing device than the flapper switch. The angle of attack sensor is used for this purpose, which consists of a synchro attached to an aerodynamic freely rotating vane protruding from the side of the fuselage near the nose of the aeroplane.

When the sensor is initially installed the vane and the synchro are aligned against the datum marked on the fuselage, at the most efficient angle of attack for the aeroplane in the clean configuration. The synchro is electrically connected to the stall-warning system and the stick shaker and/or the stick pusher are activated when the aerodynamic vane attains a preset angle of attack just before the stalling angle is reached. The electrical activation circuit includes a compensation device to allow for the changed attitude of the aeroplane when flap is deployed. There is also a weight-on microswitch that prevents the system from operating when the aeroplane is on the ground.

8.11.3 Stick Shakers

On some aeroplanes, where the buffet preceding a stall is absent or could be confused with turbulence, in particular heavy aeroplanes with powered controls, a stick shaker is incorporated to simulate the prestall buffet effect of turbulent airflow over the elevators. A sensor, on the angle of attack vane, that detects the angle of attack and the rate of change of the angle of attack, feeds a signal to the stick-shaker motor when either parameter indicates an approach to the stall and makes the control column vibrate at a similar frequency to that of the aerodynamic buffet. It operates when the IAS for a given configuration is at 1.05Vs. The vane is a small aerodynamic wing protruding from the side of the fuselage that indicates the angle of attack visually on an appropriately etched scale. *CS 25.207(c)*.

8.11.4 Stick Pushers

Some systems include a stick pusher that initiates the stall recovery automatically if the pilot fails to respond to the stall warning. The system is activated at a predetermined angle of attack, which is greater than that which activates the stick shaker and pushes the stick forward. Aircraft prone to succumbing to a deep stall or superstall condition or excessive wing drop, such as those with swept wings, a high-speed wing section or a 'T' tail configuration, have such a device always included in the design. This is because on those aeroplanes the airflow over the tailplane, which would assist the stall recovery for a normal aeroplane, is absent; turbulent airflow virtually covers the whole tailplane, making it ineffective. The system prevents the pilot from increasing the angle of attack any further.

8.12   Stalling Speed

By definition, the total lift of an aeroplane in level flight is equal to the mass of that aeroplane. From the total lift formula, if the altitude and wing area remain constant and the engine(s) are throttled back then the total drag will increase, consequently the speed and the total lift both decrease. The stalling speed (IAS) for the same mass and configuration of a particular aeroplane type would be the same at all altitudes were it not for the effect of compressibility increasing its value at high altitudes.

As the speed decreases, to maintain level flight the lift must be increased, which can be achieved by increasing the angle of attack. If this procedure is continued then eventually at a particular IAS the lift developed will be insufficient to sustain the aeroplane's mass in level flight and the aeroplane's wings will stall. The level-flight speed at which this phenomenon occurs can be determined by rearranging the total lift formula as follows.

If C_L = the coefficient of lift; ρ = atmospheric density; V = speed and S = wing area then the formula is:

$$(1)\ \textbf{Total Lift} = C_L . \tfrac{1}{2}\rho . V^2 . S$$

$$(2)\ V^2 = \frac{\textbf{Total Lift}}{C_L . \tfrac{1}{2}\rho . S}$$

$$(3)\ V^2 = \frac{2 \times \textbf{Total Lift}}{C_L . \rho . S}$$

$$(4)\ V = \sqrt{[(2 \times \textbf{Total Lift}) \div (C_L \times \rho \times S)]}$$

This same formula can be used to determine the value of the level-flight stalling speed if C_{Lmax} is used instead of C_L. Hence:

$$(5)\ V_s = \sqrt{[(2 \times \textbf{Total Lift}) \div (C_{Lmax} \times \rho \times S)]}$$

Because total lift in level flight is equal to total mass then it can be substituted in the formula and it becomes:

$$(6)\ V_s = \sqrt{[(2 \times \textbf{Mass}) \div (C_{Lmax} \times \rho \times S)]}$$

The stalling speed is equal to the square root of twice the aeroplane mass divided by the maximum coefficient of lift multiplied by the air density and the wing area.

8.13   Factors Affecting Stalling Speed

The effect of each of the factors affecting the stalling speed are described in the following paragraphs.

8.14   Centre of Gravity (CG)

The stalling speed and the ease with which the recovery can be achieved are both influenced by the position of the centre of gravity. Although the CG is required to remain in the CG envelope its position within that envelope both forward or aft will affect the manner of the stall and will produce quite different advantages and disadvantages. The *stalling angle is unaffected by the position of the CG.*

8.14.1   Forward CG

The stick force required to maintain level flight with a forward CG is considerably more than it is with a neutral CG, because the aeroplane is nose-heavy; effectively it is the same as a mass increase. For any given aeroplane mass the further forward the CG, the greater is the stick force *required* to maintain level flight at any particular airspeed.

8.14.1.1   Disadvantage

The disadvantage of a forward CG position is that the aeroplane at the same mass will stall at a higher airspeed than it would for a neutral CG position. This is because the stalling angle of attack is reached at a higher speed due to increased wing loading. The aeroplane may not even reach the stalling angle but instead just mush down.

8.14.1.2   Advantage

This CG position has the advantage that the pilot is assisted in the recovery from the stall because the aeroplane is nose-heavy.

8.14.2   Aft CG

With CG in an aft position, within the CG envelope, the aeroplane is tail-heavy, which causes the stick force to be much less than it would be for a neutral CG position.

8.14.2.1 Disadvantage

A disadvantage of this CG location is that because of decreased longitudinal stability the aeroplane may well stall with very little backward movement of the control column; however, the major disadvantage of this CG position is that the recovery from the stall will be quite difficult, although not impossible provided the CG is within the CG envelope. If the CG is aft of the CG envelope then recovery from the stall may well be impossible because the pilot could reach the maximum limit of the elevator angular deflection before recovery has been achieved.

8.14.2.2 Advantage

The only advantage of the CG in this position is that the stalling speed will be lower than it would be for a neutral CG position.

8.15 Mass

An aeroplane will usually stall at an angle of attack of between 14° and 16°, depending on the shape of the wing and other factors but, *it will always stall at the same angle of attack irrespective of the mass.* The speed at which this occurs varies in direct proportion to the aeroplane mass.

The IAS of the stalling angle increases with increased mass and/or a forward movement of the centre of gravity. Because of the relationship between the stalling speed and the square root of the aeroplane mass in formula (6) above then the revised stalling speed for a mass change is equal to the value of the stalling speed at the original mass multiplied by the square root of the new mass divided by the original mass.

For small mass changes up to 25% of the original total mass, the percentage change to the stalling speed is approximately equal to half of the percentage mass change. Other speeds, such as V_{IMD}, that are multiplicands of Vs can be treated in a similar manner to Vs for mass changes. The stalling speed after a mass change can be calculated by using the formula:

$$\text{Vs (new mass)} = \text{Vs (original mass)} \times \sqrt{(\text{new mass/original mass})}.$$

Example 8.1

An aeroplane of mass 150 000 kg has a stalling speed of 120 kt. Fuel usage decreases the mass of the aeroplane to 120 000 kg. Determine the value of the revised stalling speed.

Solution 8.1. By Formula:
Vs at 120 000 kg = Vs at 150 000 kg × $\sqrt{(120\,000/150\,000)}$
Revised Vs = 120 × $\sqrt{(120\,000/150\,000)}$ = 107.3 kt.

Solution 8.1. By approximation:
% mass decrease = $(30\,000/150\,000) \times 100 = 20\%$
% Speed decrease = 10%. Revised Vs = $120 - 12 = 108$ kt.

Example 8.2

An aeroplane of mass 360 000 N has a stalling speed of 130 kt. Fuel usage decreases the mass of the aeroplane to 240 000 N. Determine the value of the revised stalling speed.

> **Solution 8.2. By Formula:**
> Vs at 240 000 N = Vs at 360 000 N × $\sqrt{(240\,000/360\,000)}$
> Revised Vs = 130 × $\sqrt{(240\,000/360\,000)}$ = 106.1 kt.
>
> **Solution 8.2. By approximation:**
> % mass decrease = (120 000/360 000) × 100 = 33.3%
> % Speed decrease = 16.7%. Revised Vs = 130 − 21.7 = 108.3 kt.

8.16 Altitude

The formula to determine the total lift at the stall is:

$$\textbf{Total Lift} = \textbf{C}_{\textbf{Lmax}}.^{1}\!/_{2}\boldsymbol{\rho}.\textbf{V}^{2}.\textbf{S}$$

At the stalling angle C_{Lmax} and S are constant and the indicated airspeed is given by the dynamic pressure $^{1}\!/_{2}\rho.V^{2}$. Therefore, for the same aeroplane mass and configuration in straight and level flight, if the slight difference between EAS and IAS caused by compressibility is ignored, an aeroplane will *stall at the same IAS at all altitudes*. If compressibility is not ignored then the stalling IAS value slowly increases with increased altitude but the change only becomes significant at very high altitudes.

The V in the formula may be a TAS or an IAS. If it is a TAS then the stalling speed will decrease with increasing altitude because of the decreasing atmospheric density. As the effect of compressibility is negligible below 300 kt TAS, then the stalling speed as an IAS may be assumed to be the same as the EAS value. However, because of the relationship of IAS to TAS then the stalling speed as an IAS remains the same for all altitudes. For all practical purposes, the stalling speed for a particular aeroplane mass and configuration is the same IAS for all altitudes.

8.17 Configuration

The use of both trailing-edge flaps and/or leading-edge slats increase the effective angle of attack and decrease the stalling speed. The greater the angle of the trailing-edge flap deployed the lower is the stalling speed. Extension of the undercarriage does not enhance the amount of lift generated but does increase the total amount of drag and therefore increases the stalling speed.

To ensure that the stalling speed during the approach to land is as low as possible; it is common practice to use a combination of either leading-edge flap or slat together with trailing-edge Fowler flaps. Despite the effect of the undercarriage the stalling speed in this configuration remains relatively low.

8.18 Ice Accretion

A layer of ice on the surface of a wing changes its shape and its lifting ability. The rougher surface increases the surface friction and reduces the energy of the boundary layer. The shape of the aerofoil is changed, resulting in modified airflow and pressure gradients. Together, all of these factors increase the total drag and decrease the maximum lift coefficient (C_{Lmax}). The increased mass caused by the ice accretion will result in the stalling speed being increased by up to 25%.

8.19 Wing Planform

It is the airflow at right angles to the leading edge of a wing that determines the amount of lift that a wing will generate. Thus, the lift for a straight-winged aeroplane is produced by the full airspeed passing over the wing but for a swept-wing aeroplane it is created by a speed equal to the speed of the free airflow multiplied by the cosine of the angle subtended between the direction of the free airflow and the normal to the leading edge of the wing.

In other words, the lift of a swept-wing aeroplane is produced by a speed equal to the free-airflow speed multiplied by the cosine of the sweepback. Thus, the maximum lift created by a swept-wing aeroplane is less than that of a straight-winged aeroplane having the same wing area and wing loading.

Although the stalling speed is affected by many different factors, it follows then that the greater the sweepback the higher is the stalling speed. Thus, if a straight-winged aeroplane stalls at 100 kt, then the approximate theoretical stalling speed of a swept-wing aeroplane having the same wing area and wing loading would be equal to 100 kt × inverse cosine sweep angle. For example:

$$20° \text{sweep angle} = 100 \times 1.06 = 106 \, \text{kt}.$$

$$30° \text{sweep angle} = 100 \times 1.15 = 115 \, \text{kt}.$$

$$40° \text{sweep angle} = 100 \times 1.31 = 131 \, \text{kt}.$$

8.20 Summary of Factor Effects on Stalling Speed

The magnitude of the stalling speed is affected by many diverse factors:

Stalling Speed is	
Increased by:	**Decreased by:**
a. Increased mass.	a. Decreased mass.
b. Decreased flap angle.	b. Increased flap angle.
c. Forward CG position.	c. Deployment of slats.
d. Increased sweep angle.	d. Decreased sweep angle.
e. Increased load factor.	e. Decreased load factor.
f. Turning or pulling out of a dive.	f. Aft CG position.
g. Turbulence penetration.	g. High Aspect ratio.
h. Ice accretion.	
i. Heavy rain.	

8.21 The Speed Boundary

The difference between the high-incidence angle stalling IAS and the critical Mach number IAS decreases with increasing altitude and is known as the speed boundary, the speed margin or the buffet speed range. At high altitude, where most jet aeroplanes operate, the difference between these speeds is relatively small. The low-speed margin is the separation of the cruising IAS from the stalling IAS and the high-speed margin is the separation of the cruise IAS from the critical Mach number IAS. Clear-air turbulence and jet streams are prevalent at these altitudes, it is therefore extremely important to ensure that the aeroplane cruises at a stable speed to prevent the occurrence of any unwanted instability or oscillations.

The triangular shape at the top of the flight envelope chart at the altitude at which the low-speed stall IAS and M$_{\text{CRIT}}$, the high-speed stall, IAS are equal is colloquially known as 'coffin corner' or 'Q-corner.' This is because to fly at such an altitude demands a great deal of skill from the pilot; any increase or decrease of speed will cause the aeroplane to lose lift. The specific values of altitude and speed at this point vary with mass, load factor and ambient temperature. See Chapter 15. Transport jet aeroplanes fly at a much lower altitude than this because of the inherent danger.

8.22 The Effect of a Gust on the Load Factor

The load factor is equal to the total lift divided by the total mass. The load factor may change in flight as a result of:

a. decreased mass, due to the usage of fuel;
b. increased or decreased effective mass due to turbulence;
c. increased effective mass due to the forces acting on the aeroplane during a turn.

The effect that a gust of wind experienced during any turbulence has on the load factor is often expressed, as a change to angle of attack and the resulting change to the load factor has to be calculated.

Example 8.3

In straight and level flight an aeroplane has a C_L of 0.42. A 1° increase/decrease of angle of attack increases/decreases the C_L by 0.1. If a gust causes a 3° increase in the angle of attack calculate (i) the revised load factor caused by this event (ii) the change to the load factor (iii) the percentage change to the load factor.

Solution 8.3
Change to the $C_L = 3 \times 0.1 = +0.3$. New $C_L = 0.42 + 0.3 = 0.72$.

(i) New Load factor $= 0.72 \div 0.42 = \mathbf{1.71}$
(ii) Change to the load factor $= 1.71 - 1 = +\mathbf{0.71}$
(iii) The percentage change to the load factor $= \mathbf{71\%}$

A change of load factor will affect the stalling speed in the same manner as a change of mass. The formula for calculating the revised stalling speed following a load factor change is:

Vs (new load factor) = Vs (original load factor) × $\sqrt{}$(new load factor/original load factor).

Example 8.4

An aeroplane with load factor 1 has a stalling speed of 60 kt. Fuel usage increases the load factor of the aeroplane to 2. Determine the value of the revised stalling speed.

Solution 8.4
Vs at load factor 2 = Vs at load factor 1 × $\sqrt{}$(2/1). Revised Vs $= 60 \times \sqrt{}(2) = 84.9$ kt.

8.23 Turn Stalling Speed

The stalling speed in steady coordinated level turns increases to a value greater than that for level, unbanked flight because of the increased load factor. Its precise value is equal to the basic level-flight unbanked stalling speed multiplied by the square root of the load factor. The load factor during a turn is equal to one divided by the cosine of the bank angle ø. For a 60° banked turn the load factor is 2; therefore the stalling speed is equal to $\sqrt{2}$ or 1.4142 times the basic stalling speed. See Table 8.2.

Lift in a turn = Turn load factor × cosø

Turn stalling speed = Unbanked stalling speed × $\sqrt{}$(Load factor) or

Turn stalling speed = Unbanked stalling speed × $\sqrt{}$(1/cosø) or

Turn stalling speed = Unbanked stalling speed × $\sqrt{}$(Lift/Mass).

Example 8.5

Given: Stalling speed in a 20° banked turn is 65 kt. What is the stalling speed for the same aeroplane at the same mass in a 50° banked turn?

Solution 8.5
Since the basic stalling speed has not been given it is necessary to calculate the 50° banked turn stalling speed using the comparative load factors.

Vs (new load factor) = Vs (original load factor) × $\sqrt{}$ (new load factor/original load factor).
Load factor for 20° banked turn = 1/ cos 20° = 1.0641778
Load factor for 50° banked turn = 1/ cos 50° = 1.5557238
Vs (new load factor) = 65 × $\sqrt{}$(1.5557238 ÷ 1.0641778) = **78.6 kt.**
OR Simply using the cosines of the angles of bank (AoB) as follows could have shortened this calculation: cos 20° = 0.9396926; cos 50° = 0.6427876

Vs (new AoB) = Vs (original AoB) × $\sqrt{}$(cosine original AoB/cosine new AoB)

Vs (new angle of bank) = 65 × $\sqrt{}$(0.9396926 ÷ 0.6427876) = **78.6 kt.**

Table 8.2 shows that compared with the stalling speed of the lesser bank angle a 10° increase of bank angle from 20° to 30° increases the stalling speed by 5%, however, the same bank angle increase of 10° from 50° to 60° increases the stalling speed by 13% and from 60° to 70° increases the stalling speed by 21%. Thus, the increase to the stalling speed is greatest for changes of bank angle in turns that already having a large bank angle.

8.24 Stalling-Speed Definitions

All stalling speeds defined in CS 25 are calibrated airspeeds. The configuration used to determine the stalling speeds with (i) the power off and with (ii) the power set to maintain level flight at 1.5V_{SR} is specified in CS 25 as:

a. The most unfavourable position of the CG for recovery from the stall.
b. Representative masses within the normal operating range of masses.

Table 8.2 Example Turn Stalling Speeds.

Basic Stalling Speed 60 kt.					
Bank Angle ø	Cosine ø	1/Cosine ø Load Factor	$\sqrt{1/Cosine\ ø}$	Vs Multiplier [Increase]	Vs in a Turn
15°	0.9659	1.0353	1.0174	1.02 [+2%]	61.2 kt
20°	0.9397	1.0642	1.0316	1.03 [+3%]	61.8 kt
30°	0.8660	1.1547	1.0746	1.08 [+8%]	64.8 kt
45°	0.7071	1.4142	1.1892	1.19 [+19%]	71.4 kt
50°	0.6428	1.5557	1.2473	1.25 [+25%]	75.0 kt
60°	0.5000	2.0000	1.4142	1.41 [+41%]	84.6 kt
70°	0.3420	2.9238	1.7099	1.71 [+71%]	102.6 kt
75°	0.2588	3.8640	1.9657	2.00 [+100%]	120.0 kt

c. The aeroplane trimmed for straight flight at not less than $1.13 V_{SR}$.
d. The flaps, landing gear and deceleration devices in all likely combination of positions approved for operation. *CS 25.201(a) and (b).*

8.24.1 V_{CLmax}

This is the calibrated airspeed when the C_L is maximised (i.e. maximum lift is attained) and is obtained with:

a. the engines developing zero thrust at the stall;
b. the pitch controls (if applicable) in the take-off position;
c. the aeroplane in the configuration being considered;
d. the aircraft trimmed for level flight at not less than $1.13 V_{SR}$ and not greater than $1.3 V_{SR}$. *CS 25.103(b); AMC 25.103(b).*

8.24.2 V_{MS}

The lowest possible stalling speed, Vs, for any combination of AUM and atmospheric conditions with power off, at which a large, not immediately controllable, pitching or rolling motion is encountered.

8.24.3 V_{MS0}

The lowest possible stalling speed, V_{S0} (or if no stall is obtainable, the minimum steady-flight speed), in the landing configuration, for any combination of AUM and atmospheric conditions.

8.24.4 V_{MS1}

The lowest possible stalling speed, V_{S1} (or if no stall is obtainable, the minimum steady-flight speed), with the aeroplane in the configuration appropriate to the case under consideration, for any combination of AUM and atmospheric conditions.

8.24.5 Vs

This is the stalling speed or minimum steady-flight speed at which the aeroplane is controllable. The stalling speed is the greater of:

a. The minimum CAS obtained when the aeroplane is stalled (or the minimum steady-flight speed at which the aeroplane is controllable with the longitudinal control on its stop).
b. A CAS equal to 94% of the one-g stall speed (V_{S1g}). *CS Definitions page 20.*

8.24.6 Vso

This is the stalling speed or the minimum steady-flight speed for an aeroplane in the landing configuration. V_{S0} at the maximum mass must not exceed 61 kt for a Class 'B' single-engined aeroplane or for a Class 'B' twin-engined aeroplane of 2 722 kg or less. *CS 23.49(c). CS Definitions page 20.*

8.24.7 Vs1

This is the stalling speed or the minimum steady-flight speed for an aeroplane in the configuration under consideration, e.g. Flaps extended. *EU OPS 1 page 1-15.*

8.24.8 Vs1g

This is the 'one-g stalling speed' which is the minimum CAS at which the aeroplane can develop a lift force (normal to the flight path) equal to its mass whilst at an angle of attack not greater than that which the stall is identified. *CS 25.103(c).*

8.24.9 Vsr

This is the reference stalling speed, which is a calibrated airspeed selected by the manufacturer and is used as the basis for the calculation of other speeds. The reference stalling speed (V_{SR}) is approximately 6% greater than Vs and is:

a. Never less than the one-g stalling speed. *CS 25.103(a).*
b. Equal to or greater than the calibrated airspeed obtained when the maximum load factor-corrected lift coefficient, (V_{CLmax}), is divided by the square root of the load factor normal to the flight path. *CS 25.103(a).*
c. A speed reduction using the pitch control not exceeding 1 kt per second. *CS 25.103(c).*
d. Never less than the greater of 2 kt or 2% above the stick-shaker speed, if a stick shaker is installed. *CS 25.103(d).*

8.24.10 Vsro

This is the reference stalling speed in the landing configuration.

8.24.11 Vsr1

This is the reference stalling speed for the configuration under consideration.

8.25 The Deep Stall

The deep stall sometimes called the superstall is the same as the low-speed stall but, because of the aeroplane design features, when the CP moves forward it has a much more dangerous consequence. The predominant design feature most likely to cause an aeroplane to superstall is a swept wing when taken past the normal low-speed stalling angle because it tipstalls causing the aeroplane to pitch-up. A T-tailplane simply intensifies the depth of the stall because turbulent air completely enfolds the tailplane making it very difficult to recover.

The main difference to the normal low-speed stall is that because it has a much larger critical angle, the turbulent wake of the stalled main wing covers the horizontal stabiliser, rendering the elevators ineffective and preventing the pilot from recovering easily from the stall. The aeroplanes likely to suffer such an effect are fitted with stick shakers and stick pushers.

8.26 The Accelerated Stall

The accelerated stall is another variation of the low-speed stall but this type of stall is brought about by a sudden increase of the wing loading or load factor. If an aeroplane attains the critical angle of attack at a load factor greater than 1 then the aeroplane will stall at a speed much higher than the normal stalling speed at that mass.

The accelerated stall can occur during the pull up from a steep dive, during a strong vertical gust in the cruise or during a turn with excessive bank angle coupled with excessive backpressure on the control column. Any one of these manoeuvres results in the angle of attack being suddenly increased. In a dive, for instance, pulling back on the control column causes the elevators to raise the nose of the aeroplane but due to inertia the aeroplane continues to descend on its original downward path. This results in the angle of attack exceeding the critical angle of attack and the aeroplane stalls.

During these manoeuvres the load factor increases to a value greater than 1, as a result the speed at which the accelerated stall will occur is equal to the normal 1g stalling speed for the aeroplane mass multiplied by the square root of the load factor. Easing the backpressure on the control column in this situation can achieve a recovery from this type of stall.

8.27 The Power-On Stall

If during take-off in the initial part of the climb or during the go-around procedure following an abandoned landing attempt, when the aeroplane has full take-off or go-around power set, the control column is pulled back into the last half of its rearward movement from its neutral position then the aeroplane will most likely stall. This type of stall is often called a 'departure stall.'

Because the thrust produced by the engines is inclined to the horizontal a component of the thrust assists the lift in balancing the aeroplane mass. This enables the aeroplane to attain an extremely high angle of attack before it stalls. The stall will occur at a lower speed than would be normal for the mass and because of the high angle of attack the stall will be more violent than would normally be the case. It is sometimes referred to as a 'hammer stall.'

Usually, the stalling speed quoted in the Aeroplane Flight Manual (AFM) for any particular mass and/or configuration assumes that the engines are throttled back to idle. If any amount of thrust/power is set, with the throttles not being fully closed, then with regard to the stall it has the following advantages:

a. The stalling speed is lower for the same mass and configuration than the power-off stall due to the large vertical component of the thrust experienced at high angles of attack.
b. The height loss experienced during recovery is less than that of a power-off stall.

c. For propeller-driven aeroplanes the propeller slipstream over the wing will further delay the stall to an even lower speed.

The power-on stall has the following disadvantages:

a. When the aeroplane does stall it is likely to be more violent.
b. There is a strong possibility of a wing drop during the stall; this feature is particularly likely with single-engined propeller-driven aeroplanes.
c. Multi-engined propeller-driven aeroplanes will suffer from tip stalling because the wing area behind the power units will have a delayed stall due to the slipstream.

8.28   The Shock Stall

A shock stall is the separation of the boundary layer behind a shockwave. It is caused by the airflow over the aeroplane's wing being disturbed by the shockwaves when the aeroplane is flying at or close to MCRIT for the aeroplane. This is described in greater detail in Chapter 15.

8.29   Stall Recovery

The method of recovery from a stall is dictated by the manner in which the airflow became separated from the upper surface of the aerofoil and is as follows:

8.29.1   The Low-speed Stall

The recovery procedure from a high angle of attack stall is to apply full thrust then push the stick fully forward in the nose-down position and in a roll-neutral position. Judiciously correct for any bank angle with rudder, coarse use of the rudder could result in the aeroplane entering a spin. When the nose has lowered and the airspeed has increased to a normal cruise value then the control column should be eased back to raise the nose to the horizon. Care must be exercised not to pull back too sharply because the aeroplane could then enter an accelerated stall. Should a wing drop during the stall then lateral correction is not applied until the wings are unstalled.

8.29.2   The Deep Stall

To unstall an aeroplane in a deep-stall condition it is essential to attempt to smooth the airflow over the tailplane. This can only be achieved by increasing the thrust and when the elevators begin to respond with the control column in the neutral roll position and the stick fully forward in the nose-down position the nose should commence to go down and a normal recovery can be instigated. Care must be exercised not to pull back too sharply because the aeroplane could then enter an accelerated stall.

8.29.3   The Accelerated Stall

The accelerated stall is caused by the aeroplane's attitude to the relative airflow changing too rapidly and this may happen in a steep turn, during a recovery from a steep dive or as a result of severe turbulence. To unstall the wings simply relax the backpressure on the control column and use a lesser position of the lateral control to continue the manoeuvre.

8.29.4 The Power-On Stall

The power-on stall normally occurs during take-off and climb-out or during the go-around procedure. It is usually due to the pilot failing to maintain positive pitch control due to a nose-up trim setting or because the flaps have been retracted prematurely. To recover from a power-on stall it is necessary to apply full thrust as soon as possible and simultaneously lower the nose. The recovery procedure is the same as the low-speed stall recovery with the stick roll-neutral nose-down, correcting for angle of bank with rudder. When the nose has been lowered and the airspeed increased to a normal cruise/climb value then gently move the control column back to raise the nose-up to the horizon.

8.29.5 The Shock Stall

To enable the boundary layer to reattach itself to the upper surface of the wing after a shock stall the speed must be reduced to below M$_{CRIT}$ and hence eliminate the shockwave.

8.30 The Spin

Should a wing drop occur during a stall it may develop into a spin; this is because the angle of attack of the downgoing wing is increased to an angle well above that of the stall as a result of its downward vertical velocity. The upgoing wing reacts in just the opposite manner and may even be unstalled due to its upgoing vertical velocity.

The downgoing wing develops very little or no lift and will continue to drop. Any attempt to correct the attitude by using aileron will worsen the situation because it will increase the angle of attack of the outboard part of the downgoing wing further. Simultaneously as the aeroplane is rolling it is also yawing due to the increased drag of the downgoing wing, this is known as autorotation or the incipient spin.

The aim of the recovery from a spin is to decrease the rolling moment in the direction of the spin and/or increase the antispin yawing moment. The sequence of control movements for recovery actions is:

a. Thrust/power to idle.
b. Full opposite rudder.
c. Control column fully forward until the spin stops.
d. Maintain ailerons in the neutral position.
e. Ease the control column back to recover from the ensuing dive.

Self-Assessment Exercise 8

Q8.1 One of the disadvantages of a swept wing is it's stalling characteristics. At the stall
(i)............occurs first, which produces (ii)...........
- (a) (i) wing root stall; (ii) a rolling moment
- (b) (i) tip stall; (ii) a nose-down moment
- (c) (i) leading edge stall; (ii) a nose-down moment
- (d) (i) tip stall; (ii) a pitch-up moment

Q8.2 For any given configuration the stick shaker is activated when the IAS is:
- (a) 1.3 Vs
- (b) 1.12 Vs
- (c) Greater than Vs
- (d) 1.2 Vs

Q8.3 An aeroplane in level flight has a stalling speed of 100 kt. In a level turn with a load factor of 1.5 the stalling speed is:
- (a) 150 kt
- (b) 122 kt
- (c) 141 kt
- (d) 82 kt

Q8.4 Compared with level flight prior to the stall, the lift (i) and the drag (ii) at the stall change as follows:
- (a) (i) increases; (ii) decreases
- (b) (i) decreases; (ii) increases
- (c) (i) decreases; (ii) decreases
- (d) (i) increases; (ii) increases

Q8.5 Which of the following statements regarding the stalling speed is correct?
- (a) The stalling speed is decreased with a smaller angle of sweep.
- (b) The stalling speed is increased with a smaller angle of sweep
- (c) The stalling speed is decreased with T-tail.
- (d) The stalling speed is decreased with increased anhedral.

Q8.6 The critical angle of attack:
- (a) decreases as the CG moves aft
- (b) alters with an increase of gross mass
- (c) remains the same for all masses
- (d) increases as the CG moves forward

Q8.7 Which of the following design features is most likely to cause an aeroplane to superstall?
- (a) A T-tail.
- (b) A canard wing.
- (c) Swept wings.
- (d) A low horizontal tail.

Q8.8 The stalling speed IAS changes in accordance with the following factors:
- (a) It increases in a turn, with increased mass and an aft CG location.
- (b) It decreases with a forward CG, lower altitude, increased mass.
- (c) It increases with an increased load factor, in icing conditions and with increased flap angle.
- (d) It increases in a turn and may increase in turbulence.

Q8.9 All other factors remaining constant, the stalling speed increases when:
- (a) spoilers are retracted
- (b) pulling out of a dive
- (c) the mass decreases
- (d) altitude is increased

Q8.10 The percentage increase of the stalling speed for a level 45° banked turn is:
   (a) 31%
   (b) 41%
   (c) 52%
   (d) 19%
Q8.11 The angle of attack at which an aeroplane stalls may be increased by the use of:
   (a) flaps
   (b) spoilers
   (c) speed brakes
   (d) slats
Q8.12 Which of the following statements regarding the spin is correct?
   (a) in a spin the airspeed continually increases
   (b) all aeroplanes are designed so that they can never spin
   (c) during spin recovery the ailerons should be in the neutral position
   (d) an aeroplane is most likely to spin when the stall starts at the wing root
Q8.13 Which of the following devices together provide stall warning?
   (a) Stick shaker and angle of attack indicator
   (b) Angle of attack indicator and airspeed indicator
   (c) Angle of attack sensor and stallstrip
   (d) Stick shaker and angle of attack sensor
Q8.14 The IAS of the stall will change in accordance with the following factors. It will:
   (a) increase during a turn, increase with increased mass and increase with a forward CG
   (b) decrease with a forward CG, decrease with higher altitude and decrease with a forward-mounted engine
   (c) increase with an increased load factor, increase with ice accretion and increase with an aft CG
   (d) increase with an increased load factor, increase with a greater flap angle and decrease during a turn
Q8.15 The effect on the stall that the downwash from a swept wing contacting a horizontal tail has is that it causes:
   (a) a nose-up tendency and/or lack of elevator response
   (b) nose-down tendency
   (c) increased sensitivity of elevator inputs
   (d) a tendency to accelerate after the stall
Q8.16 On entering a stall the CP of a straight-winged aeroplane will (i) .... and of a swept-wing aeroplane will (ii) ....
   (a) (i) move aft; (ii) not move
   (b) (i) move aft; (ii) move forward
   (c) (i) move aft; (ii) move aft
   (d) (i) not move; (ii) move forward
Q8.17 The boundary layer of a wing is caused by:
   (a) a layer of highly viscous air on the wing surfaces
   (b) the normal shockwave at transonic speeds
   (c) a turbulent stream pattern around the wing
   (d) negative pressure above the upper surface
Q8.18 The vane of a stall-warning system with a flapper switch is activated by the movement of the:
   (a) point of lowest pressure
   (b) stagnation point
   (c) centre of pressure
   (d) centre of gravity

Q8.19 The percentage lift increase in a 45° banked turn compared with straight and level flight is:
   (a) 41%
   (b) 19%
   (c) 31%
   (d) 52%

Q8.20 At high altitudes, the stalling speed (IAS):
   (a) increases
   (b) decreases
   (c) remains constant
   (d) decreases up to the tropopause and then increases

Q8.21 An aeroplane in straight and level flight has a stalling speed of 100 kt. If the load factor in a turn is 2 then the stalling speed is:
   (a) 70 kt
   (b) 200 kt
   (c) 141 kt
   (d) 282 kt

Q8.22 The pitch-up effect of an aeroplane with swept wings in a stall is due to:
   (a) aft movement of the CG
   (b) forward movement of the CG
   (c) the wing root stalling first
   (d) the wing tip stalling first

Q8.23 The function of the stick pusher is:
   (a) beyond a predetermined angle of attack to activate and push the stick forward
   (b) to activate and push the stick forward before the stick shaker
   (c) to vibrate the controls
   (d) to pull the stick, to avoid a high-speed stall

Q8.24 During an erect spin recovery:
   (a) the control column is moved to the most aft position
   (b) the ailerons are held in the neutral position
   (c) the control column is moved sideways in the opposite direction to the bank
   (d) the control column is moved sideways in the same direction as the bank

Q8.25 The normal stall recovery procedure for a light single-engined aeroplane is:
   (a) idle thrust and stick roll neutral and no other corrections
   (b) idle thrust and stick roll neutral waiting for the natural nose-down tendency
   (c) full thrust and stick roll-neutral nose-down, correcting for angle of bank with rudder
   (d) full thrust and stick-roll neutral nose-down, correcting for angle of bank with stick

Q8.26 The type of stall that has the largest associated angle of attack is:
   (a) accelerated
   (b) low speed
   (c) deep
   (d) shock

Q8.27 At low speed, the pitch-up phenomenon experience by swept-wing aeroplanes:
   (a) is caused by a boundary-layer fence on the wings
   (b) is caused by a wing tip stall
   (c) never occurs, because swept wings prevent pitch-up
   (d) is caused by the extension of trailing edge lift augmentation devices

Q8.28 Vortex generators on the upper surface of a wing will:
   (a) increase the magnitude of the shockwave
   (b) decrease the intensity of the shockwave induced air separation
   (c) increase the critical Mach number
   (d) decrease the spanwise flow at high Mach numbers

Q8.29 If the stalling speed in straight and level flight is 100 kt the Vs in a 45° banked turn will be:
  (a) 119 kt
  (b) 100 kt
  (c) 80 kt
  (d) 140 kt

Q8.30 The cause of swept-wing aeroplanes pitching up at the stall is:
  (a) the rearward movement of the CP
  (b) separated airflow at the wing root
  (c) negative camber at the wing root
  (d) spanwise flow

Q8.31 The effect on the stalling speed of a constant IAS climb as the altitude increases is that it:
  (a) is initially constant and then increases at high altitude due to compressibility
  (b) is initially constant and decreases at high altitude due to compressibility
  (c) remains constant throughout the climb
  (d) increases throughout the climb

Q8.32 At which of the following speeds should the stick shaker operate?
  (a) 1.15 Vs
  (b) 1.2 Vs
  (c) 1.5 Vs
  (d) 1.05 Vs

Q8.33 In a 15° banked turn the stalling speed is 60 kt. For the same aeroplane the stalling speed in a 60° banked turn is:
  (a) 60 kt
  (b) 83 kt
  (c) 70 kt
  (d) 85 kt

Q8.34 In straight and level flight an aeroplane has a C_L of 0.40. A 1° increase/decrease of angle of attack increases/decreases the C_L by 0.15. A gust of wind causes the angle of attack to decrease by 2°. What is the change to the load factor caused by this event?
  (a) +1.65
  (b) +0.65
  (c) −0.75
  (d) −0.35

Q8.35 In straight and level flight an aeroplane has a C_L of 0.40. A 1° increase/decrease of angle of attack increases/decreases the C_L by 0.15. A gust of wind causes the angle of attack to decrease by 2°. What is the revised load factor caused by this event?
  (a) 1.65
  (b) 0.25
  (c) −0.35
  (d) 0.65

Q8.36 In straight and level flight an aeroplane has a C_L of 0.40. A 1° increase/decrease of angle of attack increases/decreases the C_L by 0.15. A gust of wind causes the angle of attack to increase by 2°. What is the revised load factor caused by this event?
  (a) 1.75
  (b) 0.7
  (c) 1.4
  (d) 1.0

Q8.37 An aeroplane mass 250 000 N stalls at 140 kt. The stalling speed at a mass of 350 000 N is:
  (a) 172 kt
  (b) 108 kt
  (c) 88 kt
  (d) 166 kt

Q8.38 If the stalling speed of an aeroplane in straight and level flight is 120 kt. The stalling speed in a 1.5g turn will be:
  (a) 81 kt
  (b) 147 kt
  (c) 100 kt
  (d) 150 kt

Q8.39 The stalling speed in a turn is proportional to:
  (a) Lift
  (b) Mass
  (c) $\sqrt{}$Load Factor
  (d) TAS^2

Q8.40 The percentage increase to the straight and level stalling speed in a 55° banked turn is:
  (a) 32%
  (b) 10%
  (c) 41%
  (d) 45%

Q8.41 The percentage increase to the straight and level load factor in a 55° banked turn is:
  (a) 45%
  (b) 74%
  (c) 19%
  (d) 10%

Q8.42 The V speed that indicates the straight and level stalling speed in the landing configuration is:
  (a) Vs_{1g}
  (b) Vs_1
  (c) Vs_0
  (d) Vs_L

Q8.43 Which of the following design features makes an aeroplane more likely to superstall?
  (a) Swept forward wing
  (b) Swept-back wing
  (c) Engines mounted below the wing
  (d) Swept-back wing and a 'T' tail

Q8.44 When an aerofoil stalls the lift (i) . . . . and the drag (ii). . . . .
  (a) (i) decreases; (ii) decreases
  (b) (i) remains constant; (ii) decreases
  (c) (i) decreases; (ii) remains constant
  (d) (i) decreases; (ii) increases

Q8.45 The effect that tropical rain has on the drag is (i) . . . . and on the stalling speed is (ii) . . . .
  (a) (i) increase; (ii) decrease
  (b) (i) decrease; (ii) decrease
  (c) (i) increase; (ii) increase
  (d) (i) decrease; (ii) increase

Q8.46 The forward movement of the CP on a swept-wing aeroplane is caused by:
  (a) boundary-layer fences and spanwise flow
  (b) tip stalling of the wing
  (c) flow separation at the wing root due to spanwise flow
  (d) a change in the angle of incidence of the wing

Q8.47 The recovery procedure from a stall in a light single-engined aeroplane is:
(a) maximum power; stick forward roll neutral; correct for bank with rudder
(b) maximum power; stick neutral and forward, correct for bank with stick
(c) idle power; stick neutral; wait for normal nose-down tendency
(d) idle power; stick-roll neutral and forward

Q8.48 When entering a stall the CP of a straight wing aeroplane will (i) .... and for a highly swept wing will (ii) ....
(a) (i) move aft; (ii) move aft
(b) (i) move aft; (ii) not move
(c) (i) move aft; (ii) move forward
(d) (i) not move; (ii) move aft

Q8.49 When an aeroplane wing stalls:
(a) A swept wing will stall from the wing root and the CP will move aft.
(b) A swept wing will stall from the tip and the CP will move forward.
(c) A straight wing will stall from the wing tip and the CP will move aft.
(d) A straight wing will stall from the wing root and the CP will move forward.

Q8.50 If the critical angle of attack is exceeded then lift will (i) .... and drag will (ii) ....
(a) (i) decrease; (ii) decrease
(b) (i) increase; (ii) increase
(c) (i) decrease; (ii) remain constant
(d) (i) decrease; (ii) increase

Q8.51 Where does flow separation commence at high angles of attack?
(a) upper surface towards the trailing edge
(b) upper surface towards the leading edge
(c) lower surface towards the trailing edge
(d) lower surface towards the leading edge

Q8.52 The stall behaviour of a swept-wing aeroplane as a result of the turbulent downwash contacting the horizontal tail is:
(a) nose-down
(b) nose-up
(c) tendency to accelerate after the stall
(d) nose-up and/or elevator ineffectiveness

Q8.53 Stalling speed increases when:
(a) recovering from a steep dive
(b) the mass decreases
(c) the flaps are deployed
(d) there are minor height changes in the lower atmosphere

Q8.54 The combination of characteristics most likely to make an aeroplane more susceptible to a deep stall are:
(a) swept wing and wing-mounted engines
(b) swept wing and 'T' tailplane
(c) straight wing and wing-mounted engines
(d) straight wing and 'T' tailplane

Q8.55 The IAS of a stall:
(a) increases with increased altitude; increased flaps; slats
(b) increases with altitude; forward CG and icing
(c) aft CG and increasing altitude
(d) altitude does not affect the stall speed IAS

Q8.56 An aeroplane at low subsonic speed will never stall if the:
(a) CAS is kept above the power-on stalling speed
(b) IAS is kept above the power-on stalling speed
(c) critical angle of attack is not exceeded
(d) pitch angle is negative

Q8.57 The angle of attack at the stall:
(a) increases with a forward CG
(b) increases with an aft CG
(c) decreases with a decrease of mass
(d) is not affected by mass changes

Q8.58 The stall-warning device is activated by movement of:
(a) CG
(b) CP
(c) stagnation point
(d) AC

Q8.59 Which of the following increases the stalling angle?
(a) flaps
(b) slats
(c) spoilers
(d) ailerons

Q8.60 The stalling speed increases when:
(a) mass decreases
(b) spoilers are retracted
(c) altitude is decreased
(d) pulling out of a dive

Q8.61 Vortex generators are fitted to:
(a) de-energise the boundary layer
(b) prevent tip stalling
(c) prevent spanwise airflow
(d) reduce the severity of the shock-induced airflow separation

Q8.62 The method by which vortex generators achieve their purpose is that they:
(a) transfer energy from the free airstream to the boundary layer
(b) decrease the kinetic energy to delay the separation
(c) diminish the adverse pressure gradient
(d) redirect the spanwise flow

Q8.63 The stalling speed in a 60° banked turn increases by a factor of:
(a) 1.30
(b) 2.00
(c) 1.41
(d) 1.07

Q8.64 The stalling speed in a 60° banked turn at a constant altitude will be .... times greater than the level flight stalling speed.
(a) 1.60
(b) 1.19
(c) 1.41
(d) 2.00

Q8.65 At high altitude an aeroplane encounters severe turbulence but no high-speed buffet. In these conditions if the aeroplane decelerates the type of stall most likely to be experienced first is:
(a) low-speed
(b) accelerated
(c) deep
(d) shock

Q8.66 Vortex generators:
(a) transfer the energy from the free airflow to the boundary layer
(b) change the turbulent boundary layer into a laminar boundary layer
(c) reduce the spanwise flow of a swept wing
(d) take kinetic energy out of the boundary layer to reduce the separation

Q8.67 Vortex generators on the upper surface of the wing will:
  (a) decrease the stalling speed by increasing the tangential velocity of the swept wing
  (b) increase the effectiveness of the spoiler due to the increase of parasite drag
  (c) decrease the shock-wave-induced separation
  (d) decrease the interference drag of the trailing-edge flaps

Q8.68 Vortex generators:
  (a) induce laminar flow by using the free-stream flow
  (b) prevent spanwise flow
  (c) increase the energy of the turbulent boundary layer by using the free-stream flow
  (d) induce boundary-layer separation using the energy of the laminar flow

Q8.69 An aeroplane at take-off has a mass of 10 000 kg and a basic stalling speed of 150 kt. On approach to land the aeroplane has a mass of 7500 kg, 30° angle of bank and land flap selected that doubles the value of C_{Lmax}. The stalling speed in this configuration is:
  (a) 112 kt
  (b) 99 kt
  (c) 130 kt
  (d) 158 kt

Q8.70 In steady horizontal flight the C_L of an aeroplane is 0.35. A one-degree increase in the AoA increases the C_L by 0.079. A vertical gust of air instantly changes the AoA by three degrees. The increase of the load factor is:
  (a) 1.9
  (b) 1.68
  (c) 0.9
  (d) 0.68

Q8.71 The cause of swept-wing aeroplanes having a tendency to pitch nose-up is:
  (a) using devices that augment lift
  (b) wing-tip stalling
  (c) wing fences
  (d) vortex generators

Q8.72 The reason that swept-wing aeroplanes are more likely to deep stall is:
  (a) the CP moves aft
  (b) the CP moves forward
  (c) the wing root stalls first
  (d) The upper surface spanwise flow is from tip to root

9 Thrust and Power in Level Flight

9.1 Thrust

The force produced by the engine(s) of an aeroplane in a forward direction is thrust. The unit of measurement of thrust is the Newton, which is mass (kg) multiplied by acceleration (m/s^2). The total thrust developed by the engine(s) is the thrust available and can be depicted on a graph for any specific aeroplane mass against the thrust required, which is the total drag. The graph is normally drawn with thrust on the vertical axis and the speed as equivalent airspeed (EAS) on the horizontal axis. *Any graph is valid only for the gross mass of the aeroplane for which it was drawn.*

9.2 Analysis of the Thrust Curves

9.2.1 Thrust Available

At the commencement of the take-off run, when the aeroplane is stationary with the wheel-brakes applied and the throttles/thrust levers set at the maximum setting permitted for take-off, both the piston/propeller and jet engine are producing thrust but no power. This is shown for each type of aeroplane on the thrust-available curves in Figure 9.1(a) as point A for the piston/propeller aircraft and point G for the jet aeroplane at a speed of 0 kt in Figure 9.1(b).

The thrust developed by a fixed-pitch propeller will decrease as the aeroplane accelerates because the propeller blade effective angle of attack decreases, which diminishes propeller efficiency. A jet aeroplane experiences a similar decrease but this is due the change from static thrust to net thrust.

The total thrust produced by a jet aeroplane is shown in Figure 9.1(b) as an almost straight line, between points G and H and the recovery produced by the 'ram effect,' which restores the thrust available to the line H to J. As the airspeed increases more air is forced into the engine intake, effectively increasing its density; at approximately 300 kt TAS and above, the increase is sufficient to regain the thrust lost; this is 'ram-effect recovery.' The jet net thrust available in level flight is:

a. directly proportional to engine RPM;
b. directly proportional to air density;
c. inversely proportional to air temperature;
d. directly proportional to airspeed at constant RPM.

The Principles of Flight for Pilots   P. J. Swatton
© 2011 John Wiley & Sons, Ltd

9.2.2 Thrust Required

The thrust-required curve is an exact replica of the total-drag curve for the aeroplane type at the same mass and configuration. The shape of the thrust-required (drag) curve is the same for both the piston/propeller and jet aeroplanes and is labelled as such in both Figures 9.1(a) and 9.1(b). At low speeds it can be seen that the thrust required increases as the speed decreases because of the rapidly increasing effect of induced drag. This area is colloquially called 'the area of reversed command' because the increasing thrust requirement is usually associated with increasing speed. Speeds less than V_{IMD} are considered unstable and those speeds greater than V_{IMD} are considered stable.

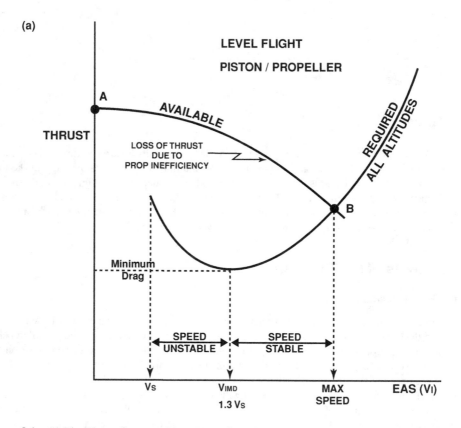

Figure 9.1   (a) The Thrust Curves – Piston/Propeller Aircraft.

9.2.2.1 Maximum Speed (EAS)

If the thrust available exceeds the thrust required in level flight then the aeroplane will climb unless the throttles/thrust levers are adjusted to maintain level flight. If level flight is maintained using the control column and trimmers without adjusting the throttles/thrust levers the aeroplane will accelerate until it reaches its maximum EAS, which is shown as the intersection of the 'available' and 'required' curves. This intersection is shown as point B on Figure 9.1(a) for piston/propeller aircraft and point J on Figure 9.1(b) for jet-engined aeroplanes. At these points the maximum thrust available equals the thrust required.

(b)

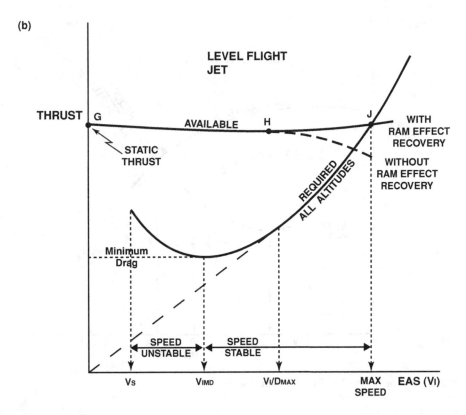

Figure 9.1   (b) The Thrust Curves – Jet Aeroplanes.

The EAS at which this occurs is equivalent to the same TAS as the intersection of the power-available curve and the power-required curve for the same aeroplane gross mass being considered. *For all masses, the highest maximum TAS will be obtained at the lowest practicable altitude for a piston/propeller aeroplane but for a jet-engined aeroplane it will be attained at the highest possible altitude.*

9.3   The Effect of the Variables on Thrust

9.3.1   Altitude

In a normal atmosphere, an increase of altitude causes the air density to decrease. At a constant aeroplane mass, the thrust developed decreases with increased altitude as a result of the decreased air density; however, the total drag, thrust required, and EAS remain constant. Because the EAS is constant at all altitudes so also is the C_L. *The thrust required (drag) is the same at all altitudes for the same EAS and gross mass; but the thrust available decreases with increased altitude.*

This effect is shown, for a piston-engined aeroplane in Figure 9.2(a) and for a jet-engined aircraft in Figure 9.2(b), by the thrust-available curves reducing, for the same mass with increased altitude. It will be seen from this diagram that the excess thrust available over thrust required and the maximum EAS are both reduced with increased altitude, i.e. the maximum EAS is obtained at the lowest practical altitude.

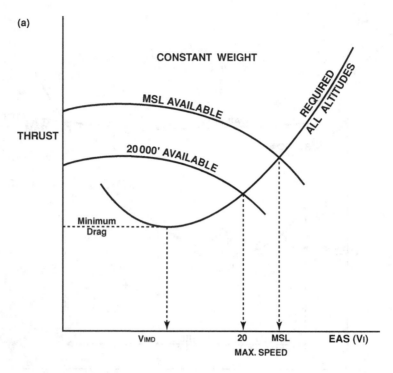

Figure 9.2   (a) The Effect of Altitude on Thrust – Piston Aircraft.

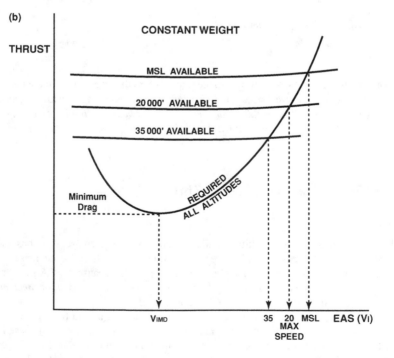

Figure 9.2   (b) The Effect of Altitude on Jet Thrust v EAS.

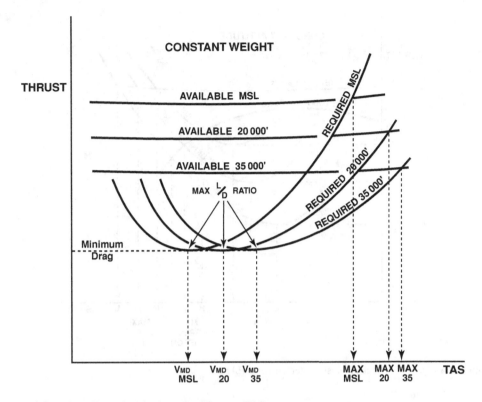

Figure 9.3   The Effect of Altitude on Jet Thrust v TAS.

Figure 9.3 depicts the same change for a jet aeroplane plotted against TAS. It can be seen that both V_{MD} and the maximum cruise speed increase in value with increasing altitude.

9.3.2 Mass

For a constant altitude, temperature and throttle/thrust lever setting the thrust available remains constant irrespective of the mass. However, the thrust required (drag) is directly affected by mass changes. If the mass is increased, the EAS must be increased to maintain level flight and this results in increased drag. This is shown graphically for a jet aeroplane in Figure 9.4(a) by the thrust-available curve remaining the same for all masses but the thrust-required curve moving upward and to the right at an increased EAS with increased mass. The same effect is depicted against TAS in Figure 9.4(b) and shows that the maximum EAS and maximum TAS both increase with decreased mass, whereas the values of V_{IMD} and V_{MD} both decrease with decreased mass. The result of these changes is that the range of stable speeds increases with decreased mass.

9.3.3 Asymmetric Flight

Beside the effect that asymmetric flight has on the control of the aeroplane the loss of thrust also has a dramatic effect. The decreased excess thrust available over that required diminishes the ability of the aeroplane to accelerate or to climb. It is possible that in conditions of heavy mass, high ambient temperature and high altitude an aeroplane may be unable to maintain altitude after suffering an engine

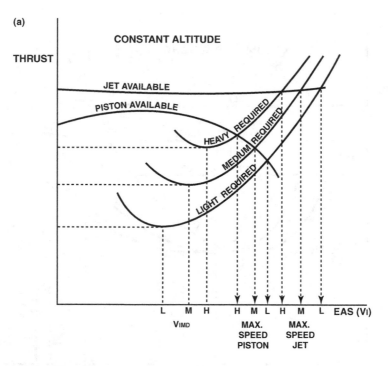

Figure 9.4   (a) The Effect of Mass on Jet Thrust v EAS.

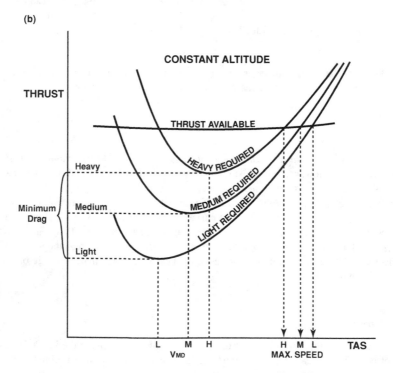

Figure 9.4   (b) The Effect of Mass on Jet Thrust v TAS.

failure. The excess thrust available over that required is considerably reduced and may be insufficient to maintain the cruise altitude. If such is the case the aeroplane will then drift-down until the thrust available can sustain level flight. The altitude at which this occurs is the stabilising altitude. See Figure 9.5.

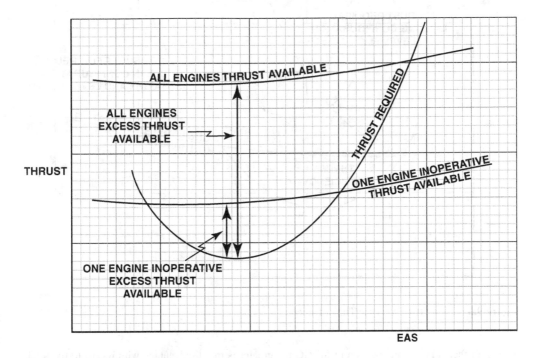

Figure 9.5 Twinjet Asymmetric Thrust Available.

If an engine has failed there is an increase in total drag due to the windmilling engine; any additional drag for any reason will exacerbate the situation. Figure 9.6 depicts the thrust required for various manoeuvres in level flight and shows the deficit of thrust available with one engine inoperative makes it impossible to maintain height.

9.3.4 Centre of Gravity

In level flight it is essential for the tailplane to be downloaded to nullify the nose-down tendency caused by the lift of the wings operating through the CP, which is located aft of the CG, and produces a turning moment about the CG. The elevator trim tab achieves this equalisation of the turning moments but in so doing causes an increase in the total drag and consequently an increase in the thrust required.

An alternative method of attaining the same result as that obtained using the trim tabs is to position the CG in such a manner that it imposes a download on the tailplane. To decrease the amount of elevator trim required balancing the nose-down moment the load and/or fuel could be positioned such that the CG is located on the aft limit of the CG envelope.

This position of the CG is advantageous not only to balance the forces acting on the aeroplane but also to decrease the fuel consumption because the decreased trim drag diminishes the thrust required, which results in increased range and endurance using the fuel available. An additional benefit of the CG at this location is that the aeroplane stalling speed is reduced.

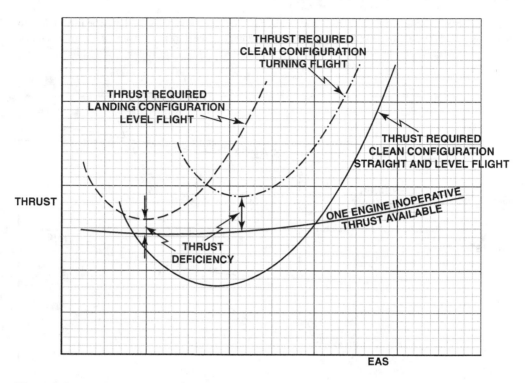

Figure 9.6   Twinjet Asymmetric Thrust Deficit.

The further forward the location of the CG the greater is the disadvantage with regard to the total drag and the thrust required. However, it does have the advantage that it assists recovery from a stall.

9.4   Power

The rate of doing work is referred to as power; it is a measure of the work done by an engine(s). The power available generated by a jet-engined aeroplane is equal to the product of the thrust available and the TAS, but for a piston/propeller aeroplane it is a measure of the effort produced at the propeller shaft measured by its braking effect on the engine, i.e. brake horsepower (BHP). The power required is the thrust required multiplied by the TAS. Power curves can be drawn, for any given mass and altitude, depicting the power available and the power required. The graph is usually drawn with power on the vertical axis and true airspeed (TAS) on the horizontal axis. Examples of these can be seen in Figure 9.7(a) for a piston-engined aeroplane and Figure 9.7(b) for a jet aircraft.

9.5   Analysis of the Power Curves

At the commencement of the take-off run, when the aeroplane is stationary with the wheel-brakes applied and the throttles/thrust levers at the maximum permitted take-off setting, both types of engine are producing thrust but no power. A comparison of the thrust produced by each engine type at the brakes release point (BRP) has no meaning. The curves of the power available for both piston/propeller engine and jet engine are those that should be compared.

At the BRP, at speed 0 kt, point A in Figure 9.7(a), the power-available curve for the piston/propeller aeroplane commences at approximately 60% of the maximum BHP possible, from this point, the power available increases with increased TAS. It reaches a maximum, in the speed range determined by the thrust available, and is shown as point B on the diagram. At speeds above this range the power available decreases with increased speed. For jet aircraft shown in Figure 9.7(b) the power available at the BRP begins at 0 Newtons and is depicted as point G on the diagram and increases linearly with increased TAS through point H to point J.

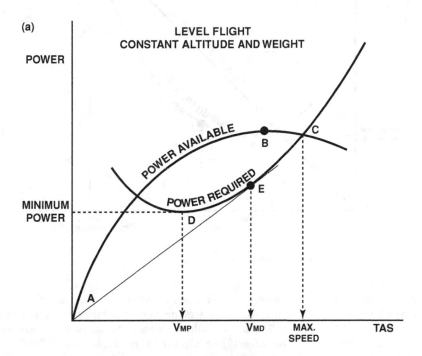

Figure 9.7   (a) The Power Curves – Piston/Propeller Aeroplanes.

9.5.1 Maximum TAS

The maximum speed in level flight, in terms of TAS, can be determined from the power curves and is the point at which the maximum power-available and maximum power-required curves intersect. This speed will be the same as the maximum EAS determined from the thrust curves converted to a TAS for the same conditions. This is shown as point C on the piston/propeller (Figure 9.7(a)) and by point J on the jet aeroplane (Figure 9.7(b)).

The lowest speed at which it is possible to fly is not normally defined by the left-hand intersection of the thrust available/drag curve or the power available/ power-required curve but by conditions of stability, control or stall. The range of speeds available is determined by the power available.

9.5.2 V_MP and V_MD

The power-required curve has a similar shape to that of the thrust-required curve. The lowest point on this curve is the position at which the least power is required. The speed at which this occurs is referred to as the velocity of minimum power (V_MP); its value is dependent on the mass and pressure altitude. It is depicted as point D on the piston/propeller diagram (Figure 9.7(a)) and as point K on the jet diagram (Figure 9.7(b)).

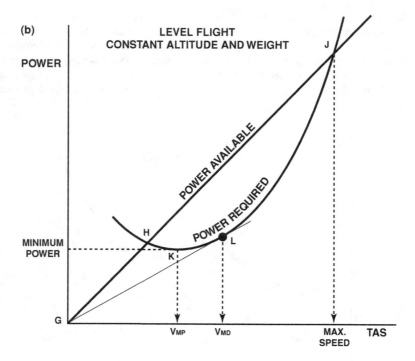

Figure 9.7   (b) The Power Curves – Jet Aircraft.

V$_{MP}$ can also be shown as V$_{IMP}$ on the drag curve and on which it is the smallest area enclosed by a horizontal line from the left vertical axis to the total-drag curve and the vertical from that point downward. (See Figure 7.6) V$_{MD}$ is the TAS attained at the point where the tangent from the origin meets the power-required curve. Shown as point E in Figure 9.7(a) and as point L in Figure 9.7(b).

From these graphs it can be determined that V$_{IMP}$ is a lower speed than V$_{IMD}$ and V$_{MP}$ is less than V$_{MD}$. Despite the fact that this means flying at an EAS less than V$_{IMD}$ and will require more thrust, it requires less power. V$_{IMP}$ is the speed to be flown to obtain the maximum endurance for piston/propeller aircraft for a given quantity of fuel. *At speeds below V$_{IMP}$, both the thrust required and the power required increase dramatically and the speed becomes more unstable.*

9.6   The Effect of the Variables on Power

9.6.1   *Altitude*

In a normal atmosphere, at constant mass, constant throttle/thrust lever setting and constant EAS, as altitude increases the total drag remains constant but the TAS increases. Thus, to maintain level flight the power required, i.e. total drag × TAS, increases with increased altitude. This is shown in Figure 9.8(a) for a piston-engined aeroplane, by the power-required curve moving up and to the right. As altitude increases the thrust available decreases and despite the increased TAS, for the same EAS, the resulting power available, i.e. thrust available × TAS, decreases with increased altitude.

This is depicted in Figure 9.8(a) by the power-available curve moving down to the right with increased altitude. It can be seen from the diagram that the excess power available over power required decreases with increasing altitude, as does the maximum cruising TAS. Thus, the range of cruising speeds available in level flight diminishes with increasing altitude.

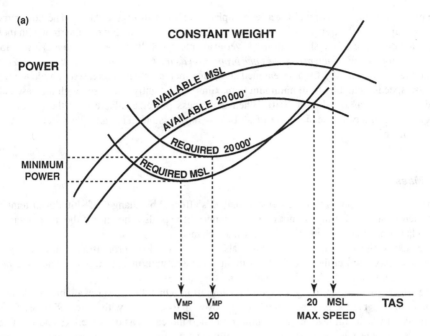

Figure 9.8   (a) The Effect of Altitude on Power – Piston Aircraft.

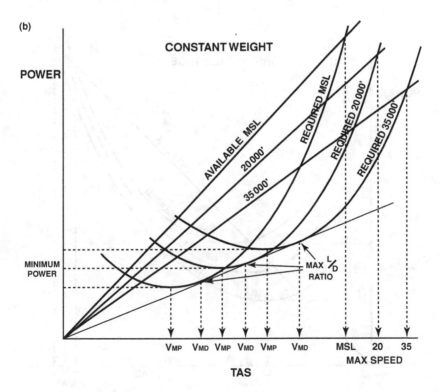

Figure 9.8   (b) The Effect of Altitude on Power – Jet Aircraft.

The effect of increased altitude for a jet aeroplane is shown in Figure 9.8(b). The power-required curve moves up and to the right and the power-available diagonal moves down to the right with increased altitude. However, due to the shape of the power-available graph line, *the maximum cruise TAS for a jet-engined aeroplane will increase in the same proportion as the power required with increased altitude.* Although the maximum TAS for a jet-engined aeroplane increases with increased altitude the range of level cruise speeds available, from minimum to maximum, is slightly decreased with increased altitude because of the increased value of V_{MD}. When the excess power available over the power required is reduced to zero, any further increase of altitude is not possible. This altitude is referred to as the 'absolute ceiling.'

9.6.2   Mass

It has been previously stated that the thrust available is affected by changes of altitude, air temperature and throttle/thrust lever setting but not by mass changes. It was also shown that the thrust required and the EAS change is directly proportional to any mass change.

Consequently, the TAS and power required also increase in direct proportion to any mass increase. However, the maximum cruise TAS in level flight decreases with increased mass. Thus, the range of level cruise speeds available diminishes with increasing mass.

These effects can be seen in Figure 9.9 for jet aircraft in which the power-available graph line is the same for all masses but the power-required curve moves up to the right with increased mass. Therefore, at high masses to maintain level flight the power required increases, as does the IAS and the TAS. The result of this is that as mass increases the excess of power available over power required decreases.

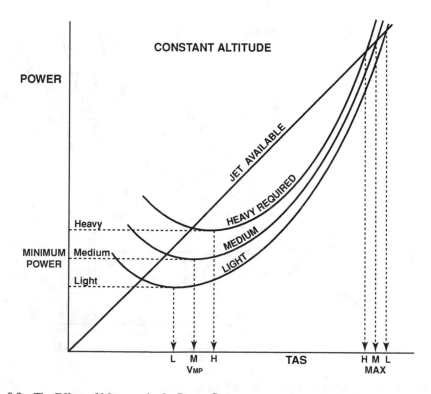

Figure 9.9   The Effect of Mass on the Jet Power Curves.

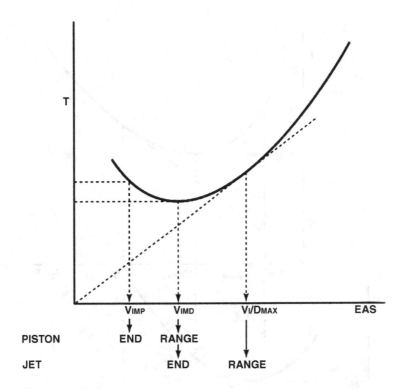

Figure 9.10   Range and Endurance Speeds.

Figure 9.10 shows a summary of the speeds flown to obtain the maximum range and the maximum endurance for both the piston/propeller and jet aeroplanes.

The relationship of the thrust required and power-required curves is illustrated in Figure 9.11.

9.7 Summary

- The aerodynamic centre is at 25% of the chordline from the leading edge.
- Total Lift $= C_L \frac{1}{2} \rho V^2 S$. Total Drag $= C_D \frac{1}{2} \rho V^2 S$.
- TAS $=$ EAS $\div \sqrt{}$relative density.
- New Vs $=$ Old Vs $\times \sqrt{}$(New mass $\div$ Old mass).
- For every angle of attack there is a corresponding IAS and C_L.
- For a constant Mach number and mass, an increase of altitude will decrease the IAS and will require an increased angle of attack to maintain level flight.
- Parasite drag is directly proportional to V^2.
- Induced drag is inversely proportional to V^2 and the aspect ratio.
- Induced drag is directly proportional to the aeroplane mass^2 and $C_L{}^2$.
- V_{IMD} is directly proportional to the mass.
- V_{IMD} approximately equals 1.3Vs for a piston and 1.6Vs for a jet.
- $V_I/Dmax = 1.32V_{IMD}$ and 1.7Vs for a piston and 2.1Vs for a jet.
- Higher masses require proportionately higher speeds.
- Higher flap settings require lower speeds; the greater the flap setting angle the lower is the speed.

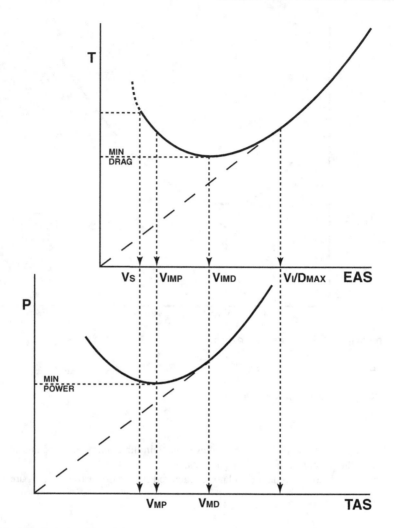

Figure 9.11   The Relationship of the Thrust Required and Power Required Curves.

- The maximum lift/drag ratio occurs at an angle of attack of 4° and at a speed of V_IMD. Lowering flap decreases the lift/drag ratio.
- If the thrust available exceeds the thrust required the aeroplane will accelerate.
- IAS is directly proportional to the square root of the aeroplane mass.
- The maximum TAS in level flight is attained at the lowest practicable altitude for a piston-engined aeroplane.
- The maximum TAS for a jet-engined aeroplane in level flight is attained at the highest practical altitude.
- Thrust available decreases with increased altitude for a constant mass.
- Thrust available is unaffected by the changes of mass.
- V_MP is always less than V_MD.

Self-Assessment Exercise 9

Q9.1 The effect of asymmetric thrust is to:
(a) decrease the ability to climb
(b) improve the turn performance
(c) decrease the fuel consumption
(d) increase the gradient of climb

Q9.2 The speed at the point where the tangent from the origin to the power-required curve for a jet aeroplane intersects is:
(a) the minimum power speed
(b) the maximum range speed
(c) V_{MD}
(d) the long-range cruise speed

Q9.3 Which of the following variables will not affect the shape or position of the Total Drag v IAS curve for speeds below the critical Mach number?
(a) configuration
(b) mass
(c) aspect ratio
(d) altitude

Q9.4 As speed is reduced below V_{IMD} total drag:
(a) increases and speed stability decreases
(b) decreases and speed stability increases
(c) increases and speed stability increases
(d) decreases and speed stability decreases

Q9.5 As speed is reduced from V_{MD} to V_{MP} power:
(a) required decreases and drag increases
(b) required decreases and drag decreases
(c) required increases and drag increases
(d) required increases and drag decreases

Q9.6 The speed to attain the minimum power required for a turbojet aeroplane is:
(a) less than the speed for the minimum drag
(b) higher than the speed for the minimum drag
(c) slower in a climb and faster in a descent
(d) the same as the minimum drag speed

Q9.7 The region of speed instability is:
(a) the same as the region of reversed command
(b) the region in which manual control is not possible
(c) at speeds below the low buffet speed
(d) the region above the thrust available and drag curve intersection

Q9.8 With increasing altitude, the power-required curve moves:
(a) up and to the right
(b) down and to the right
(c) up and to the left
(d) down and to the left

Q9.9 For a piston-engined aeroplane at a constant mass, angle of attack and configuration an increase of altitude will require:
(a) increased power but the same TAS
(b) increased power and TAS
(c) lower power but an increased TAS
(d) the same power but an increased TAS

Q9.10 For a piston-engined aeroplane at a constant altitude, angle of attack and configuration, an increased mass will require:
(a) more power and speed
(b) more power but the same speed
(c) the same power but more speed
(d) more power but less speed

Q9.11 The region of 'reversed command' means:
(a) a lower airspeed requires more thrust
(b) the thrust required is independent of the airspeed
(c) a thrust reduction results in acceleration
(d) a momentary increase of airspeed has no influence on the thrust required

Q9.12 At speeds below V_{IMD}:
(a) the aeroplane can only be controlled in level flight
(b) a lower speed requires more thrust
(c) a higher speed requires more thrust
(d) the aeroplane cannot be controlled manually

Q9.13 In which of the following conditions is thrust available equal to drag?
(a) in a descent at constant IAS
(b) in level flight at constant IAS
(c) in level flight accelerating
(d) in a climb at constant IAS

Q9.14 How does C_L vary with altitude?
(a) It increases with increased altitude.
(b) It decreases with increased altitude.
(c) It remains constant with increased altitude.
(d) It decreases with decreased altitude.

Q9.15 Maximum horizontal speed occurs when:
(a) thrust = minimum drag
(b) thrust does not increase with increasing speed
(c) maximum thrust = total drag
(d) thrust = maximum drag

Q9.16 The Maximum IAS for a piston-engined aeroplane in level flight is attained at the:
(a) lowest practical altitude
(b) optimum cruise altitude
(c) service ceiling
(d) practical ceiling

Q9.17 The speed of V_{MP} for a jet aeroplane is:
(a) lower than V_{MD} in a climb and higher than V_{MD} in a descent
(b) the same as V_{MD}
(c) always lower than V_{MD}
(d) always higher than V_{MD}

Q9.18 The tangent from the origin to the power-required curve touches the curve at the speed of:
(a) V_{MP}
(b) where the Lift/Drag ratio is at a maximum
(c) where C_D is minimum
(d) where the Lift/Drag ratio is minimum

Q9.19 If a piston-engined aeroplane is to maintain a given angle of attack, configuration and altitude at an increased gross mass it will require:
(a) an increased airspeed and constant power
(b) an increased power and a decreased airspeed
(c) an increased airspeed and increased power
(d) a higher coefficient of drag

Q9.20 At a constant mass, angle of attack and configuration, the power required at a higher altitude:
- (a) increases and the TAS increases by the same percentage
- (b) increases and the TAS remains constant
- (c) decreases slightly because of the decreased density
- (d) remains unchanged but the TAS increases

Q9.21 At a constant mass, altitude and airspeed, moving the CG from the forward limit to the aft limit:
- (a) decreases the induced drag and reduces the power required
- (b) increases the power required
- (c) affects neither the drag nor the power required
- (d) increases the induced drag

Q9.22 To maintain a given angle of attack, configuration and altitude at a higher gross mass the:
- (a) lift/drag ratio must be increased
- (b) airspeed must be increased but the drag remains constant
- (c) airspeed and drag will both be increased
- (d) airspeed will be decreased and the drag increased

Q9.23 In level unaccelerated flight, if the aircraft mass is decreased:
- (a) minimum drag and V_{IMD} both decrease
- (b) minimum drag increases and V_{IMD} decreases
- (c) minimum drag decreases and V_{IMD} increases
- (d) minimum drag increases and V_{IMD} increases

Q9.24 If the thrust available exceeds the thrust required in level flight the aeroplane:
- (a) will accelerate
- (b) will descend if the airspeed remains constant
- (c) decelerates if it is in the region of reversed command
- (d) will decelerate

Q9.25 If the altitude and angle of attack remain constant, an increased mass will require:
- (a) increased airspeed and increased power
- (b) increased airspeed and decreased power
- (c) decreased airspeed and increased power
- (d) decreased airspeed and decreased power

Q9.26 In level flight, the power-required curve would move . . . . if the altitude is decreased:
- (a) up and to the right
- (b) down and to the right
- (c) down and to the left
- (d) up and to the left

Q9.27 To fly at a speed between V_{IMD} and Vs it would require:
- (a) increased flap
- (b) increased thrust
- (c) decreased thrust
- (d) decreased flap

Q9.28 The most important aspect of the 'backside' of the power curve is the:
- (a) elevator must be pulled to lower the nose
- (b) speed is unstable
- (c) aeroplane will not stall
- (d) altitude cannot be maintained

Q9.29 The maximum horizontal speed occurs when the:
- (a) thrust is equal to the minimum drag
- (b) thrust does not increase further with increasing speed
- (c) maximum thrust is equal to the total drag
- (d) thrust is equal to the total drag

Q9.30 The critical engine inoperative:
   (a) does not affect the aeroplane performance since it is independent of the power plant
   (b) decreases the power required because of the lower drag caused by the windmilling engine
   (c) increases the power required and decreases the total drag due to the windmilling engine
   (d) increases the power required because of the greater drag caused by the windmilling engine and the compensation for the yaw effect

Q9.31 If the angle of attack, mass and configuration are constant for a reciprocating-engined aeroplane the effect of increasing the altitude on the power required is that it will be:
   (a) unchanged but the TAS will increase
   (b) increased and the TAS will increase by the same percentage
   (c) increased but the TAS remains constant
   (d) decreased slightly because of the lower air density

Q9.32 The relationship of V_{MP} to V_{MD} is:
   (a) $V_{MP} > V_{MD}$
   (b) $V_{MP} < V_{MD}$
   (c) $V_{MP} = V_{MD}$
   (d) cannot be compared with V_{MD} because they come from different graphs

10 Advanced Control

10.1 Wing Torsion and Flexing

It is important to differentiate between strength and stiffness. Strength is defined as the ability of an object to withstand a load, whereas, stiffness is its ability to withstand distortion, deformation or twisting. There is an axis that is a significant characteristic of a wing, about which the wing will twist or flex known as the torsional axis or elastic axis. See Figure 10.1. Any force applied along the chordline of the axis will not cause the wing to twist but it may flex or bend (move at right angles to the chordline) because of the force. Torsional or flexural vibrations are usually damped by the structural rigidity of the airframe and are by themselves fairly innocuous. However, if an external force acts together with the vibration the reaction may be such that it causes structural failure.

10.2 Wing Flutter

Incorrect balancing or slack control runs may cause a control surface to vibrate in the airflow this is not a torsional vibration or flexural vibration. The wing is subject to the interaction of the aerofoil mass and the aerodynamic load imposed on the aerofoil causing distortion by bending and torsion of the structure that can result in a violent vibration in the resonance frequency, which if excessive can cause structural failure. This is 'flutter' and the aeroelastic coupling will affect its characteristics. The risk of experiencing flutter increases with IAS and may be one of the following three types:

a. torsional flexural flutter;
b. torsional aileron flutter;
c. flexural aileron flutter.

10.3 Torsional Flexural Flutter

The result of the wing flexing **and** bending when an aerodynamic load is applied to it is torsional flexural flutter. Figure 10.2 depicts the sequence of events that constitutes this type of flutter that are as follows:

a. The wing is in steady cruising flight with the CG aft of the torsional axis. The reactive force that is created by the mass bending the wing balances the lift generated.

The Principles of Flight for Pilots   P. J. Swatton
© 2011 John Wiley & Sons, Ltd

(a) WING TWIST OR TORSION

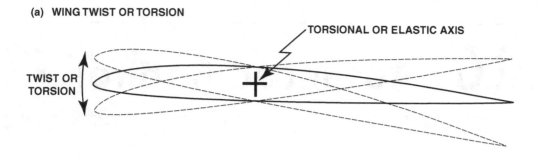

(b) WING BENDING OR FLEXING

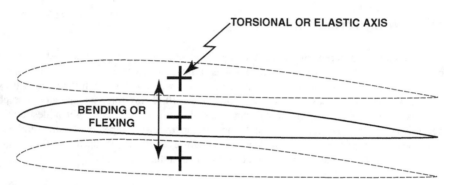

Figure 10.1   The Torsional or Elastic Axis.

b. The angle of incidence is momentarily increased by an outside disturbance that increases the lift. At that instant the wing flexes upward because the increased lift exceeds the reactive force. The situation is exacerbated because due to inertia the CG lags behind the torsional axis and further increases the angle of attack and the lift.

c. As a result, of the inherent rigidity of the wing it reaches its bending limit, which arrests the torsional axis but the CG continues to move due to inertia. This decreases the angle of incidence and reduces the lift produced to less than the reactive force and results in the wing descending.

d. The wing ceases its descent when it reaches the aeroelastic-bending limit. This halts the movement of the torsional axis but the CG again continues to travel because of inertia. Once again, this increases angle of attack and the lift generated which exceeds the reactive force once more and the cycle commences to repeat itself.

A full cycle of one flutter appears as a sine wave due to the movement of the aeroplane. See Figure 10.3. Resistance to flutter increases with increasing wing stiffness. The designers can reduce an aeroplane's sensitivity to flutter by using stabiliser trimmers instead of traditional elevator trimmers and can further prevent its predisposition to flutter by employing either one of the two following methods:

a. Designing the aeroplane so that the CG is always ahead of the torsional axis. This is achieved by constructing the aeroplane with wing-mounted engines ahead of the wing, which act as mass balances counterbalancing lift and reducing the bending stress, or

b. By designing the rigidity of the aeroplane wings such that the critical flutter speed exceeds the maximum permissible speed at which the aeroplane may be operated. Wings without wing-mounted

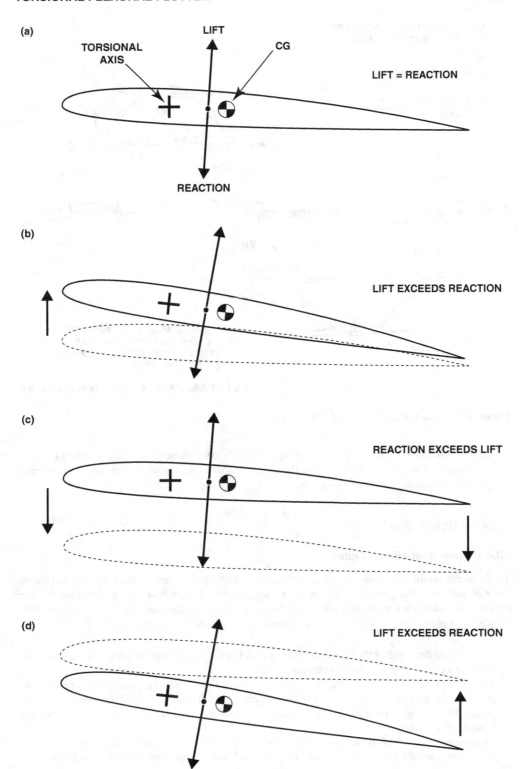

Figure 10.2   The Sequence of Torsional Flexural Flutter.

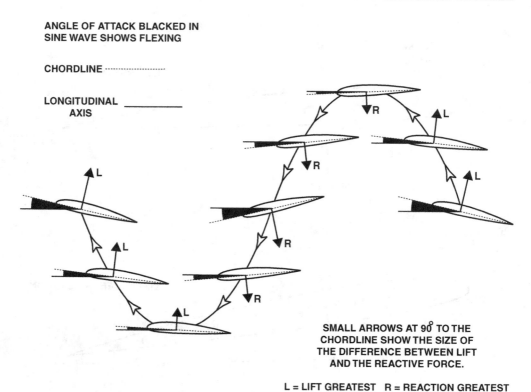

ANGLE OF ATTACK BLACKED IN
SINE WAVE SHOWS FLEXING

CHORDLINE ·······················

LONGITUDINAL _____
    AXIS

SMALL ARROWS AT 90° TO THE
CHORDLINE SHOW THE SIZE OF
THE DIFFERENCE BETWEEN LIFT
AND THE REACTIVE FORCE.

L = LIFT GREATEST R = REACTION GREATEST

Figure 10.3   One Cycle of Torsional Flexural Flutter.

engines have a more rigid construction and therefore are heavier. However, they do have the advantage that changes of thrust have less affect on the longitudinal trim than would be the case if the engines were mounted at the rear of the aeroplane.

10.4   Aileron Flutter

10.4.1   Torsional Aileron Flutter

Torsional aileron flutter is caused by cyclic deformations generated by aerodynamic, inertial and elastic loads on the wing. The speed at which this phenomenon occurs is adversely affected by excessive free play or backlash of the control runs. Aileron deflection produces aerodynamic loads that cause the wing to twist. The sequence of such an event is as follows and is depicted in Figure 10.4:

a. A downward deflection of the aileron results in the lift force at the hinge of the aileron being increased. This induces the wing to twist about the torsional axis.
b. The trailing edge of the wing at the aileron hinge point and the attached aileron rise, but because the aileron CG is behind the hinge the upward movement of the aileron lags behind the movement of the wing. This induces an even greater upward force from the aileron and further increases the twisting moment of the wing.
c. The twisting motion of the wing is stopped by its torsional reaction. However, the upward impetus of the aileron together with its aerodynamic load and the stretched control runs result in the

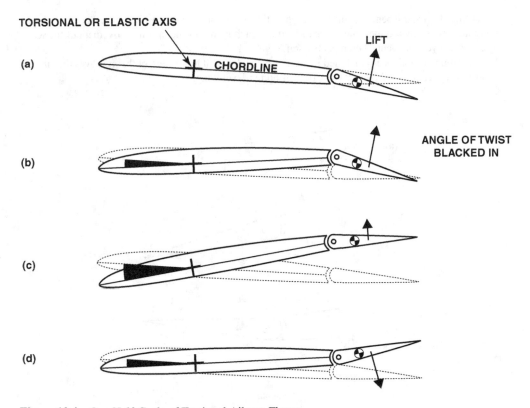

TORSIONAL OR ELASTIC AXIS

(a)   CHORDLINE   LIFT

(b)   ANGLE OF TWIST BLACKED IN

(c)

(d)

Figure 10.4   One Half-Cycle of Torsional Aileron Flutter.

continued upward movement of the wing. This now imposes a downward load on the trailing edge of the wing.

d. Resulting from the stored energy in the twisted wing and aileron aerodynamic load the wing now twists in the opposite direction and the cycle begins over again.

Torsional aileron flutter can be prevented at the design stage by making the controls irreversible or by mass balancing the ailerons (described later in this chapter) provided that the aeroplane does not have fully powered controls and has a means of reversion to manual control.

10.4.2   Flexural Aileron Flutter

Similar to torsional aileron flutter but caused by the inertia of the aileron, flexural aileron flutter results in the aileron movement lagging behind that of the wing. This causes the wing-flexing movement to be amplified. The sequence of events described below is shown in Figure 10.5.

a. The wing and aileron in equilibrium before the balance is upset by a disturbance.
b. A disturbance causes the wing to rise but the aileron lags behind due to inertia. This aggravates the situation by driving the wing further up.
c. The wing reaches its bending limit and starts to move down but the aileron continues to move upward due to inertia increasing the downward movement of the wing.

d. When the downward bending limit is reached the wing commences to rise again and the aileron continues its downward movement due to inertia, which increases the upward movement of the wing again. This cycle of events then repeats itself.

e. If an aileron mass balance is included in the aeroplane design it has the effect that it moves the aileron CG to the hinge-line and prevents this type of flutter developing.

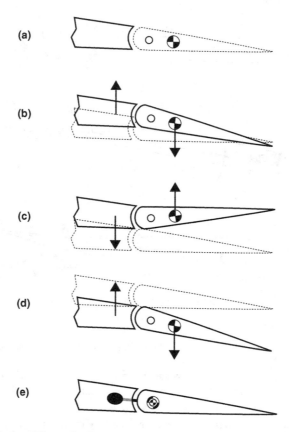

Figure 10.5   One Cycle of Flexural Aileron Flutter.

10.4.2.1   The Mass Balance

To dampen the effect of flutter on a control surface a balance mass attached to the leading edge of the main control surface is positioned in front of the hinge-line. If an aileron mass balance is included in the aeroplane design, it is important that the weight be positioned as near to the end of the wing as possible because then the size of the balance does not need to be very large.

However, this type of balance mass has the disadvantage that it could cause torsional vibration of the aileron itself. To prevent such an occurrence the mass balance can be designed as part of the leading edge of the aileron.

It is possible for aeroplanes not fitted with powered controls to experience flexural flutter from the tailplane or the elevators. This is a rare occurrence since generally the tailplane is torsionally stronger than the wings because of its smaller size but this may not be so if the aeroplane has a large fin.

10.5 Divergence

In extreme cases of weak torsional rigidity when the CP is situated in front of the torsional axis of the wing its angle of incidence increases so much that it results in structural failure. The twisting action of the wing consequential to these conditions is divergence.

If a disturbance momentarily increases the angle of incidence the lift will increase and the CP will move forward. Because the torsional axis is behind the CP this twists the wing against the torsional rigidity of the wing. The movement of the CP forward further increases the lift, which exacerbates the situation. Should the disturbance cause the wing angle of incidence to decrease and the CP to move aft then the reduction of lift so caused twists the wing in the opposite direction.

The *total lift of a wing increases in direct proportion to the square of the speed* but the torsional rigidity is the same for all speeds. Because of this there is a limiting speed beyond which the twisting motion of the wing will result in structural failure. In other words, the aerodynamic moment overcomes the wing torsional rigidity at divergence speed.

There are two methods available to the aeroplane designer to ensure that structural failure due to the twisting of the wing does not occur. They are to:

a. Design the torsional rigidity of the wing so that the divergence speed is well in excess of the maximum operating speed.
b. Design the wing so that the torsional axis is ahead of the aerodynamic axis, this measure prevents divergence at any speed.

10.6 Control Secondary Effects

Movement of the aeroplane around one primary axis induces a secondary reaction about another axis. For example, if the rudder is moved to yaw the aeroplane about its normal axis the outside wing of the aeroplane travels slightly faster than the inner wing and therefore develops more lift. As a consequence, the outer wing will rise and the inner wing will fall. Therefore, the secondary effect of applying rudder is for the aeroplane to roll into the turn.

A further example of a control secondary effect is when aileron is used to bank the aeroplane it causes the aeroplane to sideslip into the turn. The aerodynamic force of the keel surface acting behind the CG results in the aeroplane yawing towards the lower wing. The secondary effect of the aileron is the yaw.

Recent innovations to reduce the control secondary effects have been the **flaperon** that simultaneously uses the flaps and ailerons and the **elevon** that simultaneously uses the elevators and ailerons. Some fast jet aeroplanes combine the function of the stabilator with that of the ailerons and are known as **tailerons**.

10.7 Adverse Yaw

During a bank the upgoing wing effectively experiences a decreased angle of attack whilst the effective angle of attack of the downgoing wing increases. Consequently, the slope of the total lift vector generates a rearward component on the upgoing wing and a forward component on the downgoing wing. These are yaw moments but in the opposite direction to that of the roll caused by the difference of the induced drag of each wing. This is adverse yaw.

When banking, the induced drag created by the upgoing aileron is less than that of the downgoing aileron and produces a yawing moment in the opposite direction to the roll. This is adverse aileron yaw, which is particularly significant at low airspeeds because of the large aileron deflection necessary to achieve the same rate of roll as at higher speeds with less deflection.

10.8   Counteraction Devices

Adverse yaw may be counteracted by any one of the following devices:

a. rudder/aileron coupling;
b. slot/aileron coupling;
c. spoiler/aileron coupling;
d. differential aileron deflection;
e. frise ailerons.

10.8.1   Rudder/Aileron Coupling

The deflection of the ailerons on aeroplanes fitted with rudder coupling causes the rudder to move an amount proportional to the angular displacement of the aileron and to oppose adverse yaw.

10.8.2   Slot/Aileron Coupling

The slot/aileron coupling method is of particular benefit to light aeroplanes that have to execute large bank angles at low airspeeds such as crop-spraying aeroplanes. The ailerons are coupled to leading-edge slots directly in front of the aileron that open automatically when the aileron is downgoing.

10.8.3   Spoiler/Aileron Coupling

When the spoilers and ailerons are coupled at low speed to overcome adverse yaw they are called roll spoilers. Their deployment is coupled with that of the upgoing aileron and destroys the lift of the wing on which they are extended. Their operation is asymmetric and the loss of lift experienced by their extension on the downgoing wing causes the wing to drop in the direction of the roll and the increased drag results in a yaw being made in the same direction. The spoilers on the upgoing wing remain retracted and flush with the surface of the wing.

10.8.4   Differential Aileron Deflection

The differential aileron deflection system ensures that the drag generated by both ailerons when rolling into or out of a turn is equal, thus eliminating adverse yaw. This is achieved by constructing the aileron system such that the angular deflection of the upgoing aileron is always proportionately greater than the deflection of the downgoing aileron. The downgoing aileron produces yaw opposing the turn and roll of that demanded, thus reducing the drag difference. An example of this for a left-banked turn would be that the left aileron goes up 5° and the right aileron goes down 2°. *The upgoing aileron always has a greater deflection than the downgoing aileron.*

10.8.5   Frise Ailerons

To counteract adverse yaw the Frise ailerons are designed so that their leading edge produces sufficient profile drag on the upgoing aileron to equal the induced drag of the downgoing aileron and thus eliminate adverse yaw. See Figure 10.6.

The slipstream of propeller-driven aeroplanes rotates in the same direction as the propeller and imparts an asymmetric airflow over the fin and rudder. The side aerodynamic force so induced causes the aeroplane to yaw, which, being dependent on the direction of rotation, may reduce or increase the effect of adverse aileron yaw.

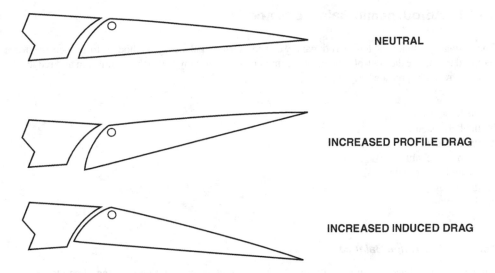

NEUTRAL

INCREASED PROFILE DRAG

INCREASED INDUCED DRAG

Figure 10.6   The Frise Aileron.

10.9   Control-Surface Operation

The airflow over a deflected control surface will attempt to move the control surface back to a neutral position. The perpendicular distance of the control surface centre of pressure (CP) from the hinge-line is the arm, which when multiplied by the lifting force of the control surface produces the hinge moment of the control surface. See Figure 10.7.

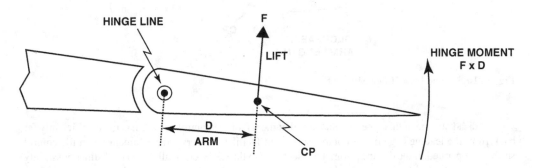

Figure 10.7   The Hinge Moment.

The magnitude of the lift force generated by a control surface is directly proportional to the square of the EAS (V^2). In a manual control system the pilot has to provide the force necessary to overcome the hinge moment. At the lower airspeeds assistance is provided to the pilot by a counterbalance device. Control balancing is achieved either by decreasing the hinge moment or by setting up a force that acts against the hinge moment.

10.10   Aerodynamic Balance Methods

The primary function of any aerodynamic control surface balancing method is to reduce the force required to operate the control. The most common methods employed to achieve this aim and reduce the manoeuvre stick force are:

a. the hinge balance;
b. the horn balance;
c. the internal balance;
d. the balance tab;
e. the antibalance tab;
f. the spring tab;
g. the servo-tab.

10.10.1   The Hinge Balance

An aerodynamic balance can be built into the control surface by insetting the hinge-line closer to the CP of the control surface. The 'hinge balance' is mounted in such a manner that when air strikes the deflected control surface ahead of the hinge-line it decreases the amount of force required to move the control because it partially balances the aerodynamic force aft of the hinge-line. See Figure 10.8.

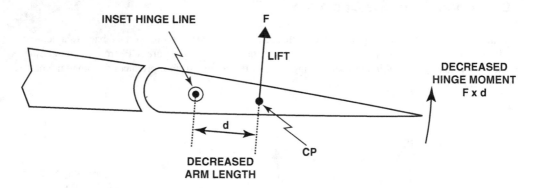

Figure 10.8   The Inset Hinge Moment.

The disadvantage of this type of balance is that the CP of the control surface has to be sufficiently far back from the leading edge of the control surface to facilitate its forward movement when the control surface is operated. If the distance from the inset hinge to the CP is too small there is a distinct possibility that when the control surface is moved to its maximum deflection that the CP could move forward of the hinge-line, thus reversing the hinge moment; this is overbalance. See Figure 10.8.

10.10.2   The Horn Balance

Any of the control surfaces can be aerodynamically balanced by designing a portion of the control surface to be ahead of the hinge-line; this is the horn balance. When the control surface is deflected the horn balance will produce an opposing moment to the hinge moment, which reduces the amount of force required to operate the control surface. See Figure 10.9.

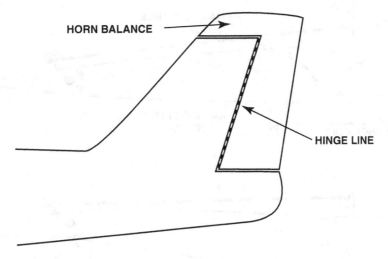

Figure 10.9   The Horn Balance.

10.10.3   The Internal Balance

The aerodynamic balance of a control surface can be achieved internally without incurring any additional drag. The leading edge of the control surface of an internal balance is shaped such that it has a projection, called the balance panel, which is connected by a flexible diaphragm within a sealed chamber to a fixed part of the aeroplane structure, usually the spar and vented to atmosphere. The diaphragm divides the sealed chamber in two. See Figure 10.10.

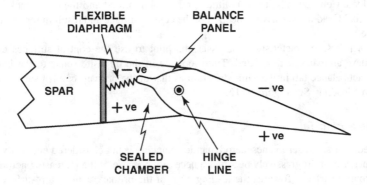

Figure 10.10   The Internal Balance.

10.10.4   The Balance Tab

A balance tab is a small ancillary hinged flap attached to the trailing edge of a primary control surface. Its purpose is to reduce the force required to operate the control surface at a given airspeed by deflecting in the opposite direction to that of the primary control surface. See Figure 10.11.

Although the balance tab is relatively small when compared with the primary control surface it is a considerable distance from the hinge-line of the primary control surface and therefore produces a relatively large opposing moment. Its operation slightly reduces the effectiveness of the primary control surface and causes a small drag penalty.

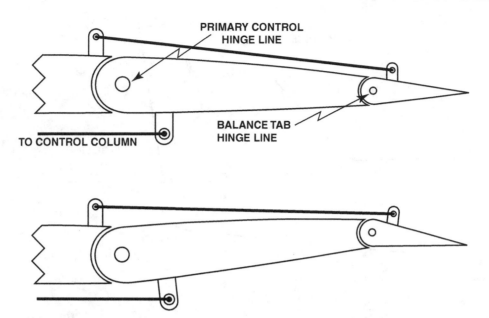

Figure 10.11   The Balance Tab.

10.10.5   The Antibalance Tab

Because the hinge moment on some aeroplanes is too small, often as the result of the CP being too close to the hinge-line of the control surface, it is too easy to deflect the control surface against the aerodynamic load. Consequently, there is little control column load and there is a lack of feel to the controls. This could lead to excessive deflection of the control surface and result in serious overstressing of the airframe.

To counteract these characteristics and assist the pilot to use the control surfaces correctly it is necessary to have antibalance tabs fitted. These operate in the opposite sense to the balance tab by deflecting the antibalance tab in the same direction as that of the primary control surface and increase the aerodynamic loading. See Figure 10.12.

10.10.6   The Spring Tab

At low airspeeds, for some aeroplanes, aerodynamic balancing is not considered necessary, however, as the airspeed increases it progressively becomes a necessity because of the increasing aerodynamic load on the main control surface. Towards the leading edge of the primary control top surface is pivoted a lever. The top of the lever is connected to the control column and the balance tab is connected to the same lever just above the pivot. Thus, movement of the control column is transmitted to the primary control surface but does not directly operate it. To ensure that the balance tab is only effective when it is required, the operating lever of the balance tab system is pivoted through a spring mounted on the elevator main control surface. This is the 'spring tab.' See Figure 10.13.

When the elevators are operated with no aerodynamic load, i.e. at low airspeeds, the control column movement is fed directly to the primary control surface with no deflection of the balance tab so that there is no geometric movement between them. However, when there is an aerodynamic load on the primary control surface, i.e. at higher speeds, one side of the spring is compressed and changes the geometric position of the balance tab relative to the main elevator control surface.

The spring causes the tab to move in the opposite direction to the main control surface at an ever-increasing angle, which is directly proportional to the increasing airspeed. Thus, the spring tab assists

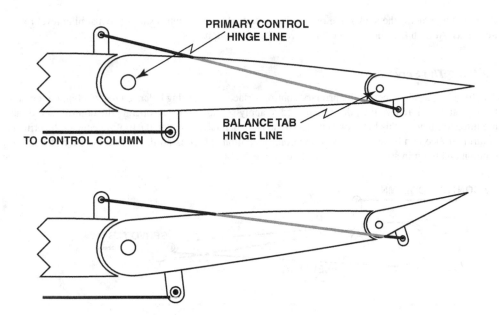

Figure 10.12   The Antibalance Tab.

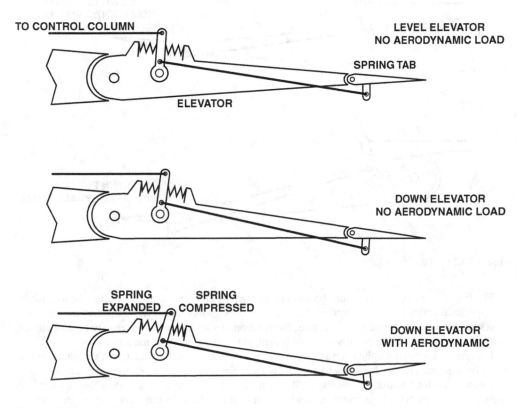

Figure 10.13   The Spring Tab.

the pilot by reducing the stick force required to operate the main control surface to a similar level for all airspeeds. At high IAS the spring tab behaves like a servo-tab.

10.10.7 The Servo Tab

The force required to operate the primary control surfaces, even using balance tabs, is often too great to be acceptable. To relieve the pilot of the need to apply so much force to operate the controls servo-tabs are fitted to carry out this function. The servo-tab is a small tab hinged on the trailing edge of the primary control surface and linked directly to the control column. There is no direct link between the control column and the primary control surface. See Figure 10.14.

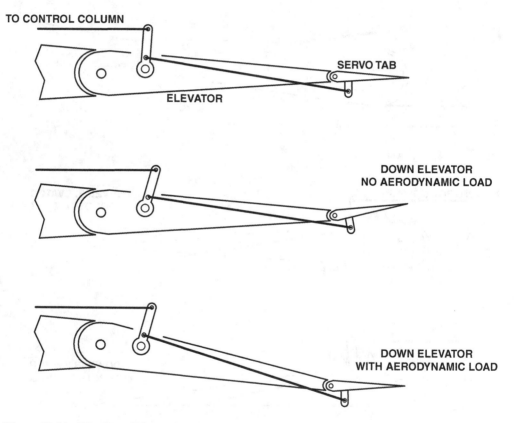

Figure 10.14   The Servo Tab.

The function and operation of the servo-tab is the same as the balance tab and moves the trim tab in the opposite direction to the movement of the primary control surface. However, at low IAS the servo-tab makes the main control surface less effective. The position of a servo-controlled primary control surface cannot be determined until a positive pressure is applied to the servo-tab by the airflow.

In flight, if a servo-controlled elevator jams the servo-tab reverses the direction of the pitch control input, because the elevator becomes part of the fixed horizontal stabiliser and the servo-tab now acts as a small elevator because despite its small size it has a longer moment arm than the elevator and consequently generates a large moment. Under normal circumstances an up servo-tab position would drive the elevator to a downward deflection causing the aeroplane's nose to go down, but if the elevator is fixed the servo-tab in an up position will cause the aeroplane's nose to move up.

10.11 Primary Control-Surface Trimming

Trim tabs are fitted to enable the pilot to set a required angle of deflection of the primary control surface and maintain that deflection. However, they are not suitable for jet aeroplanes because of their large range of operating speeds, preference, for this type of aeroplane, is given to the trimmable horizontal stabiliser. There are two types of trim tab in common use; the variable and the fixed.

To maintain a state of equilibrium for an aeroplane in flight the moments about each of the three axes must balance. If they do not balance then the pilot must intervene and apply additional force to the appropriate main control surface deflecting that control sufficiently to maintain equilibrium. To sustain the required position of the controls for the entire flight would place a great physical strain on the pilot. Trim tabs are provided to relieve the strain on the pilot by maintaining the control surface at the necessary angle.

For example, if an aeroplane has a tendency to fly continuously with a nose-down attitude it is necessary for the elevators to deflected upward and be maintained in this position to counteract this trend. To achieve this requirement the trim tab attached to the trailing edge of the main control surface is deflected downward and the cockpit indication is nose-up. See Figure 10.15.

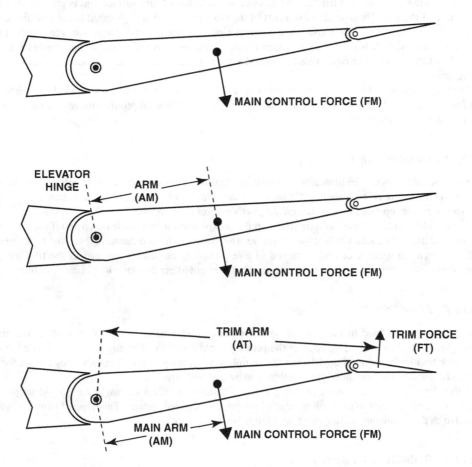

Figure 10.15 The Trim-Tab Moments.

The moment required to move the primary control surface to the required position (Mm) is equal to the force (Fm) multiplied by the length of the arm (Am) from the hinge of the elevator to the centre of pressure of the elevator main control surface. The same moment can be applied to the elevators by a smaller force (Ft) in the opposite direction generated by a trim tab multiplied by the longer arm distance (At) from the centre of pressure of the trim tab to the elevator hinge-line. In other words, (Fm × Am) = (Ft × At). The trim tab thereby does the work required to maintain the elevator out of the neutral position and there is no need for any force from the pilot's controls. See Figure 10.15.

There are two disadvantages to the use of trim tabs. They are:

a. To maintain balanced flight the primary control surface is continuously deflected and the trim tab decreases its overall effectiveness.
b. The trim tab when required to maintain balanced flight is continuously deflected in the airstream passing over the elevators thereby causing additional drag, generally known as 'trim drag.' This causes an increased thrust requirement, which increases the fuel flow and decreases the range and endurance. For this reason, the variable-incidence tailplane was introduced to eliminate the need for trim tabs on the elevators of aeroplanes having a high-speed flight profile.

The effect attained by using a trim tab is the same as any other control surface and is proportional to the square of the IAS. Thus, small movements of the tab are required at high speed but at low airspeed particularly during the landing phase when configuration changes are made the movement required to maintain balanced flight is significantly greater. It produces an aerodynamic force without the associated control surface moving. The pilot holds the control at the required deflection and removes the stick force by trimming.

In straight and level flight as speed increases the elevator is deflected further downward and the trim tab further upward. After trimming for a speed increase the stick-neutral position moves forward or for a speed decrease it moves aft.

10.11.1   Variable Trim Tabs

The variable trim tabs may be manually or electrically controlled. If manually controlled then trim wheels located on the centre console are used. However, if they are electrically controlled then they are moved by electric motors operated by a switch on the pilot's control column.

For the elevator trim if the tab is deflected up the cockpit indication is nose-down and if the trim tab is deflected down the cockpit indication is nose-up. To remain in trim and maintain level flight following a deceleration the trimmer control is moved to give a nose-up movement that causes the trim tab to move down, which makes the elevator move up. This does not affect the static longitudinal stability.

10.11.2   Fixed Trim Tabs

Fixed trim tabs are fitted to many small light aeroplanes and usually consist of a small aluminium rectangle attached to the trailing edge of the primary control surface. The angle of a fixed trim tab is manually bent to the required angle when the aeroplane is on the ground. Normally, they are set for a relatively low airspeed and the angle is determined by trial and error.

Other light aeroplanes instead of being fitted with an aluminium rectangle have rigid strips of cellulosed cord attached to the trailing edge of the primary control surface. This type of trim is not as effective as the aluminium rectangular plates but is better than nothing.

10.11.3   Stabilizer Trim Setting

The stabiliser trim setting method of trimming the pitch control is more suitable than the elevator trim tab method for jet aeroplanes because of its large range of operating speeds. The *elevator deflection required for balanced flight is inversely proportional to the IAS*, during the take-off phase a large deflection of the elevator upward is required. To maintain the stick force required within acceptable limits during this phase of flight then a large amount of trim is necessary.

The position of the CG directly affects the amount of trim necessary for the angle of attack required. A forward CG adds to the inherent longitudinal stability of the aeroplane; the pilot may run out of elevator up movement and be unable to maintain the required attitude. An aft CG decreases the longitudinal stability and makes the aeroplane more difficult or impossible to control in pitch.

The CG must remain within the CG safe envelope at all times to ensure that the aeroplane remains controllable in all phases of flight. The elevator trim or stabiliser trim setting must be preset to a value for take-off and initial climb that will ensure a minimum stick force during this critical stage of flight. Usually, there is a graph in the aeroplane performance manual from which the setting appropriate to the total mass and CG position for take-off can be determined.

In comparison with a correctly balanced aeroplane during take-off, the position of the stabiliser for a nose-heavy aeroplane requires nose-up trim from a decreased stabiliser angle of incidence.

10.12   Powered Controls

The high stick force required when manually operating the primary control surfaces of large aeroplanes, at high speeds, is impossible for most pilots to apply. Complete control can only be accurately maintained by power-assisted controls or by fully powered controls.

10.13   Power-Assisted Controls

The forces required to operate the primary control surfaces are provided by a combination of physical force by the pilot and by the power system. This system enables the pilot to 'feel' the part of the aerodynamic load imparted by the control surfaces, which is particularly important at low indicated airspeeds during take-off and landing.

The purpose of the trim system on an aeroplane fitted with power-assisted controls is to reduce the stick force to zero. When in trim the position of a power-assisted elevator is dependent on the aeroplane's speed, the position of the slats, the flaps and the location of the CG; but the neutral position of the stick does not change.

Normally, aeroplanes having power-assisted controls are fitted with an adjustable stabiliser because the effectiveness of the trim tabs is insufficient and unable to cope with large trim changes. However, if it runs away then it is more difficult to control than a runaway trim tab. There is an emergency reversion system that, in the event of a power failure, automatically enables the pilot to regain control of the aeroplane manually.

10.14   Fully Powered Controls

The primary control surfaces are operated independently and in parallel by the fully powered controls system and provide all of the operating force required. The movement of the controls by the pilot is communicated to a set of actuators, which provide the necessary physical force to attain the appropriate angular deflection.

A fail/safe system is included such that if the power fails or there is a fault, the control of one or more primary control surface reverts to manual control. Conventional trailing-edge tabs are not included in this system because the aeroplane is trimmed by adjusting the zero position of the artificial feel mechanism. Although powered controls do not require balance tabs some aeroplanes do have them fitted to relieve the servo and hinge loads. In the case of manual reversion of fully powered flight controls a servo-tab is commonly used together with control surface mass balancing to make the aeroplane easier to control.

This system provides no feel to the pilot of the aerodynamic loading of the controls because the power-operated controls are irreversible, which means that there is no feedback of aerodynamic forces from the control surfaces.

10.14.1 Artificial Feel

Feel provided by a spring that exerts a constant load at all airspeeds for the same angular deflection, has the disadvantage that at low speed the force or resistance to control movement felt by the pilot is too great and at high speed it is too small.

The feel of the controls has to be directly related to the primary input of the IAS and therefore is proportional to the difference between the pitot and static pressure, the dynamic air pressure $^1/_2\rho V^2$ (q). The system used is artificial feel commonly called 'q feel.' A resisting force directly proportional to the airspeed and the elevator deflection is applied to the movement of the control column, which if not present would make the controls much easier to move. There are two systems that are used to achieve the desired effect; the simple system and the servo-assisted system.

10.14.1.1   The Simple System

The feel of the simple system is achieved by feeding the pitot and static air pressure either side of a piston in a sealed chamber. The difference between the two pressures, the dynamic pressure, drives the piston that is attached to the control column through a suitable linkage. The simple system is shown in Figure 10.16.

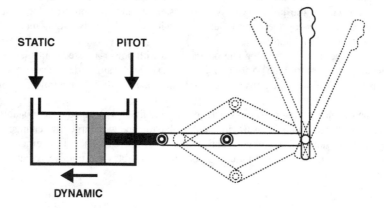

Figure 10.16 The Simple 'q' Feel System.

10.14.1.2   The Servo-Assisted Hydraulic System

The pitot and static air pressure is fed either side of a diaphragm in a hermetically sealed container. The dynamic pressure in this system causes diaphragm to operate a hydraulic servo-valve that applies pressure, equal to an amplified value of dynamic pressure, to the piston controlling the resistance to the control column movement. The servo-assisted hydraulic system is shown in Figure 10.17.

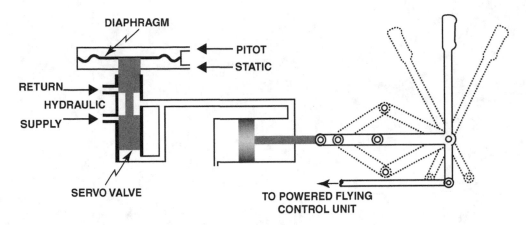

Figure 10.17   The Hydraulic 'q' Feel System.

10.15   Fly-by-Wire

Many modern aeroplane designs incorporate a method of controlling the primary control surfaces by actuators operated by electrical signals sent by a transmitter activated by movement of the pilot's control column. This is fly-by-wire. Such a system facilitates the input of complex electronic processing that modifies the pilot's input in such a manner that it prevents stalling, instability or excessive primary control-surface movement. It also has the ability to coordinate intricate movements of the control surfaces that are impossible for the pilot to achieve unaided, which improve the performance, efficiency and safety of the aeroplane.

Self-Assessment Exercise 10

Q10.1 Differential aileron deflection:
  (a) is necessary to achieve the required rate of roll
  (b) equalises the drag from both ailerons
  (c) is required to keep the total lift constant when the ailerons are deflected
  (d) increases the C_{Lmax}

Q10.2 Which of the following sets are examples of aerodynamic balancing of control surfaces?
  (a) balance tab, horn balance and mass balance
  (b) control surface leading-edge mass, horn balance and mass balance
  (c) spring tab, servo-tab and power-assisted control
  (d) servo-tab, a seal between the wing trailing edge and the control surface leading edge

Q10.3 Flutter may be caused by:
  (a) distortion by bending and torsion of the structure causing increased vibration in the reso-
      nance frequency
  (b) low airspeed aerodynamic wing stall
  (c) roll-control reversal
  (d) high-airspeed aerodynamic wing stall

Q10.4 When power-assisted controls are used for pitch control, this:
  (a) makes trimming unnecessary
  (b) makes aerodynamic balancing of the controls difficult
  (c) can only function as an elevator trim tab
  (d) ensures that partial aerodynamic forces is still felt on the control column

Q10.5 Which statement about a primary control surface controlled by a servo-tab is correct?
  (a) The servo-tab can only be used as a trim tab.
  (b) The control effectiveness is increased by servo-tab deflection.
  (c) Due to the effectiveness of the servo-tab the control surface area is reduced.
  (d) The position is indeterminate on the ground, in particular with a tailwind.

Q10.6 The purpose of a horn balance in a control system is to:
  (a) decrease the effective longitudinal dihedral of the aeroplane
  (b) decrease the required stick forces
  (c) prevent flutter
  (d) produce mass balancing

Q10.7 The type of tab most commonly used in the case of manual reversion of fully powered flight
controls is:
  (a) a balance tab
  (b) an antibalance tab
  (c) a servo-tab
  (d) a spring tab

Q10.8 Examples of the aerodynamic balancing of control surfaces are:
  (a) mass in the control surface leading edge; horn balance
  (b) Fowler flaps; upper and lower rudder
  (c) seal between the wing trailing edge and the leading edge of a control surface; horn balance
  (d) upper and lower rudder; seal between the wing trailing edge and the leading edge of a
      control surface

Q10.9 To damp flutter of control surfaces a mass balance must be positioned............ with respect
to the control surface hinge.
  (a) below
  (b) above
  (c) behind
  (d) in front

Q10.10 Regarding the spring tab, which of the following statements is correct?
  (a) At high IAS it behaves like a servo-tab.
  (b) At low IAS it behaves like a servo-tab.
  (c) At high IAS it behaves like a fixed extension of the elevator.
  (d) Its main purpose is to increase stick force per g.

Q10.11 During a left-banked turn an example of differential aileron during the initiation of the turn is left aileron (i) . . . . :right aileron (ii) . . . .
  (a) (i) 5° down: (ii) 2° up
  (b) (i) 2° down: (ii) 5° up
  (c) (i) 5° up: (ii) 2° down
  (d) (i) 2° up: (ii) 5° down

Q10.12 During entry and roll out of a turn adverse yaw is compensated by:
  (a) servo-tabs
  (b) differential aileron deflection
  (c) horn-balanced controls
  (d) antibalanced controls

Q10.13 Which of the following statements regarding control is correct?
  (a) On some aeroplanes, the servo-tab also serves as a balance tab.
  (b) Hydraulically powered control surfaces do not need mass balancing.
  (c) In general, the maximum downward elevator deflection is equal to upward.
  (d) In a differential aileron control system the control surfaces have a larger upward than downward maximum deflection.

Q10.14 Aeroplanes with power-assisted controls are usually fitted with an adjustable stabiliser because:
  (a) mechanical adjustment of trim tabs is too complicated
  (b) trim tab deflection increases M_{CRIT}
  (c) the effectiveness of the trim tabs is insufficient
  (d) the pilot does not feel the stick forces at all

Q10.15 For an aeroplane that is flying straight and level in trim, the position of the power-assisted elevator relative to the trimmable horizontal stabiliser is:
  (a) dependent on the speed, the position of the slats, flaps and CG
  (b) the elevator deflection is always zero compared to the stabiliser position
  (c) with a forward CG position the elevator is deflected upward and vice versa
  (d) the elevator is always deflected downward to ensure that sufficient authority remains for the flare on landing

Q10.16 The phenomenon counteracted by differential aileron deflection is:
  (a) adverse yaw
  (b) aileron reversal
  (c) sensitivity to Dutch roll
  (d) turn coordination

Q10.17 If the elevator trim tab is deflected up, the cockpit trim indicator shows:
  (a) nose-up
  (b) nose-left
  (c) nose-down
  (d) neutral

Q10.18 One method of compensating for adverse yaw is a:
  (a) balance tab
  (b) antibalance tab
  (c) balance panel
  (d) differential aileron

Q10.19 If the elevator jams in flight, the result for a servo-tab-controlled elevator is:
  (a) the pitch control reverses direction
  (b) the pitch control is lost
  (c) the servo-tab now operates as a negative trim tab
  (d) the pitch-control force doubles

Q10.20 If the nose of an aeroplane yaws starboard it causes:
  (a) a roll to port
  (b) a roll to starboard
  (c) a decrease in relative airspeed on the port wing
  (d) an increase of lift on the starboard wing

Q10.21 A left rudder input will cause (i) . . . . yaw about the vertical axis and (ii) . . . . roll about the longitudinal axis.
  (a) (i) left; (ii) right
  (b) (i) right; (ii) left
  (c) (i) right; (ii) right
  (d) (i) left; (ii) left

Q10.22 Servo-tabs fitted to a main control surface:
  (a) are activated by the movement of the main control surface
  (b) also act as trim tabs
  (c) make the controls less effective at low speeds
  (d) enable the control surface to be reduced in size because of the increased effectiveness

Q10.23 Relative to the main control surface hinge-line mass-balance weights are located:
  (a) in front
  (b) on the hinge-line
  (c) If the control surface has an inset hinge-line it is located aft of the hinge-line
  (d) behind the hinge-line

Q10.24 If the servo-tab-controlled elevator of an aeroplane jams in flight the effect of the servo-tab is that it will:
  (a) become a negative trim tab
  (b) become an antibalance tab
  (c) reverse the direction of the pitch input
  (d) cause the loss of pitch control

Q10.25 If it is necessary for the pilot of an aeroplane fitted with powered controls to revert to manual control in flight which of the following tabs is used?
  (a) servo-tab
  (b) antibalance tab
  (c) balance tab
  (d) spring tab

Q10.26 The purpose of a trim system on power-assisted flying controls is to:
  (a) reduce the stick force to zero
  (b) decrease the stress imposed on the trim tab
  (c) decrease the stress imposed on the hydraulic actuators
  (d) to facilitate manual control of the aeroplane in the event of a complete hydraulic failure

Q10.27 Aerodynamic balance of the ailerons is attained by:
  (a) an internal balance within a seal from the wing to the leading edge of the aileron
  (b) a seal from the wing to the trailing edge of the aileron
  (c) a weight positioned forward of the hinge-line
  (d) differential movement of the control surfaces

Q10.28 The cockpit indication when the elevator trim tab is moved up is:
  (a) dependent on the position of the elevator
  (b) neutral
  (c) nose-up
  (d) nose-down

Q10.29 To remain in trim and allow for a deceleration the trim tab moves (i) .... making the (ii) ....
  (a) (i) down; (ii) elevator move up
  (b) (i) down; (ii) adjusting the variable-incidence tailplane
  (c) (i) up; (ii) elevator move down
  (d) (i) up; (ii) making the variable-incidence tailplane increase incidence

Q10.30 A spring tab:
  (a) operates as a servo-tab at low IAS
  (b) increases the stick force per 'g'
  (c) provides a basic trim force through the spring
  (d) operates as a servo-tab at high IAS

Q10.31 The cockpit indication when the elevator trim tab is moved down is:
  (a) dependent on the position of the elevator
  (b) neutral
  (c) nose-up
  (d) nose-down

Q10.32 Stick force in artificial feel depends on:
  (a) elevator deflection and static pressure
  (b) elevator deflection and dynamic pressure
  (c) stabiliser deflection and static pressure
  (d) stabiliser deflection and total pressure

Q10.33 Stick forces provided by an elevator feel system, depend on:
  (a) stabiliser position, total pressure
  (b) elevator deflection, dynamic pressure
  (c) stabiliser position, static pressure
  (d) elevator deflection, static pressure

Q10.34 In comparison with a correctly balanced aeroplane during take-off, the position of the stabiliser for a nose heavy aeroplane requires:
  (a) increased nose-down trim from a decreased stabiliser angle of incidence
  (b) increased nose-up trim from a decreased stabiliser angle of incidence
  (c) decreased nose-up trim from an increased stabiliser angle of incidence
  (d) decreased nose-down trim from an increased stabiliser angle of incidence

Part 4
Stability

11 Static Stability

11.1 Static Stability

According to Newton's first law a body remains in a state of rest or uniform motion unless acted on by an external force. Stability is the reaction of an aeroplane after an external disturbing force to its equilibrium ceases or is removed.

The ability of an aeroplane to return to its original state following an undemanded disturbance is a measure of its stability. Too much stability is undesirable because the aeroplane is slow to respond to control inputs and is sluggish in manoeuvring. Too much instability is also undesirable because the attitude of the aeroplane requires continual correction. There are two main types of stability; they are static stability and dynamic stability:

a. The **static stability** of an aeroplane is the *immediate short-term* response of the aeroplane to a disturbance and
b. The **dynamic stability** is the *subsequent long-term* response of an aeroplane to a disturbance to its equilibrium. It is a measure of its reaction to damp out any unwanted oscillations, which depends on how fast or slow the aeroplane responds to a disturbance. The dynamic stability of an aeroplane is dependent on the design of the aeroplane, its speed and the height at which it is being flown.

A statically (short-term) stable aeroplane can be dynamically (long-term) stable, neutral or unstable. However, a statically neutral or statically unstable aeroplane can never be dynamically stable. The reaction of an aeroplane to a disturbance is conditioned by its original state of equilibrium. For instance, in straight and level flight an aeroplane is in equilibrium because the four forces are balanced and the resultant sum of these forces and their moments is zero and this will determine its reaction to a disturbance.

There are three degrees of stability – positive, neutral and negative.

a. **Positive stability**. An aeroplane that returns to its predisturbed attitude without assistance, once an external disturbing force ceases, is considered to be stable and to have positive stability.
b. **Neutral stability**. An aeroplane that remains in the attitude that it attains when an external disturbing force ceases is neutrally stable.
c. **Negative stability**. An aeroplane that continues to diverge from its predisturbed attitude after an external disturbing force ceases is an unstable aeroplane and has negative stability.

The Principles of Flight for Pilots   P. J. Swatton
© 2011 John Wiley & Sons, Ltd

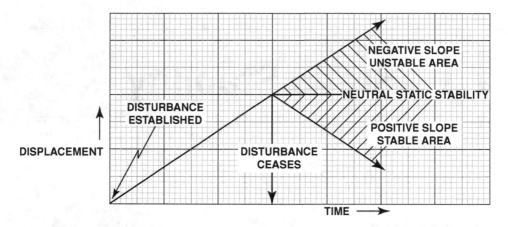

Figure 11.1   Degrees of Stability.

The degree of stability is depicted graphically in Figure 11.1 it shows the magnitude of the displacement by the vertical axis and the time period over which the disturbance was applied and the reaction to it is shown by the horizontal axis.

The graph in Figure 11.1 also indicates the point at which a disturbance is applied or established and the point at which the disturbance is removed or ceases. It shows:

a. A horizontal straight line commencing at the point where the disturbance ceases indicates neutral static stability.

b. Any reactive force in the area beneath the neutral static stability horizontal line is a positive force and indicates that the aeroplane is stable. The angle of the graph line down from the disturbance cessation point indicates the degree of static stability. The steeper the line the greater is the stability of the aeroplane.

c. Similarly, any reactive force in the area above the horizontal neutral static stability line is a negative force and indicates that the aeroplane is unstable. The angle of the graph line up from the disturbance cessation point indicates the degree of instability. The steeper the line the greater is the instability of the aeroplane.

The reaction of an aeroplane to a disturbance can be resolved into components around the three axes of the aeroplane that pass through the CG and are shown in Table 11.1. The motion is an angular velocity and the reaction to the disturbance is an angular displacement. If the total moment of the aeroplane is not zero it will rotate about the CG.

Table 11.1   The Resolution of Reactive Motion.

Axis	Control Surface	Motion (about the axis)	Positive Motion	Stability
Longitudinal (x)	Aileron	Roll (p)	Right	Lateral
Lateral (y)	Elevator	Pitch (q)	Nose-up	Longitudinal
Normal (z)	Rudder	Yaw (r)	Right	Directional (Weathercock)

11.2   The Effect of the Variables on Static Stability

The following variable factors increase the static stability of an aeroplane:

a. Decreased altitude – stability is proportional to dynamic pressure and inversely proportional to TAS therefore it is greatest at low altitude.
b. Increased IAS increases the dynamic pressure without the necessity of decreasing the altitude. Therefore, the static stability is directly proportional to the IAS.
c. Increased air density increases the dynamic pressure without having to increase the IAS. Static stability is therefore increased with decreased ambient temperature.
d. Decreased aeroplane mass increases the responsiveness of the aeroplane to the correcting forces.
e. Forward CG position increases the moment arm from the aerofoil surfaces CPs, thus increasing the moment of the correcting forces.

11.3   Directional Static Stability

The rotation of an aircraft about its normal, or vertical, axis is a yaw. If this movement is the result of an undemanded disturbance then the ability of the aeroplane to return to its original heading without the assistance of any control-surface movement is a measure of its directional static stability. It is a measure of the aircraft's ability to realign itself with the relative airflow. In a skid the aeroplane has a tendency to recover without the use of rudder. It is known as 'weathercocking' stability, alluding to the movement of a weathercock when wind is blowing.

11.4   Yaw and Sideslip

A yaw is positive to the right and negative to the left. The yawing moment, N, is calculated by using the formula:

$$N = Cn\tfrac{1}{2}\rho V^2 Sb$$

Where Cn the yawing moment coefficient; ρ = air density; V = airspeed; S = wing area and b = the wingspan.

The yaw angle is the displacement angle of the aeroplane's longitudinal axis in azimuth from a specified reference datum. It is negative when the longitudinal axis is to the left of the reference datum. See Figure 11.2(a).

Sideslip is the displacement angle of the aeroplane's longitudinal axis in azimuth from the relative airflow. It is positive when the relative airflow is to the right of the longitudinal axis and negative when the relative airflow is left of the longitudinal axis. See Figure 11.2(b). An aeroplane has directional static stability if, when in a sideslip with the airflow coming from the left initially the nose tends to yaw left. *An aeroplane that has excessive directional static stability compared to its lateral static stability is more prone to spiral dive.*

11.5   The Directional Restoring Moment

Having experienced a disturbance to its equilibrium by an outside force an aeroplane requires a restoring moment to return to its original attitude. The strength of that restoring moment is dependent on two factors, the design area of the fin and rudder and the distance of the tailplane from the CG. When these two defining factors are multiplied together the result is the fin volume, which determines the directional stability of an aeroplane.

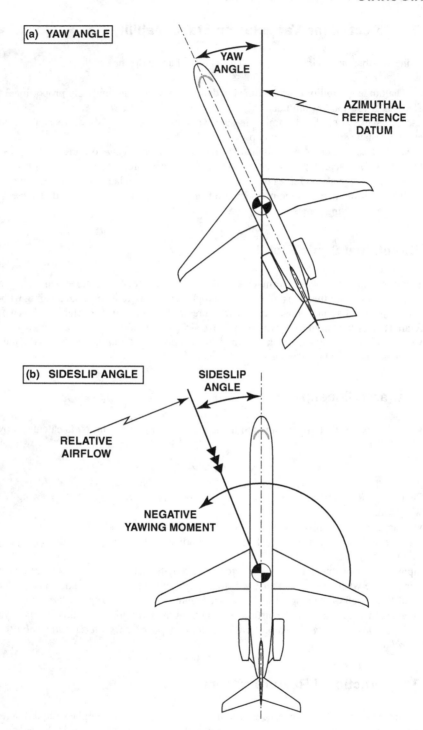

Figure 11.2   Yaw and Sideslip Angles.

11.5.1 Fin and Rudder Design

The fin, which is the vertical stabilising symmetrical aerofoil mounted on top of the fuselage at the rear, creates the correcting movement or restoring moment. The manner in which it achieves this is that the disturbing force moves the aeroplane about its normal axis causing the fin to have an angle of attack to the relative airflow. When it has a positive angle of attack to the airflow the fin produces an aerodynamic correcting force, which is lift in a sideways direction.

In a sideslip the fin has a positive angle of attack that will produce a force proportional to both the 'lift coefficient' and the area of the fin. The magnitude of the force, the sideways lift, is proportional to the area of the fin, its aspect ratio and its sweepback. To decrease the likelihood of the fin stalling (in a sideways direction) at high sideslip angles, the aeroplane design team include either a large degree of sweepback on a fin of low aspect ratio or multiple low aspect ratio fins in the design of the tailplane.

Not only is the magnitude of the sideways force proportional to the size of the fin and rudder, it is also proportional the length of the moment arm. This arm will rotate the aeroplane around its centre of gravity in the opposite direction to the disturbance. The amount of lift and the size of the restoring moment diminishes as the aeroplane returns to the direction opposite to that of the relative airflow and vanishes altogether when it is pointing exactly into the relative airflow. See Figure 11.3.

11.5.2 The Dorsal Fin

A dorsal fin is an additional fillet inserted on the top of the fuselage and forward of but joined to the fin. There are three main reasons for including a dorsal fin in an aeroplane design they are to:

a. Increase the effective surface area of the fin, thus enhancing the yawing moment and augmenting the static directional stability and the weathercocking effect.
b. Decrease the aspect ratio of the fin, therefore enlarging the sideways stalling angle of attack, which ensures that the fin continues to be effective at increased sideslip angles.
c. Maintain directional static stability when the aeroplane has a large sideslip angle.

11.5.3 The Ventral Fin

To assist the directional static stability of an aeroplane in normal flight, some aeroplanes are fitted with strakes or ventral fins. These are flat plates or strips positioned under the fuselage of the aeroplane aft of the CG and form a keel running parallel to the fore and aft axis of the aeroplane. As an alternative to this, large transport aeroplanes are often fitted with small additional fins added to the tailplane. All of these factors increase the directional static stability of the aeroplane.

11.5.4 The Moment Arm

The position of the CG affects the length of the moment arm of the fin, which directly influences the ability of the fin to achieve its purpose. The distance of the CP of the fin from the CG fixes the length of the moment arm. A large restoring moment is generated by a large fin situated a considerable distance from the CG. A small restoring moment is the result of a small fin positioned close to the CG. The combination selected by the aeroplane design team is normally determined by other factors.

Any movement of the CG will change the length of the arm of the restoring moment and either increase or decrease its effectiveness. A forward movement of the CG will lengthen the moment arm and increase the directional static stability of the aeroplane and an aft movement of the CG will shorten the moment arm and decrease the directional static stability of the aeroplane. (See Figure 11.3).

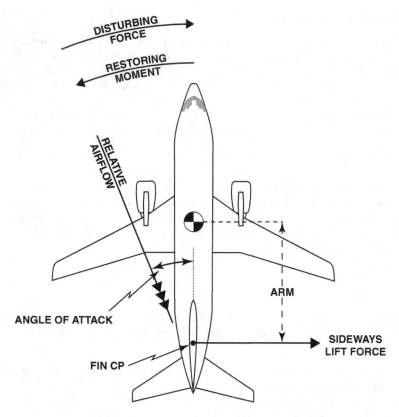

Figure 11.3   Directional Restoring Moment.

11.6   Aeroplane Design Features Affecting Directional Static Stability

11.6.1   Fuselage

When the centre of pressure (CP) of the fuselage of a normal aeroplane is well forward of the CG it causes the fuselage to be a destabilising influence on the directional static stability of the aeroplane. If such is the location of the CP relative to the CG, then in a sideslip the relative airflow exerts a larger yawing moment forward of the CG than it does aft of the CG.

In other words, the length of the fuselage ahead of the CG has an unstable static direction influence, whereas the fuselage length behind the CG has a stabilising influence. This effect is particularly noticeable at high angles of attack when the disturbed airflow around the fuselage causes the fin to stall and the aeroplane to become directionally unstable. To counteract the unstable influence of the fuselage an aeroplane requires a high vertical fin or stabiliser to produce positive directional static stability to move the CP of the fuselage to a position aft of the CG.

11.6.2   Wing

The two design characteristics that directly affect directional static stability are dihedral and sweepback.

11.6.2.1 *Dihedral*

Compared with sweepback, the effect of dihedral is insignificant; it decreases the directional static stability because the lift produced by the sloping wings (the dihedral) contributes to the yawing moment due to their inclination.

11.6.3 *Sweepback*

The drag of a swept wing is less than that of an unswept wing. The advantage of this type of wing at high speeds is gained at the expense of its poor performance at low speed. The drag experienced with a swept wing is the component of the relative airflow at 90° to the line joining the aerodynamic centres of the wing, which if the wing has parallel leading and trailing edges will be the normal to the leading edge of the wing. In Figure 11.4 the normal component of the relative airflow is greater on the port wing than the normal component on the starboard wing.

The drag on the leading wing, the port wing, in Figure 11.4 is of greater magnitude than that of the trailing starboard wing because of its decreased effective sweep angle. It is therefore a stabilising influence. *Positive sweepback has a stabilising effect on directional stability* because the CP is further aft and the stalling angle is increased. The significance and magnitude of this effect on directional static stability is directly proportional to the angle of sweepback.

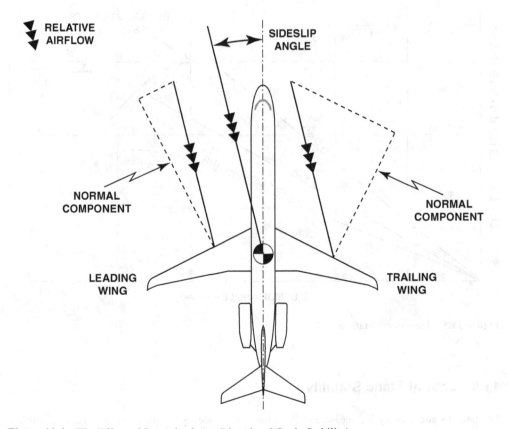

Figure 11.4 The Effect of Sweepback (on Directional Static Stability).

11.7   Propeller Slipstream

The effect caused by the slipstream of a single-engined propeller-driven aeroplane is dependent on the direction of rotation of the propeller. The corkscrew airstream produced by the propeller will strike one side of the fin more than the other. To counteract this effect the fin and rudder have to balance the asymmetric airflow to prevent sideslip, which for a single-engined aeroplane is more likely in one direction than the other. To assist in this task the fin is mounted slightly offset from the fore and aft axis into the corkscrew airstream, thus decreasing the asymmetric influence. See Figure 16.7.

11.8   Neutral Directional Static Stability

The effect that each major component has on the overall directional static stability of an aeroplane can be plotted graphically. C_N is the yawing moment coefficient and is shown on the left vertical axis of Figure 11.5. The sideslip angle is shown along the horizontal axis. The highest point of each curve is the stalling sideslip angle and is the point of neutral directional static stability; to the left of this point the aeroplane has positive directional static stability and to the right of this point the aeroplane has directional static instability.

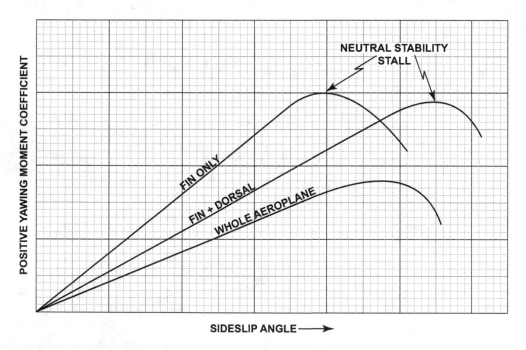

Figure 11.5   Directional Static Stability.

11.9   Lateral Static Stability

The lateral static stability is a measure of the aeroplane's tendency to return to the wings-level attitude after a disturbance has caused the aeroplane to be disturbed in the rolling plane. Positive lateral static

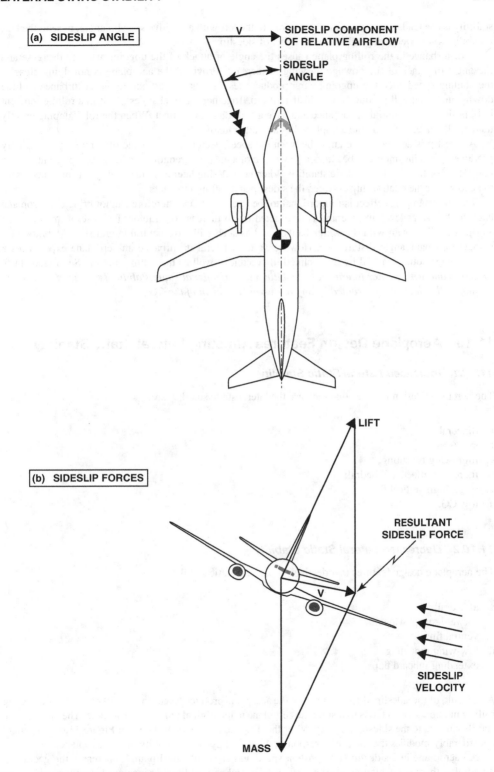

Figure 11.6 Sideslipping. (a) Sideslip Angle.

stability is the tendency of the aeroplane to roll to the left with a positive sideslip angle, nose to the left. Its ability to recover is dependent on the effect of sideslip.

A disturbance in the rolling plane causes the angle of attack of the upgoing wing to decrease and the angle of attack of the downgoing wing to increase. Provided the aeroplane is not flying close to the stalling speed, then the upgoing wing produces less lift than it did before the disturbance and the downgoing wing will produce more lift than it did. Together, these changes result in a rolling moment in opposition to the initial disturbance and have a 'roll-damping' effect. When the roll damping exactly matches the aileron torque the aeroplane has a steady rate of roll.

A sideslip is defined as the angle between the speed vector and the plane of symmetry. It not only produces a rolling moment but also a yawing moment, the strength of which is dependent on the magnitude of the directional static stability. When considering lateral static stability it is only necessary to account for the relationship between the sideslip and rolling moments.

The 'roll-damping' effect is proportional to the rate of roll and therefore cannot bring the aeroplane back to the wings-level attitude and is unaffected by an increase of altitude. Because of this then, the aeroplane will remain with the wings banked and as such will have neutral lateral static stability with respect to a bank-angle disturbance. However, after a lateral disturbance an aeroplane experiences a sideslipping motion, caused by the inclined lift vector, as well as the rolling motion. See Figure 11.6. *An aeroplane with greater lateral static stability than directional static stability is prone to developing 'Dutch' roll, which is exacerbated by any rearward movement of the CG.*

11.10   Aeroplane Design Features Affecting Lateral Static Stability

11.10.1   Increased Lateral Static Stability

The aeroplane design features that increase the lateral static stability are:

a. dihedral;
b. sweepback;
c. high-wing mounting;
d. increased effective dihedral;
e. large, high vertical fin;
f. low CG.

11.10.2   Decreased Lateral Static Stability

The aeroplane design features that decrease the lateral static stability are:

a. anhedral;
b. forward-swept wings;
c. ventral fin;
d. low-wing mounting;
e. extending inboard flaps.

As a result, of the sideslip, different parts of the aeroplane produce forces that together create a correcting rolling moment, which tends to restore the aeroplane to its original wings-level attitude. The lateral static stability reacts to the sideslip velocity 'v' or the displacement in yaw shown in Figure 11.6. This effect considerably modifies the long-term response, the lateral dynamic stability of the aeroplane.

An aeroplane in a sideslip at a constant speed and constant sideslip angle increases the geometric dihedral of the wing, which requires an increased lateral control force or increased stick force.

11.11 Sideslip Angle and Rolling Moment Coefficient

If an aircraft is subject to a negative sideslip angle (β), as shown in Figure 11.6(a), the relative airflow is to the left of the aeroplane centreline; it has lateral static stability if the resulting rolling moment is to the right. In a sideslip to the left the left wing will drop and initially the nose moves to the left but the rolling moment produces a correcting rolling moment to the right.

A roll is a rotation of the aeroplane about the longitudinal axis and is positive to the right and negative to the left. The rolling moment, L, is calculated by using the formula:

$$L = Cl \frac{1}{2}\rho\, V^2 Sb$$

where Cl = the rolling moment coefficient; ρ = the air density; V = the airspeed; S = the wing area and b = the wingspan. By transposition, the formula becomes:

$$Cl = \frac{L}{\frac{1}{2}\rho\, V^2 Sb}$$

When Cl is plotted against the sideslip angle (β) a positive line indicates lateral static instability and a negative line indicates lateral static stability. See Figure 11.7.

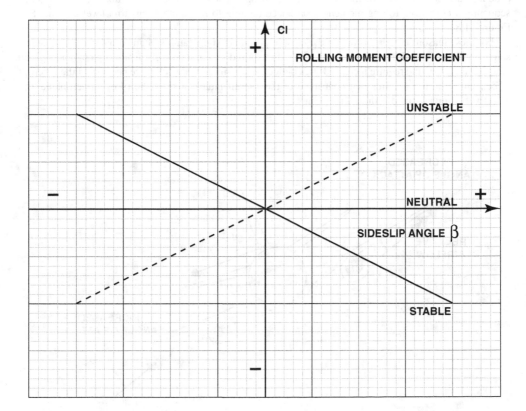

Figure 11.7 Roll Moment Coefficient Cl v Sideslip Angle β.

11.12   Analysis of Design Feature Effects

Diverse contributions of varying magnitudes are made to the overall lateral static stability by different parts of an aeroplane. The size of these contributions is dependent on the stage of flight and the configuration of the aeroplane. The largest contributions are made by:

a. Wing due to: (1) Dihedral; (2) Anhedral; (3) Sweepback.
b. Wing/fuselage interference: (1) Shielding Effect; (2) Vertical Location.
c. Fuselage/fin: (1) Fin Size; (2) Ventral Fin.
d. Handling Considerations: (1) Propeller Slipstream; (2) Crosswind Landings.
e. Flaps.

11.13   Wing Contribution

11.13.1   Dihedral

Dihedral is the upward inclination of the wings of an aeroplane from the plane passing horizontally through the lateral axis. The prime purpose of manufacturing an aeroplane with such a feature is to increase its static lateral stability. This is because if the aeroplane sideslips then the lower wing produces a greater amount of lift than the upper wing because it has a greater angle of attack; consequently a correcting rolling moment is produced tending to move the wings to a level attitude. See Figure 11.8.

For these reasons dihedral is one of the most important design features of an aeroplane. Often lateral static stability is called the 'dihedral effect.' If it is required to maintain a given sideslip angle at a given airspeed then it would require increased stick deflection and increased stick force because of the dihedral effect. Other design features that contribute to the overall lateral static stability of an aeroplane and effectively increase the dihedral of the wings are:

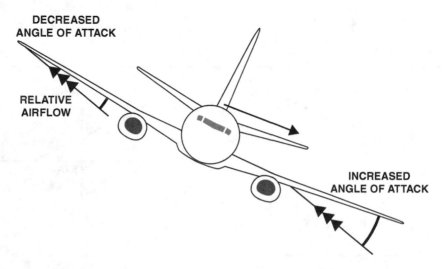

Figure 11.8   The Dihedral Effect.

a. *a 10° angle of sweepback provides the equivalent of 1° of effective dihedral;* and
b. a high-wing configuration provides approximately 5° of effective dihedral over that of a low-wing configuration.

11.13.2 Anhedral

Anhedral is the downward inclination of the wings of an aeroplane from the plane passing horizontally through the lateral axis. This design feature produces the opposite effect to that of dihedral and has a lateral static destabilising effect. This would appear to be detrimental to the performance of the aeroplane, but designers often use it to counteract other features that cause the aeroplane to become undesirably over-stable laterally and difficult to turn, such as a high wing and sweepback combination. There are many airliners currently in service with geometric anhedral, for example the Avro 146 RJ, the C5 Galaxy and the Antonov An-225. All of which have a high-mounted wing and a large degree of sweepback, which together make the aeroplane too stable hence the need for anhedral.

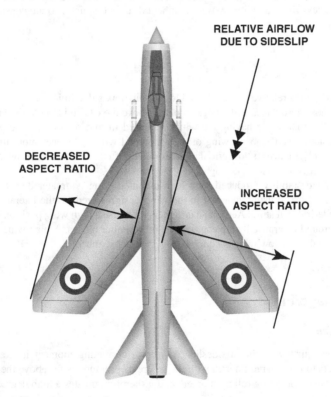

Figure 11.9 The Effect of Sweepback.

11.13.3 Sweepback

Swept wings increase lateral static stability because an aeroplane tends to sideslip in the direction of any roll. This effectively increases the wingspan of the downgoing wing and decreases its effective chord length, which together increases the effective aspect ratio. The reverse is true of the upgoing wing. The wing with the increased effective aspect ratio, the starboard wing in Figure 11.9, produces more lift than that of the low effective aspect ratio wing, and creates a corrective rolling moment that levels the wings. Sweepback produces a positive 'dihedral effect' that is not affected by altitude changes.

It is generally assumed that the stabilising effect produced by a wing having a dihedral of 1° and no sweep back is equal to that of a wing having no dihedral but a sweep back from root to tip of 10°. If the wing is straight, increasing the aspect ratio of the vertical stabiliser whilst maintaining the area as a constant enhances the static lateral stability. A swept wing not only increases the lateral static stability it also stabilises directional static stability and is least affected by turbulence. Lateral static stability also increases with increased speed.

11.14   Wing/Fuselage Interference

11.14.1   Shielding Effect

In a sideslip the fuselage causes the trailing (upgoing) wing to be shielded from the airflow, consequently the dynamic pressure over this part of the wing is less than over the rest of the wing. It therefore produces less lift and will increase the 'dihedral' effect, which is quite considerable for some aeroplane types.

11.14.2   Wing Location

The position of the wings relative to the CG and longitudinal axis determines the lateral static stability of an aeroplane. Those aeroplanes with wings mounted above the CG and longitudinal axis have greater lateral static stability than those having the wings mounted in mid or low relative positions. This is because during a sideslip a the lower wing of high-mounted wings produces more lift than the upper wing that consequently induces a flow circulation around the fuselage increasing the angle of attack of the lower wing which generates a restoring force.

If the lateral static stability produced by a high-mounted wing is greater than that required then the designer might resort to using anhedral in the wing design to reduce the lateral static stability to counteract this undesirable feature. Aeroplane designers often use a high wing position in their design to ensure adequate ground clearance by the engine pods or propeller blades. A high wing is the equivalent of 1° to 3° of dihedral and a low wing is equivalent to 1° to 3° of anhedral.

11.15   Fuselage/Fin

11.15.1   Fin Size

Because the relative airflow during a sideslip produces a correcting moment it is advantageous for aeroplane designers to use a large fin area with the centre of pressure as far above the CG as possible. This engenders a large correcting rolling moment and it therefore creates a high degree of lateral static stability. See Figure 11.10.

11.15.2   Ventral Fin

The rate of roll during a sideslip is increased by any moment caused by side surfaces mounted on the fuselage below the CG. Thus, ventral fins increase the lateral static instability of an aeroplane. Aeroplanes with a 'T' tailplane are particularly stable laterally because they necessarily have a tall fin to keep the tailplane above the turbulent downwash of the wings. Such aircraft tend to be laterally very stable, so much so that their roll response is poor and a ventral fin or keel surface is fitted quite deliberately to improve roll response.

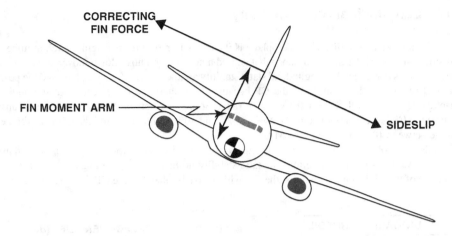

Figure 11.10 The Effect of the Fin.

11.16 Handling Considerations

11.16.1 Propeller Slipstream

In normal flight the slipstream from the propeller(s) is usually balanced symmetrically along the longitudinal axis. During a sideslip, due to a disturbance in roll, the increased dynamic pressure of the slipstream affects the upper or trailing wing. This increases the rolling moment and causes lateral static instability.

11.16.2 Crosswind Landings

During a 'crab approach' in a crosswind the aeroplane is yawed just before touchdown to align it with the runway. Sideslip is increased in the opposite direction to the yaw and if the aeroplane is too stable laterally it will roll away from the sideslip. This can create control difficulties during this critical phase of flight.

Although for most phases of flight it is desirable for an aeroplane to have a high degree of lateral static stability, on approach to land, from a handling viewpoint, it would be expedient if it were less statically stable. When making a crosswind approach and landing in an aeroplane with excessive lateral static stability the pilot has to use a lot more aileron into wind to make a stabilised approach.

11.16.3 Flaps

For an aeroplane on which the flaps are mounted close to the fuselage, when they are lowered they cause the CP to move towards the wing roots. This has a destabilising effect on the lateral static stability. Fitting drooped ailerons, which move the CP towards the wing tips, can counteract this undesirable characteristic.

Drooped ailerons have other effects that are also beneficial to the performance of an aeroplane; they increase the maximum coefficient of lift, improve the span-wise lift distribution, decrease the take-off speeds and improve the short-field performance. When they are fitted for these purposes they are called flaperons.

11.17   Longitudinal Static Stability

The longitudinal static stability of an aeroplane is its ability to return to its original pitch attitude with no outside assistance, after the disturbance that caused a nose-up or a nose-down movement has ceased. The main factors that affect the longitudinal static stability are the design of the tailplane and the position of the CG relative to the position of the CP because they determine the direction of the additional wing-pitching moment (dLX). For the working range of angles of attack for a conventional transport aeroplane the CP is located in the range between 30% and 40% of the MAC and the CG is positioned in the range between 10% and 30% of the MAC.

For other aeroplanes the CG range is normally 10% to 35%. Therefore, a conventional transport aeroplane always has positive longitudinal static stability no matter what the angle of attack because, irrespective of their individual locations, the CG will always be ahead of the CP. See Figure 11.11(b).

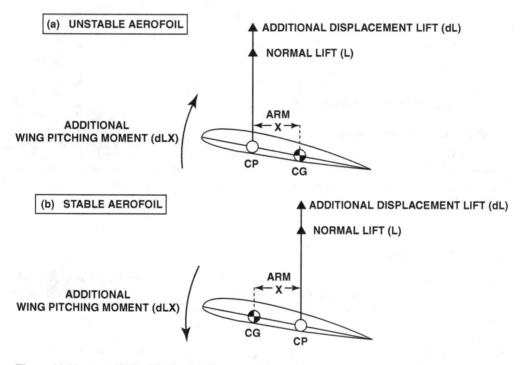

Figure 11.11   Longitudinal Static Stability.

The tailplane and elevators act as a horizontal stabiliser and the degree of positive longitudinal static stability that they provide is dependent on the size of the restoring moment they can attain; nevertheless, they do make the *largest contribution to the longitudinal static stability*. The tailplane design and the length of the moment arm from the tailplane CP to the CG condition the magnitude of the restoring moment. If the oscillations of the aeroplane caused by the disturbing force about the lateral axis of the aeroplane are of constant amplitude then the aeroplane has longitudinal static stability and longitudinal dynamic neutrality.

Although the tailplane produces very little lift compared with that produced by the wings, the pitching moment it can produce is relatively large because of the extensive distance that the tailplane CP is from the aeroplane CG. The main parameter used by aircraft designers to establish an aircraft's positive longitudinal static stability is the ratio of the tail volume to the wing volume. The tail volume is the product of the tailplane area and the tailplane moment, and the wing volume, similarly, is the product of the wing area and the wing moment. The greater the excess volume of the tail over that of the wing the greater is the longitudinal static stability.

Figure 11.12 illustrates the change to the forces and moments due to a pitch displacement and shows that the tail must be uploaded and contribute a moment of sufficient magnitude to overcome the disturbing moment to produce positive longitudinal static stability. For any given angle of attack the degree of positive longitudinal static stability is dependent on the magnitude of excess moment produced by the tailplane over that produced by the wings. This is the restoring moment.

The net pitching moment = (tail-pitching moment − wing-pitching moment). In this example the tail must produce an upward force and thus a nose-down moment. In the worst case, when the wings have an unstable influence, then the design of the horizontal stabiliser must be such as to rectify this undesirable characteristic. Some aeroplanes have the horizontal stabiliser mounted on top of the vertical tail surfaces called a 'T' tail to act as an endplate that improves the efficiency of the fin and rudder. Such a tail also increases the lateral static stability.

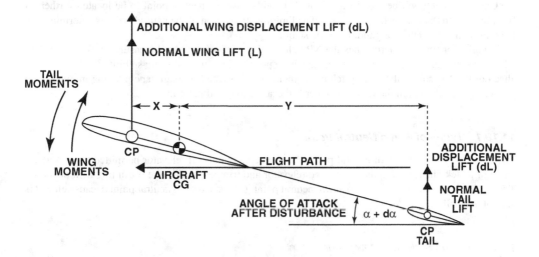

Figure 11.12   Changes to Forces and Moments Resulting from a Small Nose-up Displacement (dα).

The situation as depicted in Figure 11.12 is not the normal disposition of the pitching moments for a transport aeroplane. Usually, the CP of a transport aeroplane is aft of the CG and generates a nose-down pitching moment. To maintain level flight the tailplane has to counteract this tendency by producing a nose-up pitching moment. In other words, the tailplane must be downloaded. Thus, longitudinal static stability is ensured when the CP is aft of the CG and the tail is downloaded. A negative tail stall is the sudden reduction of this downward aerodynamic force on the tailplane and may result in an uncontrollable nose-down pitching moment.

11.18   The Centre of Pressure (CP)

The location of the CP is dependent on the angle of attack. It moves forward with increasing angle of attack until the stalling angle is attained, at which point it moves abruptly aft. The value of C_L is directly proportional to the angle of attack. It reaches its maximum value at the stalling angle.

The longitudinal static stability of an aeroplane is therefore conditioned by the relative positions of the CP and the CG. The greater the distance the CP is aft of the CG the larger is the nose-down pitching moment and the greater is the tail-down force required to counteract it. There is normally a download on the tailplane if such is the case. As an aeroplane accelerates in the transonic range the longitudinal static stability increases because as the angle of attack decreases the CP moves aft.

11.19   The Neutral Point (NP)

There is a point on the longitudinal axis at which the wing moment (increasing) is equal to the tail moment (decreasing), which results in a zero restoring moment. This is the neutral point (NP) sometimes referred to as the aerodynamic centre of the **whole** aeroplane. When the angle of attack of an aeroplane changes, the net result of the lift generated by the wings, stabiliser and fuselage act through the neutral point. This means that if the CG of an aeroplane is located at the NP it is neutrally stable.

The position of the neutral point is dependent on the size of the tailplane area and the length of the moment arm of the tailplane CP. The larger the tailplane area and/or the longer the moment arm from the CG, the greater will be the tail moment. This will cause the neutral point to be located further aft from the datum than normally is the case. The limiting aft position of the neutral point is determined by the minimum acceptable longitudinal static stability.

Usually, for transport aeroplanes the NP is located at 40% of the MAC measured from the leading edge of the wing. For an aeroplane to always have positive longitudinal static stability the CG must be ahead of the NP, so that the wing-pitching moment always exceeds the tailplane pitching moment. The further aft the NP is from the CG the greater is the static longitudinal stability.

11.19.1   Types of Static Neutral Point

There are in fact two types of neutral point, the stick-free (hands off, elevator permitted to float as the angle of attack changes) and the stick-fixed (elevator and trim tab held in the prevailing trim position). That which is commonly referred to as *the* neutral point is the stick-fixed neutral point because elevators are not normally free to float.

11.19.1.1   The Stick-Free Static Neutral Point

The stick-free static neutral point is the CG position at which the aeroplane has neutral stability with elevators free to float. This neutral point is located where *zero stick force* is required to hold the aeroplane from the trimmed angle of attack and airspeed. It is always located ahead of the stick-fixed static NP. Stick-free longitudinal static stability determines the stick force required.

11.19.1.2   The Stick-Fixed Static Neutral Point

The stick-fixed static neutral point is the CG position at which the aeroplane exhibits neutral stability with the elevators fixed. This neutral point is located where *zero stick movement* is needed to hold the aeroplane from the trimmed angle of attack and airspeed. In other words, after the control column has been moved to attain the new angle of attack, C_L or airspeed it can then be returned to its original position. Increasing the stabiliser area moves the stick-fixed neutral point aft. The stick-fixed NP is always located aft of the stick-free NP. The stick-fixed longitudinal static stability determines the degree of control and angle of elevator movement required to change the airspeed, C_L or angle of attack.

11.19.2   The Effect of the CG at the NP

If the CG is located at the NP, then:

a. The aeroplane has neutral longitudinal static stability.
b. After the disturbing force has ceased the aeroplane will remain in the attitude that it attained at the time of the cessation of the disturbance.
c. The pitching moment, C$_M$, remains constant for all angles of attack.
d. The magnitude of the lift is still dependent on the angle of attack.
e. The pilot will sense no opposing force when the control column is moved to fly at a new C$_L$ because the aeroplane does not generate a stabilising moment.

11.20   The Aerodynamic Centre (AC)

The point along the chordline of a wing at which all changes of lift take place is the aerodynamic centre (AC) of the wing and is usually located at a point 25% of the length of the chord measured from the leading edge of the wing. When the CG is located at the wing AC the magnitude of the aerodynamic pitching moment remains almost constant for all angles of attack. For good longitudinal static stability the CG should be located as close to the aerodynamic centre of the wings as possible. Aeroplanes with a delta wing normally have the CG located at approximately 10% of the chordline length ahead of the AC. A comprehensive description of the AC is given in Chapter 4.

11.21   The Centre of Gravity (CG)

In level flight, with the usual disposition of the four forces, the tailplane is downloaded to ensure longitudinal static stability. The downloading is produced by the elevator trim which, when used, causes drag that is counteracted by increased thrust. The downloading is least at high angles of attack and is greatest at low angles of attack.

On a positively cambered wing, if the CG is ahead of the wing AC it will provide a positive contribution to the longitudinal static stability. The degree of positive camber has no effect on the longitudinal static stability because irrespective of the angle of attack it produces a constant nose-down pitching moment.

11.21.1   The CG Envelope

The range within which the CG must always be located for safe operations is the CG envelope that, for transport aeroplanes, is defined as the area between the safe forward limit approximately 10% MAC and the safe aft limit approximately 30% MAC. Other types of aeroplane may have a wider range of safe CG positions. The precise limiting values vary with the aeroplane type. If for any reason the CG is located outside of the safe envelope then it is possible that the controls will have insufficient authority for the pilot to recover the aeroplane from any undesirable attitude. If the CG is positioned aft of the aft limit of the safe envelope the aeroplane will have manoeuvre instability.

11.21.1.1   CG Envelope Limitations

The safe limitations of the CG envelope are the forward and aft limits.

a. The Forward Limit

The forward limit of the envelope is determined by the amount of pitch control available from the elevators; that is the degree of manoeuvrability that an aeroplane of that type commands. For transport aeroplanes the forward limit is normally at approximately 10% of the MAC. With the CG in this position the aeroplane has the greatest longitudinal stability. In the landing configuration maximum elevator-up deflection is required when the CG is at the forward limit and full flap is selected.

b. The Aft Limit

The aft limit, which for safety reasons on a transport aeroplane is always forward of the neutral point, is confined by insufficient stick-force stability and/or excessive in-flight manoeuvrability. The minimum stick force per 'g' for the maximum permitted load factor, which is 2.5 for large transport aeroplanes, determines the aft limit of the CG envelope,. A CG position aft of this point would produce an unacceptably low value of manoeuvre stability and would make the aeroplane difficult to fly. For transport aeroplanes the aft limit is normally located at approximately 30% of the MAC.

11.21.1.2   CG Movement

The position of the CG directly affects the magnitude of the longitudinal static stability, the tailplane downforce and the stalling speed because its position determines the length of the arm of any restoring moment. If it is possible for the pilot to transfer fuel between tanks and/or pre-position the traffic load then the location of the CG can be manipulated within these limits.

Aft Movement

An *aft movement of the CG (or a forward movement of the CP) decreases the positive longitudinal static stability* of an aeroplane because the arm measured from the CG to the CP is shortened, consequently the required restorative pitching moment is diminished; however, such a movement has the advantage that it increases the longitudinal manoeuvrability of the aeroplane because it increases the elevator up effectiveness, thus decreasing the amount of deflection required to produce a particular reaction. Thus, its static stability is reduced. Both the required downforce on the tailplane and the stalling speed are decreased.

Forward Movement

A *forward movement of the CG increases the positive longitudinal static stability of an aircraft* but decreases the manoeuvrability and control response because the moment arm is increased in length and increases the nose-down pitching moment. Thus, the elevators have less nose-up authority. The same result can be achieved by decreasing the angle of attack so that the CP moves aft, thus increasing the length of the moment arm and increasing the nose-down pitching moment.

11.21.2   The Effect of CG at the Limits

11.21.2.1   CG at the Forward Limit

If the CG is located at the forward limit of the envelope it has the following effects:

a. greatest longitudinal static stability;
b. increased corrective download on the tailplane required;
c. increased corrective elevator trim causing increased trim drag;
d. decreased manoeuvrability to the minimum acceptable;
e. increased stick force required at rotation during take-off;
f. increased stalling speed (but no effect on the stalling angle);
g. increased fuel flow;
h. decreased maximum range for a given fuel load;
i. decreased maximum endurance for a given fuel load.

11.21.2.2   CG at the Aft Limit

If the CG is located at the aft limit of the CG envelope, it has the following effects:

a. decreased longitudinal static stability;
b. decreased corrective tailplane download required;
c. decreased corrective elevator trim resulting in less trim drag;
d. increased manoeuvrability to the maximum controllable;
e. decreased stalling speed;
f. decreased thrust required;
g. decreased fuel flow;
h. increased maximum range for a given fuel load;
i. increased maximum endurance for a given fuel load.

11.22   The Static Margin (SM)

The static centre of gravity margin, also known as the static margin (SM) or the CG margin, is the distance measured from the CG to the NP and is usually quoted as a percentage of the mean aerodynamic chord. This margin relates to the stick movement and the amount of force required to achieve a new airspeed and to hold it without retrimming. It is a gauge of the longitudinal static stability of an aeroplane, which is directly proportional to the length of the static margin. *The longer the static margin, the greater is the longitudinal static stability.* If the length of the static margin is zero then the CG and NP are coincident and the aeroplane is neutrally stable.

Conventional transport aeroplane designs have a static margin between 10% and 30% of the MAC ahead of the NP, which is a few feet in length; whereas for small aeroplanes the margin ranges between 5% and 15% of the MAC and will be only a few inches.

To ensure positive longitudinal static stability the CG must be ahead of the neutral point, which is a negative static margin. For good positive longitudinal static stability the CG should also be as close to the wing AC as possible. If the CG is aft of the neutral point the aeroplane has longitudinal static instability, a positive static margin and active inputs from the controls are necessary to maintain stable flight. Consequently, with the CG at any position, other than the neutral point, as the angle of attack increases the pitching moment coefficient diminishes because the length of the pitching arm decreases. This effect is shown in Figure 11.13.

11.23   The Trim Point (TP)

An aeroplane 'in trim' is in a state of equilibrium about its pitch axis. For any given aeroplane mass there is only one speed and angle of attack at which the tail moment of an aeroplane is equal to the wing-moment and the aircraft is in equilibrium. Therefore, as the flight progresses and the fuel is used the aeroplane mass decreases consequently the speed and angle of attack must be continually adjusted to maintain the equilibrium.

The trim point (TP) is the angle of attack of the aeroplane wing when the aeroplane is in equilibrium about its pitch axis. This is shown in Figure 11.14 as the point at which the wing moment equals the tail moment and the total moment is zero so the aeroplane is 'in trim.' The angle of attack at which this occurs depends on the longitudinal dihedral of the aeroplane.

11.24   Longitudinal Dihedral

The magnitude of the speed and angle of attack at the trim point is dependent on the angular difference between the chordlines of the tailplane and the wing as they were rigged during construction. This is referred to as the longitudinal dihedral. See Figure 2.2(b).

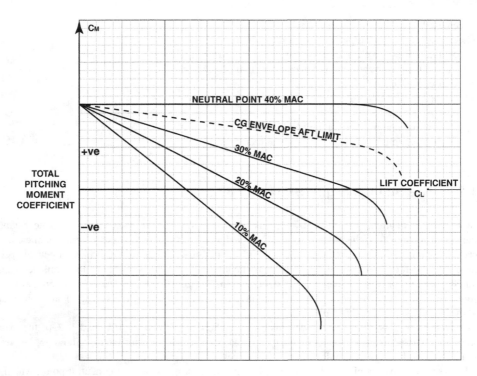

Figure 11.13   The Effect of CG on the C_M v C_L Graph.

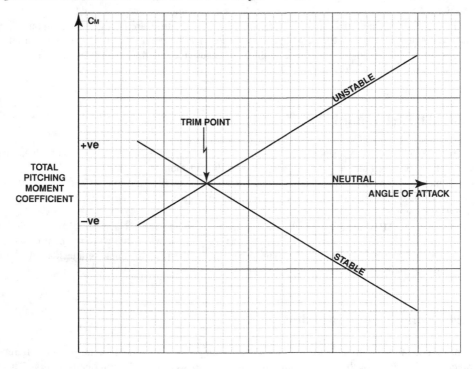

Figure 11.14   The Pitching Moment Coefficient v Angle of Attack.

Positive longitudinal dihedral is when the wing angle of incidence relative to the longitudinal axis is greater than the tail angle of incidence relative to the same datum and is a positive stabilising influence. It also ensures longitudinal stability at high angles of attack.

11.25   Aeroplane-Design Variations

On conventional aeroplanes the tailplane has a small negative angle of incidence; it has no camber and is located at the rear of the aeroplane. Some aeroplanes have the horizontal stabiliser mounted ahead of the wings, this is the canard configuration. The advantage of this configuration is that for a given speed the foreplane produces positive lift, consequently the main wings require a smaller angle of attack to develop the required wing contribution to the total lift. This results in less total drag, lower fuel consumption, increased range and endurance and a lower stalling speed when compared with a conventional aeroplane of the same mass. A similar effect can be attained if the camber of the wing is greater than the tail-plane camber.

For a conventional aeroplane the effective angle of attack of the tailplane is the angle subtended between its chordline and the airflow of the downwash from the wings. Consequently, it determines the balancing effect of the tailplane lift against the lift developed by the wings. If there is a significant change in the downwash with changing angle of attack then the stabilising effect of the tailplane will change.

For example, if a pitch-up attitude causes the angle of the downwash to increase then the effective angle of attack of the tailplane is decreased. Consequently, the restoring moment of the tailplane is decreased and the longitudinal static stability is diminished. If such were the case the aircraft designer would have to move the CG forward to regain the longitudinal static stability of the aeroplane.

11.26   The Effect of the Variables on Longitudinal Static Stability

11.26.1   Elevator Deflection

The position of the CG not only affects the positive longitudinal static stability, it also affects the pitch handling characteristics. When the elevators are deflected the pitching moment they produce must be greater than the restoring moment of the aeroplane's positive longitudinal static stability. For any given elevator deflection there is a small response if the aircraft CG is located forward (i.e. stable condition) and a large response for the same angle of deflection when the aircraft CG is located at an aft position (i.e. a less-stable condition). The elevator deflection required to counter a forward CG movement during a manoeuvre having a load factor greater than one is increased. The *neutral point defines the most aft position that the CG can be located with the aeroplane having positive longitudinal static stability.*

If the angle of attack is increased from the trim point then the natural tendency of a stable aeroplane is to produce a corrective nose-down pitching moment. To maintain this new angle of attack it is necessary to use the elevators to produce an equal and opposite pitching moment.

The tailplane is usually a symmetrical aerofoil, which is located in the downwash of the mainplane on conventional aeroplanes. Its lift is proportional to its angle of attack, however the downwash decreases the effective angle of attack, which decreases the tailplane lift and as a consequence the pitching moment is decreased, the NP moves forward and the longitudinal static stability is lessened. It is therefore necessary to increase the tailplane download. This is achieved by an upward deflection of the elevators and a downward deflection of the trim tab that effectively changes the camber of the tailplane and provides the required nose-up force and establishes a revised trim point at the increased angle of attack.

In other words, at high angles of attack of the wings the CP moves forward, decreasing the length of the moment arm but the lift force increases by a considerable amount, which increases the pitching moment. Consequently, the tail has to produce a greater nose-up moment which is achieved by changing

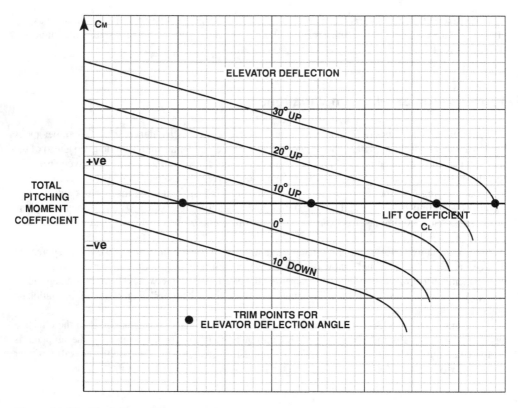

Figure 11.15   The Effect of Elevator Deflection on Trim.

the effective camber of the tailplane through raising the elevators. Normally, the longitudinal static stability is unaffected by this change. The reverse procedure is used to decrease the angle of attack of the mainplane.

In Figure 11.15 the vertical axis shows the positive and negative pitching moment coefficient and the horizontal axis shows the coefficient of lift. It can be seen that the values of C_M, the pitching moment coefficient, and C_L, the lift coefficient, are directly proportional to the angle of elevator deflection. The trim points for each of the angles of elevator deflection are shown along the horizontal axis.

The longitudinal static stability is the same for each of the elevator deflection angles shown in Figure 11.15. However, at each of the trim points the pitching moment coefficient is the same at zero, whereas the coefficient of lift increases directly in proportion to the angle of elevator deflection. Because these changes occur at fixed elevator positions it is referred to as stick-fixed stability.

11.26.2 Trim

For a stable aeroplane, the result of an increased C_L is a change to the nose-down pitching moment. To counteract this change to the trim condition it is necessary to increase the tail force, which is achieved by an upward deflection of the elevators. Consequently, the elevator hinge moment is nose-up and the trim tab has to be deflected downward to balance this effect. Therefore, a positive C_L change requires a positive trim tab angular deflection. To trim for a speed increase the elevator needs to be deflected downward to decrease the angle of attack and lower the nose of the aeroplane. This is achieved by an upward deflection of the trim tab.

11.26.3   The Fuselage

The effect that the fuselage, and to a lesser extent wing-mounted the engine nacelles, has on the aeroplane's longitudinal static stability is normally a destabilising influence, a negative contribution. If the fuselage aerodynamic centre is ahead of the wing aerodynamic centre, the aerodynamic centre of the wing–fuselage combination is further forward than it would normally be. If such is the case the aircraft CG will be further aft than is normal and the aircraft longitudinal static stability is decreased.

This destabilising influence is increased by the induced upwash of the airflow ahead of the wing. However, the downwash of the airflow aft of the wing decreases the destabilising influence caused by areas of the fuselage and nacelles aft of the wing. Rear fuselage-mounted engine nacelles have a positive effect.

11.26.4   Angle of Attack

The effective angle of attack of the tailplane is decreased with an increased angle of attack of the wings because of the increased downwash; consequently, the tailplane lift is diminished. This causes the restorative moment produced by the tail to decrease, the neutral point to move forward and the longitudinal static stability to be reduced.

11.26.5   Configuration

11.26.5.1   Trailing-Edge Flaps

Extending trailing-edge flaps increases the effective camber of the wing, which causes the downwash to increase and reduces the longitudinal static stability. However, the pressure distribution of the wing is also affected by the lowering of trailing-edge flaps and results in the CP moving aft and a change to the nose-down pitching moment; this opposes the destabilising effect caused by the increased down-wash. The overall effect of the use of trailing-edge high-lift devices is to destabilise the longitudinal static stability.

11.26.5.2   Undercarriage

Lowering the undercarriage of an aeroplane lowers the level of the dragline. If the thrust line does not pass through the CG, this causes changes to the pitching moment if the thrust alters with changes of the angle of attack of the wing. When the CG is lower than the thrust line the effect on the longitudinal static stability is stabilising and when it is above the thrust line it is destabilising.

11.27   Stick-Fixed Longitudinal Static Stability

With an aeroplane having a powered irreversible control system, the elevators and trim tabs are held in the prevailing trim position. It is the stick-fixed stability that determines the amount of elevator *movement* needed to alter the airspeed, C_L or angle of attack.

The stick-fixed longitudinal static stability for a given CG position is proportional to the rate of change of elevator angle with respect to the C_L for the entire aeroplane, wings and fuselage. The more stable the aeroplane the further the stick has to be moved to effect the change required. As the CG moves aft the amount of stick movement to produce an equivalent change of airspeed, C_L or angle of attack decreases. With the CG at the stick-fixed neutral point the change of stick position needed to sustain a change of airspeed is zero. After moving the control column to attain a new angle of attack the stick can be returned to its original position.

11.27.1   Stick-Position Stability

The amount of elevator deflection required and consequently the amount of trim is dependent on the IAS (dynamic pressure). At low indicated airspeeds it is desirable that an aft movement of the control column is required to increase the angle of attack and the amount of trim. This is particularly so during an approach to land, with a forward position of the CG, when land flap is selected and maximum up deflection is required. At high indicated airspeeds it is expedient that a forward movement of the control column is necessary to decrease the angle of attack and trim. Such is the case when the aircraft CG is positioned forward of the stick-fixed neutral point. This is stick-position stability.

The trim-indicated airspeed corresponds to a particular elevator deflection as shown in Figure 11.16. The amount of elevator deflection required for a given trim airspeed is decreased if high-lift devices are used to increase the C_L. If such is the case then the stick-stability graph line moves to the left, decreasing the trim speed for zero elevator deflection, as shown in Figure 11.16. Adjustment of the elevator trim tab has no effect on the static longitudinal stability. If the CG is in a forward position then the elevator deflection for a manoeuvre with a load factor greater than 1 will be larger.

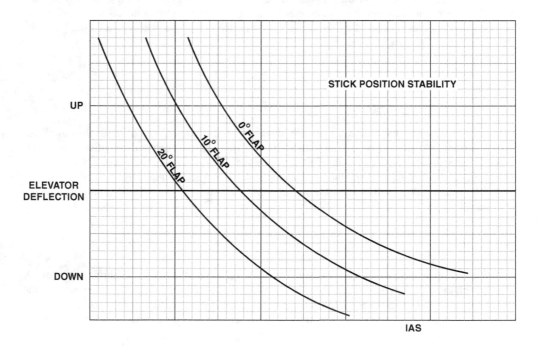

Figure 11.16   Stick-Position Stability.

11.28   Stick-Free Longitudinal Static Stability

If, when in flight, the elevators are allowed to float freely in the airstream with no manual or automatic input they will tend to align themselves with the freestream airflow. This is stick-free longitudinal static stability, which determines the *force* required to effect a change of airspeed, C_L or angle of attack. If such were the case then the tail force due to a displacement would reduce the tail moment.

The usual position adopted by the free-floating elevator is shown in Figure 11.17. The balance between the wing moment and the tail moment is changed because of the decreased tail moment. In these circumstances, the CG position at which these moments are equal, the stick-free neutral point, moves forward because the diminished tail force requires a longer arm. This reduces the stick-free CG margin and consequently the longitudinal static stability of the aeroplane. *The stick-free NP is always ahead of the stick-fixed NP*; the amount of freedom that the elevator has determines the distance that separates the two NPs.

Thus, the stick-free longitudinal static stability is less than the stick-fixed longitudinal static stability. The stick-free CG margin is smaller than the stick-fixed CG margin because the neutral points are at different positions. If a horn balance is fitted to the elevators it will trail in the relative airflow and will reverse this situation by increasing the stick-free longitudinal static stability.

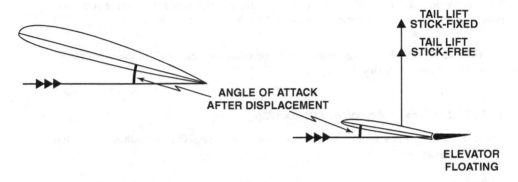

Figure 11.17   Tail Contribution Stick Free.

11.28.1   *Stick Force*

The stick force required is the amount of effort necessary to overcome the longitudinal static stability of the aeroplane. To fly the aeroplane at an airspeed different from that for which it is trimmed the stick force required is dependent on the location of the CG and the size of the tailplane. The stick force per g has to have upper and lower limits to ensure that the aeroplane has acceptable control characteristics at all times. A forward CG position increases longitudinal static stick-force stability. The magnitude of the stick force required decreases as the CG moves aft; therefore the maximum permitted aft position of the CG is limited by the minimum acceptable value of the stick force per g.

To establish the pitch change required, the function of the elevators is distinctly different to that of the elevator trim tabs. The purpose of the elevators is to produce the required pitching moment coefficient by adjusting the lift coefficient of the tail surface, whereas the function of the trim tab is to adjust the elevator hinge moment and consequently reduce the initial stick force required to deflect the elevator.

For any given attitude the stick force necessary to maintain that attitude is independent of the IAS, however, the trim-tab setting necessary to balance that stick force is dependent on the IAS (dynamic pressure). It follows then that the trim-tab setting required, maintaining level flight with a zero stick force, will also vary with the IAS. In trimmed level flight an acceleration of 10 kt affects the stick force required far more at low speed than it does at high speed. At double the speed it has half the effect.

At high speeds the down trim-tab setting required to maintain zero stick force is less than at lower speeds because the elevator-up deflection is less. If the CG is forward of the neutral point any speed increase from the trimmed position has less effect at high speeds than it does at low speeds. Such is the case with modern transport aeroplanes.

Stick-force stability can be recognised by the pilot when a pull force is required to maintain a speed less than the trimmed speed and a push force is needed to increase the speed above the trim speed. A forward CG position increases the stick-force stability. Extending the flaps during an approach to land moves the CP aft, increasing longitudinal stability, which requires a greater stick force to flare. Stick-force stability remains unchanged after trimming the aeroplane.

11.29   Certification Standard Stick-Force Requirements

The minimum acceptable standard of the stick force versus speed is specified in the certification standards promulgated by the European Aviation Safety Agency (EASA) to ensure acceptable control characteristics and they are as follows:

11.29.1   a. Class 'A' Aeroplanes CS 25.173(c)

The average gradient of the stable slope of the stick force versus speed curve may not be less than one pound for each six knots (4 Newtons for each 11.2 km/h).

(i) **AMC 25.173(c)**. The average gradient is to be taken over each half of the speed range between 0.85Vtrim and 1.15Vtrim.

11.29.2   b. Class 'B' Aeroplanes CS 23.173(c)

The stick force must vary with speed so that any substantial speed change results in a stick force that is clearly perceptible to the pilot.

11.30   The Effect of CG Position on Stick Force

The stick force required, maintaining the trim-speed in level flight, changes if the position of the CG alters during trimmed flight at a constant airspeed. A forward movement of the CG position increases the longitudinal static stability, the stick-force stability and the manoeuvre stability and an aft movement of the CG position decreases the longitudinal static stability, the stick-force stability and the manoeuvre stability. This can occur when fuel is used as the flight progresses and the aeroplane mass decreases.

As the CG moves aft it not only decreases the download required on the tailplane, but also decreases the amount of trim required to maintain level flight and increases the angle of attack. The further aft that the CG moves the less is the stick force required and the greater is the longitudinal static instability until the CG is positioned at approximately 40% of the mean aerodynamic chord (MAC), which is the stick-force neutral stability point. If the CG moves further aft, stick-position instability will exist. If such were the case, the stick would have to be pushed at lower speeds and pulled at higher speeds to maintain level flight.

This aspect is shown in Figure 11.18. If the CG is positioned beyond the aft limit of the CG envelope it leads to an unacceptably low value of manoeuvrability, stick force per 'g.' Among other things the maximum aft position of the CG envelope is limited by the minimum acceptable value of stick force per 'g.' When trimmed for zero stick force an elevator trim tab causes more drag than the horizontal stabiliser.

It is very difficult to assess the correct trim setting to apply if there is friction in the control system because it alters the feel of the controls. Friction requires additional force to overcome it, which varies with the IAS. Therefore, there is a range of forces that are needed to subjugate the friction that in turn becomes a range of trim speeds. When setting the trim for take-off due allowance must be made for the position of the CG as well as the configuration of the aeroplane.

Irrespective of the speed or the deflection of the elevators, by modifying the forces acting on the control column such that a greater pull and push force is necessary, stick-force stability is obtained. There are two devices that achieve this aim, they are the downspring and the bobweight; either of them can be introduced in the control system between the control column and the elevators.

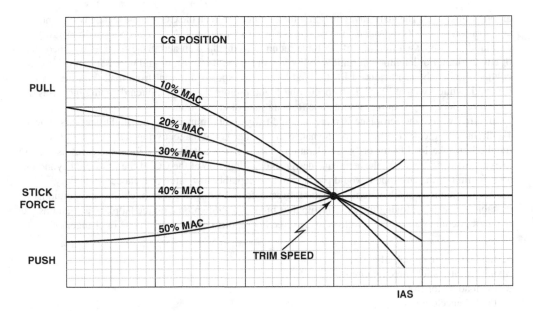

Figure 11.18   Stick Force versus IAS.

11.31   Longitudinal Static Manoeuvre Stability

The difference between longitudinal static stability and longitudinal static manoeuvre stability is that when contemplating longitudinal static stability only disturbances to the angle of attack (α) at a constant load factor (n = 1) are considered whereas longitudinal static manoeuvre stability accounts disturbances to the angle of attack **and** the load factor at a constant speed. The initial condition is always steady level flight.

Longitudinal static manoeuvre stability is really just the longitudinal static stability with the additional factor of the pitch rate and describes the behaviour of the aeroplane in more than 1-g flight. A forward CG position positively affects the longitudinal static manoeuvre stability and requires a larger deflection of the elevators. The designer of an aeroplane must ensure that there is adequate elevator control appropriate to the role of the aeroplane to enable the pilot to maintain a particular manoeuvre.

An aeroplane originally trimmed to fly straight and level when subjected to a sudden dive followed by a pull up to the initial trimmed speed and attitude encounters two disturbances both of which contribute to the overall longitudinal static stability. They are:

a. The effective angle of attack is increased and produces the additional lift necessary to maintain a curved path.
b. The nose-up rotation rate about the CG is equal to the rate of rotation about the centre of the pull-up. It rotates about its lateral axis with the nose rotating up and the tail rotating down.

11.31.1   The Manoeuvre Point

The stabilising effect of the tailplane is greater during the above manoeuvre than it is in level flight and the neutral stability CG is further aft than it is in straight and level flight. This position of the CG is the manoeuvre point and is the neutral point for the manoeuvre. The aft limit of the CG envelope is always ahead of this point. The manoeuvre margin is the distance between the CG and the manoeuvre point. For any position of the CG *the manoeuvre margin is always greater than the CG margin* (static margin).

Stability is assessed in terms of the force required to displace an aeroplane from its trimmed state of equilibrium. For longitudinal static stability it is the amount of push or pull necessary on the control column to alter the coefficient of lift C_L, and to produce speeds different from that, for which it was trimmed, while flying at "1-g". In manoeuvring flight at more than "1-g", pitch damping increases the stick force necessary to move an aeroplane from its state of equilibrium.

In other words, the aeroplane is more longitudinally stable in manoeuvring flight than it is in level 1-g flight. The greater the rate of pitch change the greater is the manoeuvring stability, the stick force required and the load factor or 'g' acceleration. The rate at which the stick force increases as the 'g' increases is conditioned by the aeroplane design and the location of the CG. If the CG were beyond the aft safe limit of the envelope it would produce an unacceptably low value of manoeuvre stability.

A longitudinally stable aeroplane requires greater force to displace it in pitch than a longitudinally unstable aeroplane. Moving the elevators up thus produces a downforce on the tailplane and attains the required pitching moment. The following are directly proportional to the pitch rate of change:

a. the manoeuvring stability;
b. the elevator stick force required;
c. the load factor (n);
d. the 'g' acceleration.

A high stick force per 'g' is a desirable quality but if required it can be reduced by the use of the elevator trim tab setting. Aeroplanes that have a small stick force per 'g' usually have a bob weight added to the elevators to increase the CG margin and so meet the certification standard requirements. Using a downspring has no effect on the manoeuvring stick-force stability and therefore cannot be used for this purpose.

11.32   Factors Affecting Stick Force

The stick force is directly related to the load factor, as shown graphically in Figure 11.19. To prevent the inadvertent overstressing of large transport aeroplanes caused by excessive control inputs, the stick force per 'g' is deliberately set high but for highly manoeuvrable aeroplanes such as fighter aeroplanes it is very low.

The stick force per 'g' has both an upper and a lower limit to ensure that the control characteristics remain acceptable. The magnitude of the stick force per 'g' is affected by the following factors. It:

a. Increases with forward movement of the CG and vice versa. The stick position per 'g' behaves in a similar manner.
b. Increases with increased load factor and vice versa.
c. Decreases with increased altitude because of reduced 'pitch damping' due to decreased air density.

11.33   Summary

The relationship of the aerodynamic centre (AC), the centre of gravity (CG) and the neutral point (NP) is depicted in Figure 11.20. They are shown at their appropriate positions on the mean aerodynamic chord (MAC) together with the minimum and maximum static margins. Also shown is the stick force required for both the stable and the unstable aeroplane.

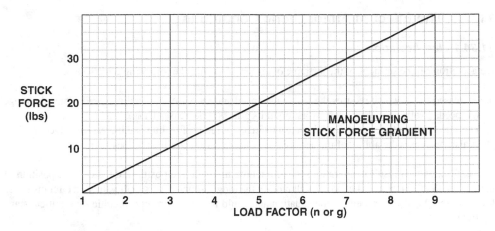

Figure 11.19   Stick Force per 'g'.

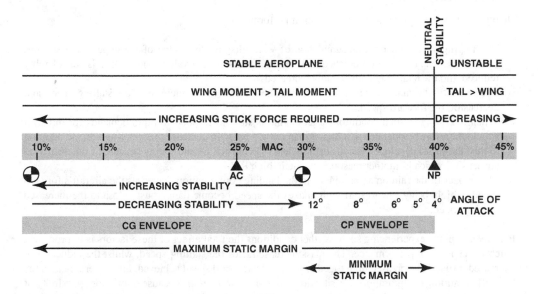

Figure 11.20   Transport Aeroplane Longitudinal Stability Summary.

Example 11.1

Given: Straight and level manoeuvre stability 150 N/g. Calculate the stick force increase required to produce a load factor of 3.

Solution 11.1
3 × 150 = 450 N.
Increase of stick force = 450 − 150 = **300 N**

11.34   The Effect of Atmospheric Conditions

11.34.1   Ice Accretion

Ice accretion on the surfaces of the airframe of an aeroplane has the following effects:

a. Increased mass, which reduces the maximum speed and increases the stalling speed. Thus, the range of speeds available is decreased. Worst of all its effects it decreases C_{Lmax}.
b. Distorted shape of the aerofoils, which results in increased profile drag and increased surface friction. It also decreases the ability of the aerofoils to generate as much lift as it does normally.

These changes may result in alterations to the pitching moment, the position of the CG, the position of the CP and the CG margin all of which will modify the longitudinal static stability. In such circumstances lowering the flaps will exacerbate the situation and could result in control difficulties at a critical stage of flight.

11.34.2   Heavy Rain

Heavy rain has the following effects on aircraft performance:

a. The impingement and rapid accumulation of water distorts the shape of the upper wing surface, causing the total lift to be diminished by up to 30% of its normal value. Thus, C_{Lmax} is considerably reduced and the total drag considerably increased.
b. The mass of the rain momentarily increases the aircraft mass, increasing the stalling speed and decreasing the forward speed.
c. When operating in such conditions a jet engine is slow to respond to rapid demands for thrust; it should therefore be operated at a higher than normal RPM to decrease the spool-up time. In addition to this, the thrust developed is considerably less than normal for the setting because of the decreased air density, which is another reason to maintain high RPM.
d. The impact of the rain on an aeroplane in the landing configuration decreases the aircraft's forward speed; which, being low to start with, makes the speed become dangerously close to the increased stalling speed.

For any aeroplane experiencing such weather conditions the total effect of these factors is to produce a much lower level of performance. The mass increase raises the stalling speed, whilst the reduced lift, decreased thrust and increased drag together reduce the forward speed achieved. For an aeroplane in the take-off or landing configuration, this situation is *extremely dangerous* because of the close proximity of the normal operating speed to the stalling speed. There is, therefore, a very high risk of the aeroplane stalling when it is close to the ground.

11.34.3   Altitude

An aircraft flying at the same IAS and load factor at two different altitudes will have decreased longitudinal static manoeuvre stability at the higher altitude. This is because the IAS at the higher altitude produces a higher TAS, which reduces the rate of pitch change and decreases the angle of attack of the tailplane, thus diminishing the tailplane contribution.

11.35   The Factors Affecting Static Stability

The individual factors affecting static stability are listed in Table 11.2:

Table 11.2   Factors Affecting Static Stability.

General Static Stability	
Positive or Increased Stability	**Negative or Decreased Stability**
(1) Decreased Altitude.	(1) Increased Altitude.
(2) Increased IAS.	(2) Decreased IAS.
(3) Increased Air Density.	(3) Decreased Air Density.
(4) Decreased Aeroplane Mass.	(4) Increased Aeroplane Mass.
(5) Forward CG Position.	(5) Aft CG Position.
(6) Aft CP Position.	(6) Forward CP Position.
Directional Static Stability	
(1) CG ahead of CP.	(1) CP ahead of CG.
(2) A long moment arm.	(2) Short moment arm.
(3) A forward movement of the CG.	(3) An aft movement of the CG.
(4) An aft movement of the CP.	(4) A forward movement of the CP.
(5) Wing sweepback.	(5) Forward wing sweep.
(6) A dorsal fin increases stability.	(6) Dihedral.
(7) A ventral fin and/or strakes.	(7) High aspect ratio fin.
(8) Large fin and rudder area.	(8) Small fin and rudder area.
(9) The vertical stabiliser design.	(9) Long fuselage ahead of CG.
Lateral Static Stability	
(1) Dihedral.	(1) Anhedral.
(2) Sweepback.	(2) Forward wing sweep.
(3) High-wing mounting.	(3) Low-wing mounting.
(4) Large, high vertical fin.	(4) Ventral fin.
(5) Low CG.	(5) Extending inboard flaps.
Longitudinal Static Stability	
(1) CG ahead of CP.	(1) CP ahead of CG.
(2) CG on forward limit.	(2) CG on aft limit.
(3) A forward movement of the CG.	(3) An aft movement of the CG.
(4) An aft movement of the CP.	(4) A forward movement of the CP.
(5) Longitudinal dihedral.	(5) High angle of attack.
(6) Long moment arm from tail CP.	(6) Extension of trailing-edge flaps.
(7) Large tailplane area.	(7) Increased wing downwash.
(8) Fuselage and engine nacelles aft of the wing.	(8) Fuselage and engine nacelles ahead of the wing.
(9) Excess tail volume.	(9) Increased altitude.

Self-Assessment Exercise 11

Q11.1 In what way is the longitudinal stability affected by the degree of positive camber of an aerofoil?
  (a) Positive, because the CP moves aft as the AoA increases.
  (b) Negative, because the lift vector slopes forward as the AoA increases.
  (c) Positive, because the lift vector slopes backward as the AoA increases.
  (d) No effect, because the camber of the aerofoil produces a constant pitch-down moment coefficient independent of the AoA.

Q11.2 One of the factors that limits the maximum aft position of the CG is:
  (a) minimum value of the stick force per 'g.'
  (b) maximum longitudinal stability of the aeroplane
  (c) maximum elevator deflection
  (d) too small an effect of the controls of an aeroplane

Q11.3 The aft movement of the CP during acceleration through the transonic flight regime will:
  (a) decrease the static lateral stability
  (b) increase the static longitudinal stability
  (c) decrease the longitudinal stability
  (d) increase the static lateral stability

Q11.4 How does the exterior appearance of an aeroplane change when it is trimmed for a speed increase?
  (a) The elevator is deflected further upward by a downward deflection of the trim tab.
  (b) The elevator is deflected further downward by a movable horizontal stabiliser.
  (c) The exterior appearance of the aeroplane will not change.
  (d) Elevator deflection is increased further downward by the upward deflection of the trim tab.

Q11.5 Aft movement of the CG will:
  (a) increase the elevator-up effectiveness
  (b) decrease the elevator-down effectiveness
  (c) not affect the elevator effectiveness
  (d) change the elevator effectiveness depending on the wing location

Q11.6 During the landing of a low-winged jet aeroplane, maximum elevator-up deflection is usually required when the flaps are (i) .... and the CG is (ii) ......
  (a) (i) up; (ii) fully aft
  (b) (i) fully down; (ii) fully forward
  (c) (i) up; (ii) fully forward
  (d) (i) fully down; (ii) fully aft

Q11.7 The effect that a ventral fin has on the static stability of an aeroplane is longitudinal (i) ......; lateral (ii) ....; directional (iii) ....
  (a) (i) none; (ii) negative; (iii) positive
  (b) (i) positive; (ii) negative; (iii) negative
  (c) (i) negative; (ii) positive; (iii) positive
  (d) (i) none; (ii) positive; (iii) negative

Q11.8 Sensitivity for a spiral dive will occur when:
  (a) the Dutch-roll tendency is too strongly suppressed by the yaw damper
  (b) the static directional stability is positive and the static lateral stability is relatively weak
  (c) the static directional stability is negative and the static lateral stability is positive
  (d) the static lateral and directional stability are both negative

Q11.9 The effect of heavy tropical rain on the aerodynamic characteristics of an aeroplane are to
(i) .... C_{Lmax} and to (ii) .... Drag:
(a) (i) decrease; (ii) increase
(b) (i) decrease; (ii) decrease
(c) (i) increase; (ii) increase
(d) (i) increase; (ii) decrease

Q11.10 In straight and level flight the value of manoeuvre stability is 150 N/g. To achieve a load factor
of 2.5 the increase of stick force necessary is:
(a) 450 N
(b) 150 N
(c) 225 N
(d) 375 N

Q11.11 The effect that the use of trim to accommodate a speed decrease has on the appearance of the
aeroplane is:
(a) the elevator is deflected further downward by a trimmable horizontal stabiliser
(b) that it has no effect
(c) the elevator is deflected downward by an upward deflected trim tab
(d) the elevator is deflected further upward by a downward deflected trim tab

Q11.12 The CG of a stable aeroplane is located:
(a) at the neutral point
(b) between the aft limit and the neutral point
(c) ahead of the neutral point by a sufficient margin
(d) aft of the neutral point

Q11.13 In a sideslip to the right, an aeroplane that has static directional stability, initially the:
(a) nose tends to move to the right
(b) right wing drops
(c) nose will maintain its direction
(d) nose tends to move to the left

Q11.14 With the CG fixed in a position forward of the neutral point if a speed change causes a departure
from the trimmed attitude; which of the following statements is correct regarding the stick-force
stability?
(a) Increased speed generates pull forces.
(b) Nose-up trim decreases the stick-force stability.
(c) Stick-force stability is not affected by trim.
(d) An increase of 10 kt has more effect on stick force at low speed than it does at high speed.

Q11.15 Static longitudinal stability is created by:
(a) the wing surface area being greater than the horizontal tail surface area
(b) a large trim-speed range
(c) the CG being located ahead of the neutral point
(d) the CG being located ahead of the wing leading edge

Q11.16 An aeroplane is in a sideslip at a constant speed and angle that increases the geometric dihedral
of the wing. Which of the following statements is correct?
(a) the required lateral control force increases
(b) the required lateral control force decreases
(c) the required lateral control force does not change
(d) the stick force per g decreases

Q11.17 The lateral static stability is decreased by:
(a) increased wing span
(b) anhedral
(c) dihedral
(d) a high wing

Q11.18 A statically unstable aeroplane is:
- (a) sometimes dynamically stable
- (b) sometimes dynamically unstable
- (c) never dynamically stable
- (d) always dynamically stable

Q11.19 An aeroplane that has positive static stability:
- (a) is always dynamically unstable
- (b) may be dynamically stable, neutral or unstable
- (c) is always dynamically stable
- (d) can never be dynamically stable

Q11.20 The effect that a positive wing sweep has on directional static stability is that it has:
- (a) a destabilising dihedral effect
- (b) a negative dihedral effect
- (c) a stabilising effect
- (d) no effect

Q11.21 The most forward allowable position of the CG is limited by:
- (a) engine thrust and location
- (b) trim system and trim-tab surface area
- (c) wing surface area and stabilise surface area
- (d) elevator capability and elevator control forces

Q11.22 The (i) stick-force stability and (ii) the manoeuvre stability are positively affected by:
- (a) (i) a forward CG position; (ii) a forward CG position
- (b) (i) a forward CG position; (ii) a nose-up trim position
- (c) (i) an aft CG position; (ii) an aft CG position
- (d) (i) a nose-up trim position; (ii) a nose-up trim position

Q11.23 The greatest positive contribution to the longitudinal static stability is provided by:
- (a) the horizontal tailplane
- (b) the engine
- (c) the fuselage
- (d) the wing

Q11.24 The effect a more aft position of the CG has on the aeroplane's longitudinal static stability (i) .... and the required deflection of the pitch control (ii) .... is:
- (a) (i) smaller; (ii) smaller
- (b) (i) larger; (ii) smaller
- (c) (i) larger; (ii) larger
- (d) (i) smaller; (ii) larger

Q11.25 A CG located beyond the aft limit leads to:
- (a) a too high pulling force during rotation for take-off
- (b) an increasing static longitudinal stability
- (c) a better recovery performance from a stall
- (d) an unacceptably low value of the manoeuvre stability (stick force per g)

Q11.26 Which of the following statements regarding stick force per g is correct?
- (a) An unacceptable value of stick force per g can only be corrected by electronic devices (stability augmentation):
- (b) If the slope of the Fe-n line becomes negative, it is not a problem for the control of an aeroplane.
- (c) The stick force per g must have both an upper and a lower limit in order to ensure acceptable control characteristics.
- (d) the stick force per g increases when the CG is moved aft.

Q11.27 If the CG is located in a forward position then the elevator deflection for a manoeuvre with a load factor greater than 1 will be:
   (a) dependent on the trim position
   (b) larger
   (c) smaller
   (d) unchanged

Q11.28 The lateral static stability of an aeroplane is increased by:
   (a) Fuselage-mounted wings, dihedral, T-tail
   (b) sweep back, under-wing-mounted engines, winglets
   (c) high wing, sweep back, large and high vertical tail
   (d) low wing, dihedral, elliptical wing planform

Q11.29 If the CG is aft of the CP straight and level flight can only be maintained when the horizontal tail loading is:
   (a) zero
   (b) downward
   (c) upward or downward depending on elevator deflection
   (d) upward

Q11.30 The manoeuvrability of an aeroplane is best when the:
   (a) flaps are extended
   (b) CG is on the aft limit of the CG envelope
   (c) speed is low
   (d) CG is on the forward limit of the CG envelope

Q11.31 The most important problem of ice accretion on an aeroplane during flight is the:
   (a) reduced C_{Lmax}
   (b) increased mass
   (c) increased drag
   (d) blocked control surfaces

Q11.32 The vertical load on the tailplane when the CG is forward of the CP is:
   (a) zero because in steady flight all loads are in equilibrium
   (b) downward because it is always negative regardless of the position of the CG
   (c) downward
   (d) upward

Q11.33 Positive static stability of an aeroplane means that once it has been displaced the:
   (a) tendency will be to move with an oscillating motion of decreasing amplitude
   (b) tendency will be to move with an oscillating motion of increasing amplitude
   (c) initial tendency to move is towards its position of equilibrium
   (d) initial tendency to move is away from its position of equilibrium

Q11.34 Which of the following statements is correct regarding lateral static stability and directional static stability?
   (a) Static directional stability is improved by installing more powerful engines.
   (b) An aeroplane with greater static directional stability than static lateral stability is prone to spiral instability.
   (c) The effects of static lateral stability and static directional stability are independent of each other because they are rotations about different axes.
   (d) An aeroplane with greater static directional stability than static lateral stability is prone to Dutch roll.

Q11.35 When the CG is at the aft limit of the CG envelope its relationship to the neutral point is:
   (a) forward of the neutral point
   (b) aft of the neutral point
   (c) at the neutral point
   (d) above the neutral point

Q11.36 The effect of the movement of the CG (i) on the down-force on the tail (ii) and on the stalling speed (iii) is:

(a) (i) rearward; (ii) reduced; (iii) Vs increased

(b) (i) forward; (ii) reduced; (iii) Vs increased

(c) (i) rearward; (ii) increased; (iii) Vs decreased

(d) (i) rearward; (ii) reduced; (iii) Vs decreased

Q11.37 The effect of the movement of the CG (i) on stability (ii) and control effectiveness (iii) is:

(a) (i) rearward; (ii) decreased; (iii) reduced

(b) (i) rearward; (ii) decreased; (iii) increased

(c) (i) forward; (ii) decreased; (iii) increased

(d) (i) forward; (ii) increased; (iii) reduced

Q11.38 With the CG at the forward limit of the envelope the effect on Vs (i) and stall angle (ii) is:

(a) (i) increases; (ii) remains constant

(b) (i) increases; (ii) increases

(c) (i) decreases; (ii) remains constant

(d) (i) decreases; (ii) decreases

Q11.39 The aft limit of the CG envelope relative to the manoeuvre point is that it is:

(a) at the manoeuvre point

(b) aft of the manoeuvre point

(c) always ahead of it

(d) furthest from it under a turn of 2g

Q11.40 Rearward movement of the CG:

(a) increase spiral instability

(b) creates a tendency to 'Dutch roll

(c) increase the elevator stick-force gradient

(d) increase the value of the minimum control speed

Q11.41 An aeroplane that is statically unstable:

(a) is always dynamically stable

(b) can never be dynamically stable

(c) is only dynamically stable at high airspeeds

(d) may or may not be dynamically stable dependent on type

Q11.42 Longitudinal static stability is provided by the:

(a) fuselage

(b) engines

(c) horizontal stabiliser

(d) wings

Q11.43 If the total moment of an aeroplane is not zero it will rotate about the:

(a) CP

(b) neutral point

(c) CG

(d) aerodynamic centre

Q11.44 The position of the CG that produces the greatest stability is:

(a) on the forward limit of the envelope

(b) on the aft limit of the envelope

(c) aft of the neutral point

(d) just forward of the neutral point

Q11.45 For take-off if the CG is towards the forward limit the effect is:

(a) the stick force required at rotation is increased

(b) VMCG is decreased

(c) the stick force required at rotation is decreased

(d) the stick force at rotation is unaffected

Q11.46 In trimmed level flight an acceleration of 10 kt has the effect on stick-force stability is:
(a) greater at high speeds
(b) greater at low airspeeds
(c) not affected by speed changes
(d) the same at all speeds

Q11.47 Positive static lateral stability is the tendency of an aeroplane to:
(a) roll to the right with a positive sideslip angle (nose to the right)
(b) roll to the left in a right turn
(c) roll to the right in a right turn
(d) roll to the left with a positive sideslip angle (nose to the left)

Q11.48 The effect of an aft movement of the CG on (i) static longitudinal stability (ii) the required control deflection for a given pitch change is:
(a) (i) decreases; (ii) increases
(b) (i) decreases; (ii) decreases
(c) (i) increases; (ii) decreases
(d) (i) increases; (ii) increases

Q11.49 A swept wing causes:
(a) adverse dihedral effect
(b) increased yaw effect
(c) positive dihedral effect
(d) adverse yaw effect

Q11.50 At a constant airspeed and sideslip angle, increase geometric dihedral would:
(a) decrease the stick force
(b) reduce the stick force to zero
(c) increase the stick force
(d) not affect the stick force required

Q11.51 If the CG is on the forward limit and the trimmable horizontal stabiliser position is maximum nose-down then:
(a) rotation will require additional stick force
(b) rotation will be unaffected
(c) early rotation is necessary
(d) the take-off warning system will be activated

Q11.52 In addition to providing positive longitudinal stability, which of the following ensures longitudinal stability at high angles of attack?
(a) longitudinal dihedral
(b) a dorsal fin
(c) a symmetrical vertical stabiliser
(d) geometric dihedral

Q11.53 When landing an aeroplane with excessive lateral stability in a crosswind it will require:
(a) less aileron into wind
(b) more rudder
(c) less rudder
(d) more aileron into wind

Q11.54 The effect that the degree of positive camber has on longitudinal stability is:
(a) positive. The CP moves aft with increasing angle of attack.
(b) negative. The lift vector slopes forward with increasing angle of attack.
(c) positive. The lift vector slopes aft with increasing angle of attack.
(d) no effect. The pitching moment is always negative irrespective of the angle of attack.

Q11.55 If the total moments of an aeroplane about an axis are not zero it will result in:
(a) a continuous rotation about that axis
(b) an angular acceleration about that axis
(c) increased stability
(d) equilibrium

Q11.56 Lateral stability is decreased by:
  (a) dihedral
  (b) winglets
  (c) anhedral
  (d) wing-tip tanks

Q11.57 To prevent excessive lateral stability an aeroplane should have:
  (a) negative dihedral
  (b) positive dihedral
  (c) winglets
  (d) wing-tip tanks

Q11.58 Roll is:
  (a) due to aileron deflection causing motion about the lateral axis
  (b) rotation about the normal axis
  (c) rotation about the lateral axis
  (d) rotation about the longitudinal axis

Q11.59 The elevator controls movement about . . . . the
  (a) longitudinal axis
  (b) lateral axis
  (c) normal axis
  (d) CP

Q11.60 The relationship between lateral static stability and directional static stability is:
  (a) dominant lateral static stability increases the tendency to spiral instability
  (b) dominant directional static stability increases the tendency to Dutch roll
  (c) dominant lateral static stability increases the tendency to Dutch roll
  (d) dominant lateral static stability decreases the tendency to Dutch roll

Q11.61 The lateral static stability of a swept-wing aeroplane:
  (a) increases with increased speed
  (b) decreases with increased speed
  (c) decreases the dihedral effect
  (d) is unaffected by speed

Q11.62 The effect of a swept wing on directional static stability of an aeroplane is that it:
  (a) increases the dihedral effect
  (b) has a stabilising effect
  (c) decreases the dihedral effect
  (d) is unaffected

Q11.63 Which of the following conditions increase (i) stick-force stability (ii) manoeuvre stability?
  (a) (i) an aft CG position; (ii) an aft CG position
  (b) (i) a forward CG position; (ii) nose-up trim
  (c) (i) a forward CG position; (ii) a forward CG position
  (d) (i) nose-up trim; (ii) nose-up trim

Q11.64 The aerodynamic centre (AC) is the point at which:
  (a) C_L is the same for all angles of attack
  (b) all aerodynamic forces are the same for any angle of attack
  (c) the lift for any angle of attack is the same
  (d) the pitching moment is constant for all normal angles of attack

Q11.65 Rotation about the lateral axis is:
  (a) pitching
  (b) rolling
  (c) yawing
  (d) spinning

Q11.66 The wing arrangement that decreases static lateral stability is:
  (a) high wing
  (b) dihedral
  (c) anhedral
  (d) large wingspan

Q11.67 Lateral stability is increased by the following combination of characteristics:
  (a) low wing; dihedral; elliptical wing
  (b) high wing; high vertical stabiliser; sweepback
  (c) high wing dihedral; straight wing
  (d) low wing; anhedral; sweepback

Q11.68 An aeroplane with positive static stability is:
  (a) always dynamically stable
  (b) always dynamically unstable
  (c) dynamically neutral, stable or unstable
  (d) always dynamically neutral

Q11.69 The elevator deflection required to counter a forward CG movement for a manoeuvre having a load factor greater than one is:
  (a) increased
  (b) decreased
  (c) constant
  (d) dependent on the position of the trim

Q11.70 The effect that a ventral fin has on static stability is (i) Directional (ii) Lateral (iii) Longitudinal.
  (a) (i) none; (ii) negative; (iii) positive
  (b) (i) positive; (ii) negative; (iii) none
  (c) (i) negative; (ii) none; (iii) none
  (d) (i) positive; (ii) none; (iii) none

Q11.71 The position that the CG should be located is:
  (a) between the aft limit and the neutral point
  (b) a sufficient margin ahead of the neutral point
  (c) aft of the neutral point
  (d) at the neutral point

Q11.72 The result of the CG being aft of the aft limit is:
  (a) excessive stick forces
  (b) increased longitudinal stability
  (c) manoeuvre instability
  (d) decreased lateral stability

Q11.73 Wing dihedral:
  (a) decreases the lateral static stability
  (b) increases the lateral static stability
  (c) is the only way in which lateral static stability can be improved
  (d) is only positive for aeroplanes with high-mounted wings

Q11.74 Which of the following statements regarding dihedral is correct?
  (a) Dihedral is necessary for slip-free turns.
  (b) The 'effective dihedral' component is its contribution to the static lateral stability.
  (c) 'Effective dihedral' is the angle between the $^1/_4$-chordline and the lateral axis of the aeroplane.
  (d) Dihedral contributes to dynamic lateral stability but not to static lateral stability.

Q11.75 The effect of having the CG on the forward limit and full nose-down trim on rotation at take-off is that the:
  (a) stick forces will decrease
  (b) stick forces will increase
  (c) the nose-wheel will lift off immediately
  (d) rotation will occur earlier in the take-off ground run

Q11.76 Which of the following produces the greatest manoeuvrability?
    (a) an aft CG position
    (b) high flap settings
    (c) low airspeed
    (d) a forward CG position

Q11.77 A high-wing aeroplane has zero dihedral. This is the equivalent to the aeroplane having:
    (a) anhedral
    (b) zero dihedral effect
    (c) positive dihedral effect
    (d) negative dihedral effect

Q11.78 The wing shape least sensitive to turbulence is:
    (a) winglets
    (b) swept wings
    (c) straight wings
    (d) wing dihedral

Q11.79 In level flight as the angle of attack, α, is decreased:
    (a) the CG will move aft and improve the longitudinal stability
    (b) the CP will move forward and improve the longitudinal stability
    (c) the stability is unaffected however the stalling speed increases
    (d) the CP movement aft of the CG improves the longitudinal stability

12 Dynamic Stability

Dynamic Stability is the ***subsequent long-term*** response of an aeroplane to a disturbance to its equilibrium. It is the change to the stability of an aeroplane with the passage of time. Any disturbance causes changes to the aerodynamic forces and moments acting on the aeroplane that are complex, particularly if the displacement is to its directional stability because it affects the aeroplane in both the yawing and rolling planes.

An aeroplane that is statically stable is not necessarily dynamically stable. However, for an aeroplane to be dynamically stable it must be statically stable. In other words it cannot have long-term stability without first having short-term stability. *An aeroplane that has positive static stability may be dynamically neutral, stable or unstable*; but an aeroplane that is statically unstable can never attain dynamic stability. The dynamic stability of an aeroplane is dependent on the aeroplane design, its speed and the altitude at which it is flying. For conventional aeroplanes it is assumed that the coupling between the pitching and directional motions of the aeroplane can be ignored. Thus, separate consideration may be given to the longitudinal dynamic stability and the lateral dynamic stability. The factors that affect the dynamic stability of an aeroplane are:

a. the linear velocity and mass (momentum);
b. the static stability in all planes;
c. the angular momentum (angular velocity and moments) about all three axes;
d. the aerodynamic damping moments due to roll, pitch and yaw.

If a disturbance to the equilibrium of an aeroplane induces an oscillation, the reaction of the aeroplane after the source of the disturbance is removed over the ensuing period of time determines its dynamic stability. A description of each state of dynamic stability derives from the form of motion assumed by the aeroplane during this period. There are five states of dynamic stability and for each of them the aeroplane must have positive static stability. They are illustrated in Figure 12.1, and are:

a. **Negative dynamic stability** – the amplitude of the oscillation increases with the passage of time.
b. **Neutral dynamic stability** – the amplitude of the oscillation remains constant with the passage of time.
c. **Positive dynamic stability** – the amplitude of the oscillation decreases with the passage of time.
d. **'Dead beat' positive dynamic stability** – the oscillations cease immediately because they are heavily damped.
e. **'Divergent' negative dynamic stability** – the oscillations cease because the motion completely diverges from the original motion.

The Principles of Flight for Pilots   P. J. Swatton
© 2011 John Wiley & Sons, Ltd

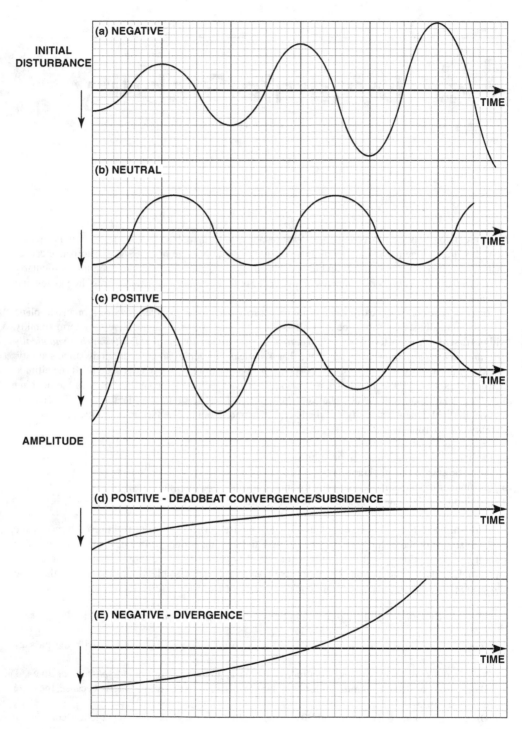

Figure 12.1   Dynamic Stability Forms of Motion.

The time taken from the commencement to the completion of one whole oscillation is dependent on the magnitude of the static stability. The greater the static stability the shorter is the time period. If the amplitude of the oscillation remains constant in equal periods of time it is simple harmonic motion. See Figure 12.1(b).

If such were the case the oscillation would continue indefinitely. In practice there is always some damping or restriction to the continued oscillation amplitude. The viscosity of the air causes the damping of an aeroplane's oscillation, which is proportional to the speed of the aeroplane. (See Chapter 1).

Damping is expressed in terms of the number of cycles necessary for the amplitude of the oscillation to diminish to one half of its original size. The diminishment of the oscillation amplitude will be attained in fewer cycles in a more viscous fluid, in this case warmer air at a lower altitude. Because the effectiveness of the aerodynamic damping is dependent on the TAS it is diminished at high altitude. Eventually the oscillations will disappear altogether so that the equilibrium of the aeroplane is slowly restored. If this is achieved without oscillation overshoot the motion is 'dead beat.'

Oscillations that occur over a lengthy time period are of no concern to the aeroplane designer because even if they are not well damped the pilot can easily control the aeroplane. However, those that occur over a time period similar to that of the pilot's response time must be heavily damped. This is because the pilot's response may become out of phase with the motion and develop pilot-induced oscillations (PIOs).

The minimum damping requirement is that the amplitude of the oscillation must diminish to at least half of its original amplitude in one complete cycle of the motion. Compliance with this requirement is often not achieved with modern aeroplanes and as a result the designer has to include autostabilisation systems such as pitch dampers or yaw dampers to enhance the basic stability of the aeroplane.

12.1 Longitudinal Dynamic Stability

In trimmed level flight a disturbance in pitch usually oscillates about its original state causing fluctuations in the values of speed, height and indicated load factor. When an aeroplane has positive longitudinal dynamic stability the amplitude of the oscillations gradually diminish as the aircraft returns to its original trimmed flight state. See Figure 12.1(c). This is a stable periodic motion. An oscillation that increases its size with the passage of time is an unstable periodic motion. See Figure 12.1(a). However, an aeroplane that oscillates about the lateral axis with constant amplitude is statically stable but dynamically neutral. See Figure 12.1(b).

12.1.1 The Phugoid

The oscillatory motion of the aeroplane in pitch subsequent to a disturbance to its trimmed state can be shown as two separate oscillations. One is a short-period oscillation and the other is a *long-period oscillation* called a *phugoid*. The phugoid is usually poorly damped; it has large variations of speed and height and has an almost uniform load factor (n). It has a constant energy motion in which the potential energy and the kinetic energy are continuously interchanging. (See Figure 12.2).

The following is an example of such an effect. If the forward speed of an aeroplane is decreased by a disturbance it will result in decreased total drag and decreased total lift, causing the aeroplane to descend but enabling it to accelerate to a speed above its predisturbed value. This results in the aeroplane starting to climb once more.

Thus, an oscillatory motion is initiated where the height loss and speed gain are followed by a height gain and speed loss, which is the phugoid. As a result, the speed and the height of the aeroplane vary considerably. The degree to which the horizontal stabiliser is able to dampen the phugoid oscillation is dependent on the drag characteristics of the aeroplane. Because of the modern design trend towards low drag, phugoid oscillation has become a greater problem because its 'damping' is very weak.

12.1.2 Short-Period Oscillation

The motion that causes short-period oscillation is normally heavily damped. It not only involves small changes of speed and height but also large changes of load factor. The speed and height of the aeroplane remain approximately constant. It is a pitching oscillation about the lateral axis with one degree of freedom. The time taken to complete one whole oscillation is the periodic time of the short-period oscillation and is between 3 and 5 s. A vertical gust can cause such an oscillation because it causes the angle of attack to increase, which increases the total lift and results in a changed pitching moment. (See Figure 12.2).

   If this oscillation is dynamically unstable then the correction required to eliminate it is continually changing in phase with the oscillation itself. It is difficult for the pilot to apply the necessary correction because of the lag time of the reaction to the oscillation. This may result in pilot-induced oscillations (PIOs) the magnitude of which can become dangerously large very quickly, especially at high speed and/or mass at low altitude. If such is the case the usual remedy is for the pilot to allow the innate dynamic stability to damp out the oscillation. The damping effect of the tailplane is decreased in a low-density atmosphere and is therefore less effective at high altitude.

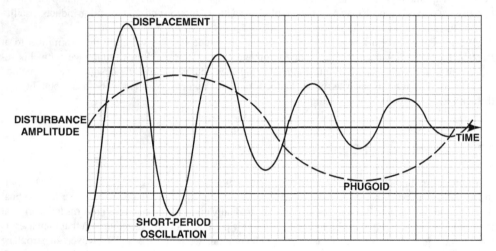

Figure 12.2   Longitudinal Dynamic Stability Basic Components.

12.1.3 Factors Affecting Longitudinal Dynamic Stability

An aeroplane cannot be dynamically stable unless it has static stability. The longitudinal dynamic stability, that is the return of an aeroplane to its state of equilibrium after the disturbing force has been removed in the long term, is dependent on:

a. longitudinal static stability;
b. aerodynamic pitch damping;
c. moments of inertia in pitch;
d. angle of pitch;
e. rate of pitch.

12.2   Lateral Dynamic Stability

Wing dihedral increases the lateral dynamic stability of the aeroplane; so also do wing sweep, a high wing mounting and a low CG. The extension of flap will adversely affect the lateral dynamic stability of an aeroplane. The motion resulting from a lateral disturbance when the aeroplane is trimmed in level flight consists of the three components that occur simultaneously. They are roll, yaw and sideslip; the aerodynamic effect of which will vary with the magnitude of each component. The usual three resulting motions are rolling, spiral and Dutch roll.

12.2.1 Sideslip

The lateral dynamic stability of an aeroplane is determined mostly by the relative effects due to sideslip of the:

a. **Yawing moment** – weathercock effect. Too much weathercock effect will cause spiral instability.
b. **Rolling moment** – dihedral effect. Too much dihedral effect will cause Dutch roll.

12.2.2 Rolling

Initially only the angle of bank will change but it is quickly damped because the downgoing wing produces an increased amount of lift as a result of its increased angle of attack and the upgoing wing produces a decreased amount of lift due to its decreased angle of attack.

12.2.3 Spiral

A gradually tightening spiral motion is the result of a combination of bank and yaw if the aeroplane is laterally dynamically unstable. It is not usually of great importance because, even if the aeroplane has negative divergent dynamic stability, the rate of divergence is slow and can be controlled by the pilot.

12.2.4 Dutch Roll

This is an oscillatory motion having a short periodic time that involves roll, yaw and sideslip. Because this motion is either undamped or only weakly damped it is more of a problem with lateral dynamic stability than it is with longitudinal dynamic stability.

12.3 Spiral Instability

Spiral instability occurs when the **directional stability** of an aeroplane is **strong** and the **lateral stability is weak**. When a wing drops, due to a disturbance, the tendency of the aeroplane to correct this departure from the balanced condition in level flight is dependent on the lateral stability of the aeroplane. Additionally, when a wing drops, it causes other secondary motions to be initiated. The fin and rudder tend to yaw the aeroplane into the airflow in the direction of the dropped wing.

Because of the yaw, the higher wing on the outside of the turn is travelling faster than the lower wing and develops a greater amount of lift. This generates a rolling moment in the opposite direction to the correcting moment of the dihedral and if the rolling moment is greater it will tend to increase the angle of bank. In a level balanced turn, if the controls are released and the bank subsequently steadily increases then the aeroplane is spirally unstable.

If the rolling moment is of sufficient magnitude to overcome the correcting force, produced by the dihedral and the damping yaw effect, the bank angle will increase and the aeroplane will enter a steadily steepening spiral dive. This is spiral instability that can be described as a combined lateral and directional aperiodic motion.

Spiral instability is relatively unimportant, nevertheless, it can be diminished at the design stage of the aeroplane by decreasing the fin area, thus lessening its directional stability and reducing its tendency to yaw into the sideslip. As a consequence, of these measures, the uppermost wing has a reduced lift gain, which improves the spiral stability. **Dihedral decreases** *the tendency to* **spiral instability**. If spiral instability remains uncorrected then spiral divergence persists. This causes the speed to increase as the pitch decreases and the roll increases, resulting in a descent.

When a high-performance aeroplane maintains a prolonged turn, or experiences asymmetric thrust as a result of engine failure, because of the interaction of directional and lateral stability, the yaw so produced causes it to rapidly develop a rolling motion in the direction of the yaw and to immediately enter a steep spiral dive.

12.4   Dutch Roll

Dutch roll is the undulating motion of an aeroplane in the directional and lateral planes, following a lateral disturbance to the equilibrium of the aeroplane. **Dutch roll** *occurs when the* **directional stability** *of an aeroplane is* **weak** *and the* **lateral stability** *is* **strong**, which may be exacerbated by an aft CG position because of the shortened corrective moment arm. If an aeroplane is disturbed laterally the combined rolling motion and yawing motion, either of which may be extreme, can cause oscillatory instability, which is much more serious than spiral instability.

This type of instability is prevalent to a varying degree for combinations of high wing loading, swept-back wings (particularly at low IAS) and high altitude. This means that if a wing drops it is rapidly corrected but the aeroplane does not turn into the sideslip or it may even turn away from the sideslip. The aeroplane therefore wallows. *The tendency and sensitivity to 'Dutch' roll is increased if the static lateral stability is increased but the directional static stability remains constant.*

The following is a simplified explanation of the sequence of events that result in the undulating motion of the aeroplane in the directional and lateral planes. If an aeroplane is yawed to port, the port wing effectively slows down and develops less lift but the starboard wing effectively speeds up and develops more lift and therefore causes the aeroplane to roll to port. *This consequence is exacerbated if the aeroplane has dihedral and/or swept-back wings because of the altered effective aspect ratio.* This decreases the tendency of the wings to dampen the rolling motion that leads to sideslip. Coincident with the production of greater lift, the starboard wing also generates an increased amount of drag, which produces a yaw to starboard resulting in the port wing developing more lift and reversing the direction of the roll.

Thus, an excessive restoring force creates the undulating motion. This can be *mollified at the planning stage by introducing a small amount of anhedral in the wing design.* If the oscillatory undulating motion maintains a constant magnitude a state of neutral Dutch roll exists. If the amplitude of the undulating oscillations decreases as time progresses without control input then the Dutch roll is stable. However, if the amplitude of the undulating oscillations increases with the passage of time then the Dutch roll is unstable.

It is difficult to manually counteract Dutch roll. If the correcting control inputs are out of phase with the yawing and rolling motion it will increase the instability to a dangerous degree. Modern jet transport aeroplanes are fitted with yaw dampers to relieve the pilot of the need to manually control Dutch roll.

The yaw-damper system is electrohydraulic. The direction and rate of yaw are sensed by gyros that feed the information to a computer that controls the hydraulic jacks operating the rudder. In this manner, the unintentional yaw is corrected before the motion becomes oscillatory. If this fails to correct the motion then decrease the altitude and speed and this will have the desired effect.

Some aeroplanes fitted with a yaw-damper system are also fitted with a roll-damper system that senses an unintentional rolling motion and corrects it in a similar manner to the yaw damper but operating the ailerons or an asymmetric spoiler to perform the correction. Although roll dampers are not fitted primarily to counteract Dutch roll they are very useful in preventing its onset.

12.5   Asymmetric Thrust

If at the time of a disturbance upsetting the equilibrium of an aeroplane it is in an asymmetric thrust condition its ability to recover is impaired because of the decreased thrust available and the increased total drag experienced caused by the failed engine. This may cause one wing tip to stall and consequently induce the aeroplane to roll and yaw to such an extent that it enters a steep spiral descent or a spin.

In normal level flight the thrust available and total drag are symmetrically disposed about the aeroplane's centreline. In the event of an engine failure a strong yawing moment towards the failed engine results from the loss of the thrust from that engine. The aeroplane will sideslip away from the failed engine but if it has a large degree of lateral static stability it will also roll towards the failed engine.

This situation is particularly dangerous in conditions of low forward speed and high thrust settings, such as during a take-off or go-around procedure because the low forward speed diminishes the authority of the controls. The yawing moment produced by the failed engine has to be counteracted by a large rudder deflection and the rolling motion has to be counteracted by a large amount of aileron deflection. Furthermore, the reduced thrust available increases the response time to any increased thrust demands.

12.6   Aerodynamic Damping

Damping is the suppression of an unwanted oscillation of the aeroplane. It may be achieved manually, by an electrohydraulic system or by aerodynamic means. Aerodynamic damping is the natural subjugation of the effect of an outside disturbance that is attained by the reaction of the airflow acting on the surface of an aerofoil. This type of damping is dependent on the viscosity of the air for its success. The denser the atmosphere the faster the air will cause the aerofoil to extinguish the unwanted movement.

An example of such damping is that of the effect of the fin in restoring directional stability. If the fin is moving sideways the airflow is causing it to have an angle of attack, which generates sideways lift in the opposite direction to the original movement. The change to the fin angle of attack is a function of the true speed of the sideways movement and the aeroplane TAS. At high altitude in a standard atmosphere for a given EAS the TAS will be high but the sideways movement will be small and the size of the damping force will be small. The damping force is a function of the fin change of angle of attack and the EAS. Because the EAS is low at high altitude then the damping force will be small.

Aerodynamic damping in all planes is dependent on the ratio of the true velocity of the disturbance to the horizontal true velocity of the aeroplane. Thus, stability is proportional to dynamic pressure and inversely proportional to the TAS. Consequently, stability decreases with increased altitude and directional stability diminishes at a greater rate than lateral stability. Therefore, *the tendency for an aeroplane to Dutch roll increases with increased altitude and the tendency to spiral instability decreases with increased altitude.*

12.7   Summary

Dynamic stability is greater than static stability in level flight and it requires a greater amount of elevator deflection to hold the aeroplane in a steady pullout than it does in level flight.

12.8   The Factors Affecting Dynamic Stability

The factors that affect the dynamic stability of an aeroplane are:

12.8.1   a. General

1. Mass and linear velocity.
2. Static stability.
3. Angular momentum.
4. Aerodynamic damping moments magnitude.

12.8.2   b. Longitudinal

1. Static longitudinal stability.
2. Aerodynamic pitch damping.
3. Pitch inertia moments.
4. Pitch angle.
5. Rate of pitch.

12.8.3   c. Lateral

1. Dihedral effect.
2. Weathercock effect.

Self-Assessment Exercise 12

Q12.1 The yaw damper corrects which aeroplane behaviour?
(a) spiral dive
(b) buffeting
(c) Dutch roll
(d) tuck under

Q12.2 Which of the following systems suppresses the tendency to 'Dutch roll'?
(a) rudder limiter
(b) yaw damper
(c) roll spoilers
(d) spoiler mixer

Q12.3 Which of the following statements is correct with regard to the phugoid of a conventional aeroplane?
(a) An aft CG position decreases the period time of the phugoid.
(b) Damping of the phugoid is usually very weak.
(c) The speed is constant throughout one period of the phugoid.
(d) The usual period-time of a phugoid is 3 s.

Q12.4 Following a disturbance about its lateral axis, an aeroplane oscillates at constant amplitude about its lateral axis. The aeroplane is statically (i) ....; dynamically (ii) ....
(a) (i) unstable; (ii) stable
(b) (i) stable; (ii) unstable
(c) (i) unstable; (ii) neutral
(d) (i) stable; (ii) neutral

Q12.5 Which motions interact in a Dutch roll?
(a) pitch and adverse yaw
(b) roll and yaw
(c) pitch and yaw
(d) pitch and roll

Q12.6 Which of the following statements is correct?
(a) A dynamically stable aeroplane is impossible to fly manually.
(b) Dynamic stability is only possible when the aeroplane is statically stable about the relevant axis.
(c) Static stability infers that the aeroplane is dynamically stable about the relevant axis.
(d) Dynamic stability means that after being displaced from its original state of equilibrium an aeroplane will return to that condition without oscillation.

Q12.7 The essential requirement for dynamic stability is:
(a) positive static stability
(b) a large deflection range of the stabiliser trim
(c) a small CG range
(d) effective elevator

Q12.8 The 'short-period mode' is an:
(a) oscillation about the lateral axis
(b) oscillation about the vertical axis
(c) oscillation about the longitudinal axis
(d) oscillation induced by the pilot

Q12.9 An aeroplane that is longitudinally statically unstable is simultaneously dynamically:
(a) stable
(b) neutral
(c) unstable
(d) positively stable

Q12.10 Dynamic longitudinal stability requires:
(a) an effective elevator
(b) a small CG envelope
(c) positive static longitudinal stability
(d) a variable-incidence tailplane

Q12.11 In a phugoid:
(a) speed is constant
(b) an aeroplane is unstable
(c) oscillations are damped out in 3 s
(d) damping is weak

Q12.12 An aeroplane that oscillates about its lateral axis with constant amplitude is statically (i) .... and dynamically (ii) ....
(a) (i) stable; (ii) stable
(b) (i) stable; (ii) neutral
(c) (i) neutral; (ii) neutral
(d) (i) unstable; (ii) unstable

Q12.13 An aeroplane that has positive static stability is dynamically ....
(a) either neutral, stable or unstable
(b) always stable
(c) always unstable
(d) always neutral

Q12.14 An aeroplane is prone to spiral divergence if it has the following conditions:
(a) weak directional stability and positive lateral stability
(b) positive longitudinal stability
(c) overactive inputs from the yaw damper when recovering from Dutch roll
(d) positive directional stability and weak lateral stability

Q12.15 The effect that increased geometric dihedral has on (i) lateral stability (ii) Dutch roll and (iii) spiral instability is:

	(i) Lateral Stability	(ii) Dutch Roll	(iii) Spiral Instability
(a)	Increase	Increase	Decrease
(b)	Increase	Increase	Increase
(c)	Increase	Decrease	Decrease
(d)	Decrease	Decrease	Increase

Q12.16 Which of the following conditions make an aircraft prone to spiral divergence?
(a) Positive lateral stability, positive directional stability.
(b) Overactive yaw damper.
(c) Positive longitudinal stability.
(d) Positive directional stability, weak lateral stability.

Q12.17 Short-period oscillation is:
(a) oscillation in yaw
(b) oscillation in pitch
(c) oscillation in roll
(d) an unstable PIO

Q12.18 The tendency to Dutch roll increases with:
(a) decreased dihedral
(b) a forward CG location
(c) increased static lateral stability
(d) increased static directional stability

Q12.19 With increasing altitude and a constant IAS (i) the static lateral stability and (ii) the dynamic lateral/directional stability of an aeroplane with sweptback wings will:
- (a) (i) decrease; (ii) decrease
- (b) (i) decrease; (ii) increase
- (c) (i) increase; (ii) decrease
- (d) (i) increase; (ii) increase

Q12.20 Sensitivity to the Dutch roll is increased by:
- (a) increased static lateral stability
- (b) increased static directional stability
- (c) a forward movement of the CG
- (d) increased anhedral

Q12.21 The effect that a swept wing has on lateral stability with increased altitude is (i) static (ii) dynamic:
- (a) (i) increased; (ii) decreased
- (b) (i) decreased; (ii) decreased
- (c) (i) unchanged; (ii) decreased
- (d) (i) increased; (ii) unchanged

Part 5
Manoeuvre Aerodynamics

13 Level-Flight Manoeuvres

13.1 The Manoeuvre Envelope

13.1.1 The Flight Load Factor

The flight load factor, generally referred to as the load factor, is defined as the ratio of aerodynamic force component acting at right angles to the longitudinal axis of the aeroplane, (i.e. the lift), to the aeroplane mass in the same units of measurement. It is therefore a nondimensional number and is considered to be positive in an upward direction. *CS 25.321(a)*.

The operating limitations of an aeroplane in terms of speed and load factor can be shown graphically and is used to specify the design requirements for new aeroplanes or to illustrate the performance of an aeroplane type for comparison purposes. The load factor is plotted against the vertical axis and the speed as an EAS is plotted against the horizontal axis. This is the manoeuvre envelope. An example is shown in Figure 13.1.

13.2 Manoeuvre-Envelope Limitations

The limitations of the basic manoeuvre envelope are, the stalling speed, the g limitations and the maximum speed.

13.2.1 The Stalling Speed

From basic theory then the Load Factor (n) = lift/mass.
If V_B = Basic Stalling Speed and V_M = Manoeuvre Stalling Speed.
Then, because Lift = Mass in level flight and Total Lift = $C_L \frac{1}{2}\rho V^2 S$.
Thus, the value of C_L at stalling speed = C_{Lmax}.

$$\text{Then, '}n\text{'} = \frac{\text{Lift at the manoeuvre stalling speed (}V_M\text{)}}{\text{Lift at the basic stalling speed (}V_B\text{)}}$$

The Principles of Flight for Pilots   P. J. Swatton
© 2011 John Wiley & Sons, Ltd

$$\text{Therefore, } 'n' = \frac{C_{Lmax} \frac{1}{2}\rho \, V_M^2 S}{C_{Lmax} \frac{1}{2}\rho \, V_B^2 S}$$

$$\text{So, by cancellation } 'n' = \frac{V_M^2}{V_B^2}$$

Then by transposition and square rooting the result $V_M = V_B\sqrt{n}$.

In other words, the Manoeuvre Stalling Speed = The Basic Stalling Speed multiplied by the square root of the load factor.

As the load factor in a turn increases so also does the stalling speed. Its precise value is equal to the basic level-flight, unbanked, stalling speed multiplied by the square root of the load factor. The load factor during a turn is equal to one divided by the cosine of the bank angle ø.

Turn stalling speed = Unbanked stalling speed × $\sqrt{}$Load factor.

Turn stalling speed = Unbanked stalling speed × $\sqrt{}$(Lift/Mass).

Turn stalling speed = Unbanked stalling speed × $\sqrt{(1/\cos ø)}$

For a 60° banked turn the load factor is 2; therefore, the stalling speed is equal to $\sqrt{2}$ or 1.4 times the basic stalling speed. See Table 13.1, in which the basic stalling was assumed to be 60 kt. The stalling speed is calculated for a series of load factors and plotted on the manoeuvre envelope as the left limiting parameter.

Table 13.1   Example Turn Load Factor and Stalling Speeds.

Bank Angle	Cosine ø	1/Cosine ø	$\sqrt{}$1/Cosine ø	Load Factor	Vs in the Turn in kt
15°	0.9659	1.0353	1.0174	1.04	**61.0**
30°	0.8660	1.1547	1.0746	1.15	**64.5**
45°	0.7071	1.4142	1.1892	1.41	**71.4**
50°	0.6428	1.5557	1.2473	1.56	**74.8**
60°	0.5000	2.0000	1.4142	2.00	**84.9**
70°	0.3420	2.9238	1.7099	2.92	**102.6**
75°	0.2588	3.8640	1.9657	3.86	**117.9**
80°	0.1736	5.7588	2.3997	5.79	**144.0**

13.2.2   The 'g' Limitation

An aeroplane is designed for the minimum strength required in its role. This ensures that the mass of the aeroplane is as low as possible compatible with its role. The minimum strength required for a transport aeroplane is such that it can safely withstand a positive force equal two and a half times the force of gravity (+2.5g) and a safe negative force equal to the force of gravity (−1g). These values are usually factorised by approximately 50% so that structural failure will not occur until +3.75g and −1.5g are reached, however, lesser damage may happen at lower force values than these. Exceeding the normal maximum g limit is *hazardous*, up to the 50% safety margin beyond which it is *catastrophic*. See Figure 13.1.

The positive limiting manoeuvre load factor (n) for Class 'A' transport aeroplanes is specified in CS 25 and for Class 'B' Normal and Commuter aeroplanes in CS 23. The maximum value for any speed up to V_D is calculated from the following formula with the overriding limitation that it may not be less than

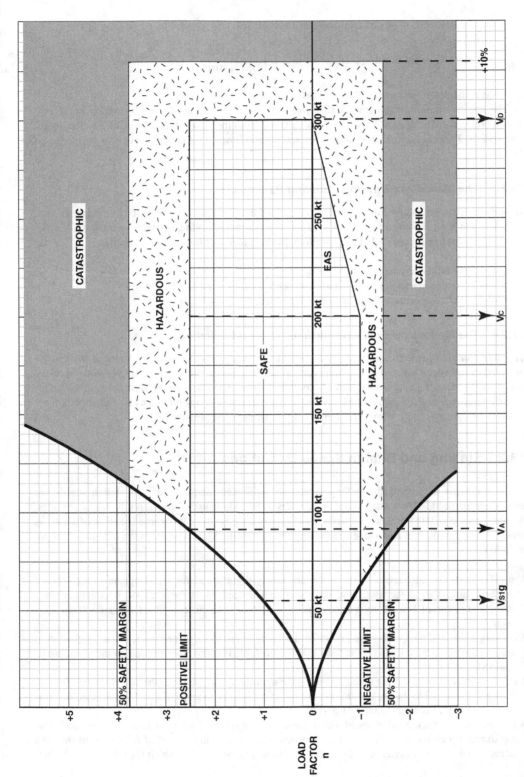

Figure 13.1 The Large Transport Aeroplane Manoeuvre Envelope. *CS 25.333(b)*

+2.5 and does not need to be greater than +3.8 and represents the upper limit of the envelope:

$$\textbf{Lowest limiting load factor} = \textbf{2.1} + \frac{\textbf{24 000}}{\textbf{(M + 10 000)}}.CS\ 25.337(b)$$

where M is the design maximum take-off mass in pounds.

The negative limiting manoeuvre load factor may not be less than −1.0 for speeds up to Vc (the design cruise speed) and must vary linearly with speed from the value at Vc to zero at VD (the design dive speed). These are drawn as straight lines and are the lower limit of the manoeuvre envelope. *CS 25.337(c).*

13.2.3   The Manoeuvre-Envelope Limiting Parameters

The limiting parameters of the manoeuvre envelope for a light aeroplane are the critical speeds VA, VNE, VNO VS1 and VSO. However, for a large transport aeroplane the limiting parameters are VS1g, VA, VC and VD, as shown in Figure 13.1. The manoeuvre limiting load factor may not be less than:

a. 4.4 for light utility aeroplanes. *CS 23.337(a)(2);*
b. 2.5 for large transport aeroplanes in the clean configuration; *CS 25.337(b).*
c. 2.0 for large transport aeroplanes with flaps extended. *CS 25.345(a)(1).*

13.2.4   The Manoeuvre-Envelope Maximum-Speed Limitation

The maximum EAS that an aeroplane is permitted to fly is VD. There is a safety factor of 5% to 10% included in this speed. Exceeding the maximum speed limitation could result in damage to the weakest panels of the aircraft's structure or possibly control reversal and is deemed as *hazardous* up to the 5% or 10% safety margin and *catastrophic* at speeds faster than this. VD is plotted as a vertical line on the manoeuvre envelope and is the right extremity of the envelope. See Figure 13.1.

13.3   Stalling and Design Speed Definitions

In the following definitions the stalling speeds are calibrated airspeeds and the design speeds are equivalent airspeeds. The values stated for the design speeds usually only apply to lower altitudes. At approximately 20 000 ft there is a steep reduction to the design speeds because the value of MCRIT is less than VD and requires the manoeuvre envelope to be modified. The speeds shown in Figure 13.1 are as follows:

VS0. This is the stalling speed or the minimum steady-flight speed for an aeroplane in the landing configuration. VSO at the maximum mass must not exceed 61 kt for a Class 'B' single-engined aeroplane or for a Class 'B' twin-engined aeroplane of 2722 kg or less. *CS 23.49(c). CS Definitions page 20.*

VS1. This is the stalling speed or the minimum steady-flight speed for an aeroplane in the configuration under consideration, e.g. flaps retracted. *JAR 1 page 1-15.* It is the lowest speed on the green arc of the ASI.

VS1g. This is the 'one-g stalling speed', which is the minimum CAS at which the aeroplane can develop a lift force (normal to the flight path) equal to its mass whilst at an angle of attack not greater than that at which the stall is identified. *CS Definitions Page 20.*

VA. Is the design manoeuvring speed and is the maximum speed at which it is possible to safely use maximum up elevator; if this speed is exceeded with the elevator deflected fully upward then the aeroplane may suffer permanent distortion. VA is the highest speed at which the aeroplane will stall before exceeding the manoeuvre maximum load factor and does not need to exceed V_C. It is not less than the stalling speed with the flaps retracted multiplied by the square root of the limiting load factor at a speed of V_C. As shown in Figure 13.1 V_A has a load factor of $+2.5$. $\mathbf{V_A = V_{S1g}\sqrt{n} = V_S\sqrt{2.5}}$. Because of this, a change of mass causes the value of V_A to alter by a percentage amount equal to **approximately half of the percentage mass change**. An alternative calculation that may be required because of a mass change is that:

$$\textbf{New } V_A = \textbf{Old } V_A \times \sqrt{\textbf{New Mass}/\textbf{Old Mass}}$$

Example 13.1

Given: V_A 240 kt; Aeroplane Mass 50 000 kg; Load Factor 2.5. Calculate V_A for an aeroplane mass of 40 000 kg.

Solution 13.1
Exact new $V_A = 240 \times \sqrt{4/5} = \mathbf{214.7kt}$.

Percentage mass decrease $= 10\,000$ kg $\div 50\,000$ kg $\times 100 = 20\%$
Approximate new $V_A = 240 - 10\% = 240 - 24 = \mathbf{216\ kt}$.

V_A is not shown on the ASI, but it should be placarded close to the ASI. For a normal category light aeroplane having a maximum load factor of 3.8g then $V_A = V_{S1g}\sqrt{3.8} = 1.95\ V_{S1g}$. If only one value of V_A is quoted then it is the speed at which the aeroplane will stall at the manoeuvring load factor at the MTOM. *CS 25.335(c)*.

V_B is the design speed for maximum gust intensity. If a gust is experienced at speeds in excess of V_B then the aeroplane may be overstressed. At altitudes where a Mach number limits V_C, V_B is chosen by the manufacturer to provide the optimum margin between the low-speed buffet and the high-speed buffet and need not be greater than V_C to ensure that there is no danger of an inadvertent stall. *CS 25.335(d)*.

V_C is the design cruising speed. The minimum value of V_C must exceed V_B by a sufficient margin to allow for inadvertent speed increases that may be caused by severe turbulence. V_C may not be less than $[V_B + (1.32 \times$ the reference gust velocity)] EAS. It need not exceed V_C or the maximum speed in level flight at the maximum continuous-thrust setting appropriate to the altitude. V_C is limited to a specific Mach number at altitudes at which a Mach number limits V_D. *CS 25.335(a)*.

V_D is the design diving speed, which is never greater than V_{NE} and at high altitude may be limited to a specific Mach number. *CS 25.335(b)*.

V_{DD} is the design drag devices speed. For each drag device the design speed must exceed the recommended operating speed by a sufficient margin to allow for probable variations in speed control. For drag devices intended for use during high-speed descents V_{DD} may not exceed V_D. *CS 25.335(f)*

V_F is the design flap speed. For each wing-flap position it must sufficiently exceed the recommended operating speed for that stage of flight to allow for probable variations in control airspeed and for the transition from one flap position to another. *CS 25.335(e)(1)*. It may not be less than:

a. $1.6V_{S1}$ with take-off flap at the maximum take-off mass;
b. $1.8V_{S1}$ with approach flap at the maximum landing mass;
c. $1.8V_{S1}$ with land flap at the maximum landing mass. *CS 25.335(e)(3)*.

13.4   Limiting Speeds

To ensure an aeroplane is not endangered structurally or suffers from control problems there are a number of limiting speeds imposed on its operation as follows: Some speeds are colour coded on the ASI as listed in Table 13.2.

V_{FE} – The maximum speed at which it is safe to fly with the flaps in a prescribed extended position. It must not exceed V_F the design flap speed. *CS 25.1511. CS Definitions Page 19.*

V_{FO} – The maximum speed at which the flaps may be operated either extended or retracted is V_{FO}. *CS 25.335(e).*

V_{LE} – The maximum speed at which an aeroplane may be safely flown with the undercarriage (landing gear) extended. *CS Definitions Page 20; CS 25.1515(b).*

V_{LO} – The maximum speed, at which the undercarriage (landing gear) may be safely operated, either extended or retracted. V_{LO} is a lower speed than V_{LE}, because damage may be caused to the undercarriage wheel-bay doors at any greater speed. *CS Definitions Page 20; CS 25.1515(a).*

V_{MO}/M_{MO} – The maximum operating IAS (or Mach number, whichever is critical at a particular altitude), which must not be deliberately exceeded in any flight condition, is referred to as V_{MO}/M_{MO}. V_{MO} is the maximum operating IAS and M_{MO} is the maximum operating Mach number. This speed is that which, allowing for moderate upsets ensures the aircraft will remain free from buffet or other undesirable flying qualities associated with compressibility. It must not exceed V_C. When climbing at V_{MO}, the TAS and the Mach number are increasing, so care must be taken to ensure that M_{MO} is not exceeded. Conversely, when descending at M_{MO}, the TAS and IAS increase, and care must be taken not to exceed V_{MO}. *CS 25.1505.*

V_{NE} – The maximum IAS that must never be exceeded and is shown as a red line across the speed arc on the ASI. *CS Definitions Page 20.*

V_{NO} – is the maximum structural cruising speed or the normal operating limit. It should be equal to or greater than V_C, which enables the aeroplane to withstand moderate to severe turbulence. It is the greatest speed on the green arc of the ASI.

Table 13.2   The ASI Speed Arc.

	Speed	
Colour	From	To
Red Line	V_{NE}	V_{NE}
Yellow Arc	V_{MO}	V_{NE}
Green Arc	V_{S1}	V_{NO}
White Arc	V_{SO}	V_{FE}

13.5   The Load Factor

The load factor of an aeroplane is defined as the ratio of its total lift to its total mass and is signified by the letter 'n.' A heavy aeroplane, therefore, has a low load factor and is better able to withstand severe turbulence than a light aeroplane. As a formula then:

$$\text{'n'} = \text{Total Lift} \div \text{Total Mass}$$

The wing loading is defined as the aeroplane mass ÷ wing area. The maximum lift attainable at any airspeed happens when the aeroplane is at C_{Lmax}.

$$\text{Maximum Lift } (C_{Lmax}) = C_{Lmax} \tfrac{1}{2}\rho V^2 S$$

Because total lift equals total mass at the stall then using the basic lift formula it becomes:

$$\text{Total Mass} = C_{Lmax} \tfrac{1}{2}\rho Vs^2 S$$

Ignoring compressibility, then the maximum load factor can be calculated as:

$$n_{max} = \frac{Lmax}{Mass} = \frac{C_{Lmax}\tfrac{1}{2}\rho V^2 S}{C_{Lmax}\tfrac{1}{2}\rho Vs^2 S} = \frac{(V)^2}{(Vs)^2}$$

From this it can be derived that if an aeroplane is flying at twice the stalling speed, i.e. 2Vs and the angle of attack is adjusted to attain the maximum lift possible the resulting load factor can be calculated as:

$$n_{max} = \frac{(2Vs)^2}{(Vs)^2} = \frac{4Vs^2}{Vs^2} = 4$$

In other words, for an aeroplane flying at a speed equal to twice the stalling speed, if the angle of attack is regulated to attain the maximum lift possible at that speed, it will experience a maximum load factor equal to four times the force of gravity or '4g'. Any aeroplane flying at a multiplicand of Vs and at the angle of attack to attain the maximum lift will experience a load factor equal to the square of the multiplicand as a 'g' force, e.g. 1.3Vs creates '1.69g,' 3Vs creates '9g,' 4Vs creates '16g' and so on.

13.6   The Gust Load Factor

A gust is a localised sudden and rapid change to the speed of the air in the atmosphere that can be either horizontal or vertical. The horizontal gust is of little importance because it causes a change to an aeroplane's dynamic pressure that results in an insignificant change to the load factor. The vertical gusts are far more important because they change the effective angle of attack, total lift and the load factor. The gust load is the extra load imparted to the aeroplane by vertical gusts or turbulence. Its magnitude is unaffected by increased altitude but is increased with increased aspect ratio and/or decreased mass.

The load factor for any given angle of attack can be derived from the basic load factor for the normal cruise angle of attack because it is increased by the same percentage as the increase of angle of attack. The vertical component of a gust increases the effective angle of attack, and therefore the lift and the load factor, by the same percentage. See Figure 13.2.

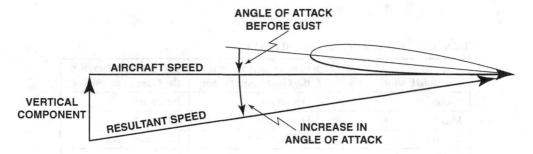

Figure 13.2   The Gust Vertical Component

The load factor deriving from a gust is determined by the magnitude of the vertical velocity of the gust and the forward speed of the aeroplane. To determine the new load factor resulting from a gust divide the new C_L by the original C_L. The speed usually used for these purposes is V_B, which is defined as the design speed for maximum gust intensity. *CS 25.335(d)*. The instantaneous load factor during a gust can be calculated as follows:

Example 13.2

Given: In steady horizontal flight the C_L of an aeroplane is 0.45. A one-degree increase in the AoA increases the C_L by 0.09. A vertical gust of air instantly changes the AoA by two degrees. Calculate the new instantaneous load factor.

Solution 13.2
Increase of $C_L = 2 \times 0.09 = 0.18$
New $C_L = 0.45 + 0.18 = 0.63$
New load factor $= 0.63 \div 0.45 = \mathbf{1.4}$

To avoid stalling and exceeding the limiting load factor an aeroplane must be operated within the limitations imposed by the gust envelope, which is graphically constructed by using the arbitrary values of gust speed against the EAS of the aeroplane originating from a load factor value of $+1.0$ and specified in *CS 25.341(a)(5)(I)*, as shown in Table 13.3.

Table 13.3 The Gust Envelope Limitations.

'g' limit	Speed	EAS	Gust Load Factor	Gust Speed at msl	Gust Speed at 15 000 ft	Gust Speed at 50 000 ft
+ve	V_B (maximum gust intensity speed)	150 kt	+2.5	+66 fps		
+ve	V_C (design cruise speed)	250 kt	+2.75	+56 fps	+44 fps	+26 fps
+ve	V_D (design dive speed)	300 kt	+2.0	+28 fps	+22 fps	+13 fps
−ve	V_B	150 kt	−0.5	−66 fps		
−ve	V_C	250 kt	−0.75	−56 fps	−44 fps	−26 fps
−ve	V_D	300 kt	−0.25	−28 fps	−22 fps	−13 fps

Table 13.4 Factors Affecting the Gust Load Factor.

Variable Factor	Factors that Increase the Gust Load Factor	Factors that Decrease the Gust Load Factor
Altitude	Decreased	Increased
Mass	Decreased	Increased
Speed	Increased EAS	Decreased EAS
Wing Loading	Decreased	Increased
Lift v Angle of Attack curve	Steep	Shallow
Turbulence	Vertical Updraught	Vertical Downdraught
Aspect Ratio	Increased	Decreased

The maximum speed for flight in turbulence or rough air (VRA) can be derived from the gust envelope. See Figure 13.3. It must be within the range of requirements imposed to determine VB but it must be low enough to ensure that any gust will not cause the speed to exceed VMO. In severe turbulence the limiting factors are the stall and the structural limitation. A heavy aeroplane has a high CL and a small load factor but unless the speed is reduced to VRA in any turbulence the limiting load factor may be exceeded and structural failure could result. The effect of variations of mass, altitude and speed must be accounted for and require separate gust envelopes to be drawn up for every combination of factors to be considered but the general effect of these factors can be seen in Table 13.4. A graph that does account all of these considerations is the buffet onset boundary (BOB) Diagram shown in Figure 13.4.

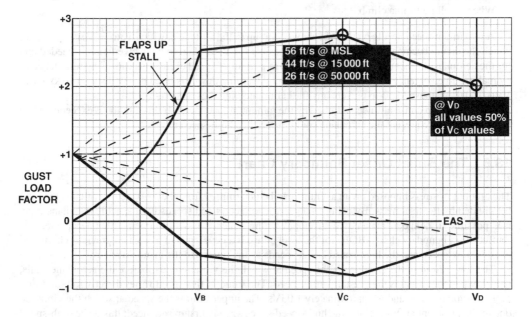

Figure 13.3   The Gust Envelope. (V–n diagram).

13.7   Buffet

The small, rapid movements of the control surfaces and vibrations of the airframe caused by turbulent airflow are 'buffet.' It occurs at very low and very high forward speeds. At high altitude, the buffet restricts the ability of an aeroplane to manoeuvre. The *buffet margin is the difference in speed between the normal cruise speed and the buffet speed,* the low-speed margin to the low-speed buffet and the high-speed margin to the high-speed buffet. Both margins decrease with increased altitude and/or increased mass.

13.7.1   Low-Speed Buffet

At high angles of attack, usually approximately 14°, and with low forward speed the airflow becomes turbulent this is the low-speed buffet. The IAS at which this occurs is just above the stalling speed (**Vs**), usually approximately 1.05Vs, at which speed the stall-warning device should activate *CS 25.207(c)*. The reference stalling speed (**VSR**) is 2 kt above the speed that activates the stall-warning device and is approximately 6% above Vs.

13.7.2   High-Speed Buffet

At very high Mach numbers, the shockwave above the upper surface of the wing causes the airflow to become turbulent towards the trailing edge of the wing, which results in small, rapid movements of the control surfaces; this is the high-speed buffet. The speed at which it occurs is the maximum operating speed, V_{MO}/M_{MO}. Although it is not permitted it is possible to inadvertently exceed M_{MO} in the cruise because of:

a. Gust upsets, or
b. Unintentional control movements, or
c. Passenger movement, or
d. When levelling from a climb.

A maximum speed needle on the ASI shows V_{MO} up to the altitude at which $V_{MO} = M_{MO}$ and above which the datum becomes M_{MO}. Because it is possible for V_{MO}/M_{MO} to be inadvertently exceeded and for the aeroplane to encounter high-speed buffet, as a result of speed excursions caused by turbulence or by making an emergency descent, a high-speed aural warning device is fitted that will sound 10 kt above V_{MO} or 0.01M above M_{MO} to prevent any further unintentional excursion.

The maximum cruise altitude is limited by the minimum load factor. If this altitude is exceeded when there is turbulence present the aeroplane could experience high-speed buffet.

13.8   The Buffet Onset Boundary Chart

The amount of stress imposed on an airframe can be determined from the load factor, which is the total lift divided by the total mass. By using a range of load factors and speeds for a specific aeroplane type, it is possible to define a 'manoeuvre envelope' within which it is safe to operate that aeroplane. The 'buffet-onset boundary chart' (BOB) for the level cruise is a development of the two previous charts and uses the low-speed and the high-speed buffet speeds as limitations to the manoeuvre envelope.

The lower limit of the envelope is the speed at which the vibration, just before the aeroplane stalls, is first experienced and is the prestall buffet speed. The value of this speed varies directly in proportion to the aeroplane mass and is approximately $1.05V_s$. The upper limit is the speed at which the vibration due to the turbulent airflow behind the high-speed shockwave is first experienced; this is the high-speed buffet and has a maximum value of V_{MO}/M_{MO}, which is constant for all masses.

The difference between these limiting speeds decreases with increased altitude and/or mass. This causes manoeuvrability to become increasingly restricted with increased altitude and is of considerable importance when flying in bad weather or clear-air turbulence (CAT) at high altitude. It is unlikely that a pilot would allow the speed of the aeroplane to inadvertently decrease to stalling speed, but it could happen.

For any given combination of mass, centre of gravity position and airspeed, the maximum operating altitude is that altitude at which a normal positive acceleration increment of 0.3g, without exceeding the buffet onset boundary, is possible. It is, therefore, common practice to draw the 'buffet-onset boundary chart' for a range of cruise masses at 1.3g. Any manoeuvre having a load factor exceeding 1.3g will initiate the buffet onset.

Figure 13.4 shows the buffet-onset boundary chart for a typical transport aircraft. Entering the chart with the cruising altitude, cruise Mach number, CG position and aeroplane mass it is possible to determine the maximum permissible bank angle at that altitude and the load factor this imposes on the aeroplane. The buffet margins at 1g derived from the chart also provide the speeds at which both the low-speed and the high-speed buffets will occur. The buffet margin decreases with increased altitude and/or increased aeroplane mass. Transport aeroplanes are prohibited from exceeding the buffet-onset Mach number. Above the graph, to the right of the chart are the details of the example shown on the graph.

Any of the following factors reduces the margins between cruising speed and the high-speed and the low-speed buffets: increased altitude, increased mass, a forward CG position, and an increased bank

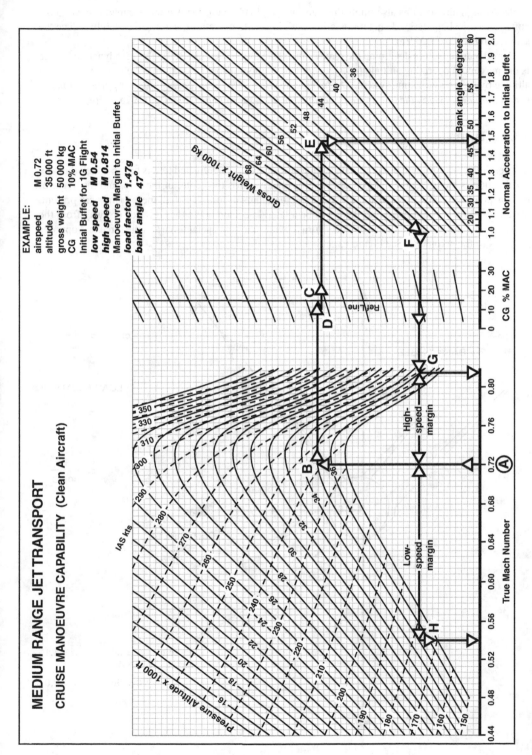

Figure 13.4 The Buffet Onset Boundary Chart.

angle. At high altitude, this also limits the manoeuvring load factor. Inadvertent excursions beyond the buffet-envelope boundaries do not necessarily result in unsafe operations, but it will certainly reduce the safety margin. *CS 25.251(e).*

Aerodynamic factors determine the maximum usable altitude on the BOB. The aerodynamic ceiling is the maximum operating altitude for a jet aeroplane although some aircraft may be limited to a lower altitude by the pressurization system. *The 'manoeuvre or aerodynamic ceiling,' for any particular mass, is that altitude at which the low-speed buffet and the high-speed buffet coincide. CS 25.251(e)2.*

13.9   Turns

In level flight, any turn will be coordinated or uncoordinated. A coordinated turn is one in which the only acceleration is along the normal axis of the aeroplane, towards the centre of the turn, and the angle of bank is precisely correct for the aircraft speed and the radius of turn. If not all these factors are present, the turn will be uncoordinated and there will be a lateral acceleration either towards or away from the centre of the turn. The total lift of a balanced turn provides the centripetal force and the force that opposes the mass of the aeroplane. See Figure 13.5.

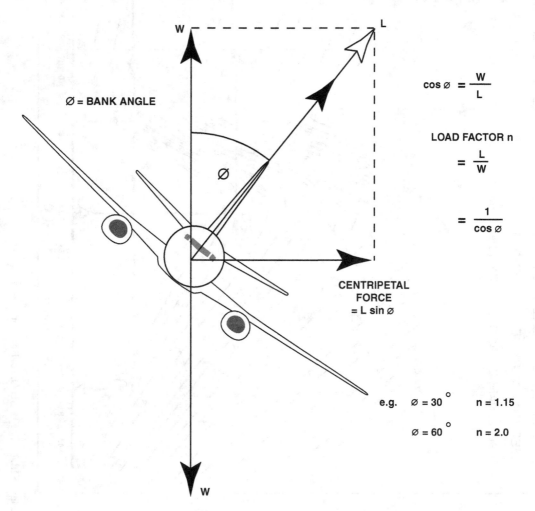

Figure 13.5   The Load Factor in a Turn.

If there is acceleration towards the centre of the turn the aeroplane is slipping and if it is away from the centre of the turn the aeroplane is skidding. Either condition will be indicated on the turn and slip indicator. An overbanked aeroplane will slip into the turn, and requires an increase of speed or a reduction in bank angle to correct this condition. An underbanked aeroplane will skid out of the turn, which requires a decreased speed or an increased bank angle to correct this condition. These are indicated on the turn and slip indicator by the needle one-way and the ball the other. (See page 306).

To maintain altitude and airspeed when executing a turn both the angle of attack and the thrust have to be increased above their level flight values. To ensure the correct response when a jet aeroplane enters or straightens out from a turn and to eliminate any adverse yaw any one of the following three devices may be employed differential ailerons, aileron/rudder coupling roll or control spoilers.

13.9.1 The Load Factor in a Turn

The load factor in level flight is 1 and is allocated the abbreviation 'n', which is equal to (Lift ÷ Mass). In a turn the mass, acting vertically downward, is equal to the total lift multiplied by the cosine of the angle of bank (Mass = Total Lift × Cosine bank angle). Therefore, the load factor 'n' in a turn is greater than in level flight and is equal to the total lift divided by the lift multiplied by the cosine of the bank angle. In other words, the load factor in a turn = 1/cosine bank angle or the secant of the bank angle. For example, the load factor in a 60° banked turn is equal to the secant of 60° = 2. See Figure 13.5. Load factors for various bank angles are shown in Table 13.5.

Total Lift in a turn = Mass ÷ cosine bank angle

Load Factor in a turn = 1/cosine bank angle = secant bank angle

Lift in a turn = Turn Load Factor × cosine bank angle.

Example 13.3

Given: In a level 40° banked turn calculate the lift developed by an aeroplane of mass 60 000 N.

Solution 13.3
Load Factor = Lift ÷ Mass
By transposition Lift = Load factor × Mass
Load factor in a turn = Lift ÷ cosine bank angle

By substitution then Lift = Mass ÷ cosine bank angle = 60 000 N ÷ cos 40°
Lift developed in the turn = 60 000 N ÷ 0.766 = **78 324 N**

13.9.2 The Turn Radius

When a body turns around a fixed point at a constant speed there is **acceleration towards the centre of the turn** that can be calculated by the formula: $V^2 \div r$, where V is the true airspeed in metres per second and r is the radius of the turn in metres and gravity is 9.81 metres per second per second. If the body has a mass M then the centripetal force can be found by the formula:

$$\text{Centripetal force} = \frac{(\text{Mass} \times \text{Speed}^2)}{(\text{gravity} \times \text{turn radius})} = \frac{MV^2}{gr}$$

$$\text{Therefore, from Figure 13.5, Lift sin } \phi = \frac{(\text{Mass} \times \text{Speed}^2)}{(\text{gravity} \times \text{turn radius})} = \frac{MV^2}{gr}$$

And, from Figure 13.5: Lift × cos ø; = Mass

$$\text{By substitution then, } \tan ø = \frac{(\text{Speed}^2)}{(\text{gravity} \times \text{turn radius})} = \frac{V^2}{gr}$$

$$\text{By transposition then, } \textbf{turn radius} = \frac{(\textbf{Speed}^2)}{(\textbf{gravity} \times \textbf{tanø})} = \frac{V^2}{g \times \tan ø}$$

Thus, the radius of a turn is determined by the airspeed and the angle of bank; *mass has no effect on the radius of turn.* Lowering flap or increasing the thrust to maintain a safe speed below VFE will decrease the radius of turn. A combination of increased bank angle together with a decreased airspeed will decrease the radius of turn whilst increasing the rate of turn.

Example 13.4

Given: An aircraft is in a 30° coordinated turn at a TAS of 150 kt. Assuming gravity is equal to 9.81 m/s² and 1 kt is equal to 0.515 mps the turn radius will be:

Solution 13.4
150 kt = 150 × 0.515 = 77.24 mps

$$\text{Turn radius} = \frac{(\text{Speed}^2)}{(\text{gravity} \times \tan ø)} = \frac{(77.24^2)}{9.81 \times \tan 30°} = \frac{5965.36}{5.6638} = \textbf{1053m.}$$

Example 13.5

Given: An aircraft is in a 45° coordinated turn at a TAS of 200 kt. Assuming gravity is equal to 9.81 m/s² and 1 kt is equal to 0.515 mps the turn radius will be:

Solution 13.5
200 kt = 200 × 0.515 = 103 mps

$$\text{Turn radius} = \frac{(\text{Speed}^2)}{(\text{gravity} \times \tan ø)} = \frac{(103^2)}{9.81 \times \tan 45°} = \frac{10\,609}{9.81} = \textbf{1081m.}$$

Example 13.6

Given: An aircraft is in a 45° coordinated turn at a TAS of 360 kt. Assuming gravity is equal to 10 m/s² and 1 kt is equal to 0.515 mps the turn radius will be:

Solution 13.6
360 kt = 360 × 0.515 = 185.4 mps

$$\text{Turn radius} = \frac{(\text{Speed}^2)}{(\text{gravity} \times \tan ø)} = \frac{V^2}{gr} = \frac{(185.42)}{10 \times \tan 45°} = \frac{34\,373.16}{10} = \textbf{3437.3m.}$$

13.9.3 Rate of Turn

The radius of turn for any combination of bank angle and speed can be determined from the above formula. To reduce the radius of turn then either the speed must be decreased or the angle of bank must be increased. However, if the turn is to remain a coordinated turn then the bank angle must be increased and the speed adjusted accordingly. Because the vertical component of lift must be maintained at the magnitude required to equal the aircraft mass, then increasing the angle of attack will increase the total lift and the load factor.

The stalling speed in a turn increases; it is therefore necessary to increase the speed to avoid stalling the aeroplane during the turn. The minimum radius of turn for an aeroplane type is determined by the thrust available to keep the speed above V_M, the manoeuvring stalling speed and the load factor limit, 'n'.

An underbanked turn will cause the aeroplane to skid out of the turn and requires the speed to be reduced or the bank angle to be increased. An overbanked turn will cause the aeroplane to slip into the turn that can be corrected by decreasing the angle of bank or by increasing the speed.

To avoid exceeding the limiting load factor the pilot has to be able to judge the angular speed of the aeroplane, which is the rate of turn. This is the ratio of the TAS to the turn radius, usually measured in radians per second. A radian is the arc of a circle equal to the radius of the circle. **One radian = 57.2958°.**

Thus: **Rate of turn = (TAS mps ÷ radius of turn in metres) radians per second**

$$= V \div r \tan \phi \text{ radians per second}$$

But, turn radius $= \dfrac{(\text{Speed}^2)}{(\text{gravity} \times \tan \phi)} = \dfrac{V^2}{g \times \tan \phi}$

Therefore, by transposition rate of turn $= \dfrac{V}{V^2/(g \times \tan \phi)}$

So, **Rate of turn** $= \dfrac{\mathbf{g \times \tan \phi}}{\mathbf{V}}$ **degrees per second.**

where, $V = $ TAS in metres per second and $g = 9.81$ m/s^2.

Example 13.7

Given: An aircraft is in a 30° coordinated turn at a TAS of 180 kt. Assuming gravity is equal to 9.81 m/s^2 and 1 kt is equal to 0.515 mps the turn radius and rate of turn will be:

Solution 13.7

180 kt $= 180 \times 0.515 = 92.7$ mps

Turn radius $= \dfrac{(\text{Speed}^2)}{(\text{gravity} \times \tan \phi)} = \dfrac{(92.7^2)}{9.81 \times \tan 30°} = \dfrac{8593.29}{5.6638} = 1517.2$ m.

Rate of turn $= \dfrac{g \tan \phi}{V} = \dfrac{9.81 \times \tan 30°}{92.7} = 0.0610982$ rad/s $\times 57.2958 = 3.5°$ **per second**

Most passenger transport aeroplanes are restricted to a maximum rate of turn of 3° per second, referred to as a rate one turn. Using a rate of turn indicator the pilot is able to adjust the bank angle to suit the TAS. In this manner, the load factor is maintained well within limits and the 'g' felt by the passengers is quite acceptable.

Examples of all turn calculations that may require to be made in the EASA examination are shown in Table 13.5. These examples assume the TAS to be 150 kt, that 'g' is 9.81 m/sec^2 and that 1 radian = 57.2958°.

Table 13.5   Example Turn Calculations.

Example Radius of Turn, Rate of Turn and Load Factor Calculations						
Assuming: 1 kt = 0.515 mps; TAS = 150 kt; V = 77.25 mps; V^2 = 5967.56 g = 9.81 m/sec^2; 1 rad = 57.2958°;						
Calc.	Turn Radius		Rate of Turn		Turn Load Factor	
Bank Angle	$g \times \tan\emptyset$	$\dfrac{V^2}{g \times \tan\emptyset}$ m	$\dfrac{g \times \tan\emptyset}{V}$ rad	Rate of Turn °/s	$\cos\emptyset$	$\dfrac{1}{\cos\emptyset}$
15°	2.6286	2270	0.0340	1.95	0.9659	1.0353
30°	5.6638	1054	0.0733	4.20	0.8660	1.1547
45°	9.81	608	0.1270	7.28	0.7071	1.4142
50°	11.6911	510	0.1513	8.67	0.6428	1.5557
60°	16.9914	351	0.2200	12.60	0.5	2.0000
70°	26.9528	221	0.3489	19.99	0.3420	2.9238
75°	36.6114	163	0.4739	27.15	0.2588	3.8637
80°	55.6352	107	0.7202	41.26	0.1736	5.7588

13.10   Turn and Slip Indications

For a twin-engined aeroplane in straight and level flight with the left engine inoperative and wings level, the turn indicator will be centred because it is not turning and the slip indicator will also be centred because the wings are level and there is no lateral acceleration. The instrument measures the resultant acceleration, which is vertical. However, in practice the aeroplane will be flying with a constant sideslip towards the dead engine or yaw, which is not shown on the slip indicator. In these circumstances it is normal practice to fly level with approximately 5° of bank towards the live engine and, although the aeroplane is not sideslipping, the slip indicator will be displaced to the right toward the live engine because of the bank.

Self-Assessment Exercise 13

Q13.1 In steady horizontal flight the C_L of an aeroplane is 0.35. A one-degree increase in the AoA increases the C_L by 0.079. A vertical gust of air instantly changes the AoA by two degrees. The new load factor is:
(a) 1.9
(b) 1.45
(c) 0.9
(d) 0.45

Q13.2 In a gust, which combination of speeds is applicable for the structural strength of an aeroplane in the clean configuration at MSL?
(a) 28 fps and V_D
(b) 66 fps at all speeds
(c) 66 fps and V_B
(d) 56 fps and V_C

Q13.3 The load factor that determines V_A is the:
(a) gust load factor at 66 fps
(b) manoeuvring flap limit load factor
(c) manoeuvring limit load factor
(d) manoeuvring ultimate load factor

Q13.4 To maintain altitude and speed when an aeroplane is rolled into a turn the pilot must (i) . . . . .the thrust: (ii) . . . . .the angle of attack.
(a) (i) maintain; (ii) increase
(b) (i) increase; (ii) increase
(c) (i) increase; (ii) maintain
(d) (i) increase; (ii) decrease

Q13.5 The data that may be obtained from a buffet-onset boundary chart is:
(a) M_{MO} for various masses and altitudes
(b) M_{CRIT} for various masses and altitudes
(c) the low-speed and the high-speed stall Mach numbers for various masses and altitudes
(d) the low-speed and the high-speed buffet Mach numbers for various masses and altitudes

Q13.6 The manoeuvring load factor positive limit for a utility category light aeroplane in the clean configuration is:
(a) 6.0
(b) 4.4
(c) 2.5
(d) 3.8

Q13.7 Two aeroplanes at different masses execute a 20° banked turn at 150 kt IAS. The heavier aeroplane will:
(a) have a greater rate of turn than the lighter aeroplane
(b) turn at the same radius as the lighter aeroplane
(c) have a larger turn radius than the lighter aeroplane
(d) have a smaller radius of turn than the lighter aeroplane

Q13.8 The limiting load factor for a large transport aeroplane in the manoeuvring diagram is:
(a) 2.0
(b) 4.4
(c) 6.0
(d) 2.5

Q13.9 Two identical aeroplanes mass 1500 kg commence a 20° banked turn. Aeroplane A flies at 130 kt and aeroplane B flies at 200 kt. Which of the following statements is correct?
(a) The load factor for A is greater than the load factor for B.
(b) The turn radius for A is greater than the turn radius for B.
(c) The C_L for A is smaller than the C_L for B.
(d) The rate of turn for A is greater than the rate of turn for B.

Q13.10 Given: TAS 300 kt; Bank angle 45°; $g = 10$ m/s². The turn radius is:
(a) 2387 m
(b) 4743 m
(c) 9000 m
(d) 3354 m

Q13.11 The manoeuvring load factor for a large jet transport aeroplane with flaps extended is:
(a) 1.5
(b) 2.5
(c) 4.4
(d) 2.00

Q13.12 The buffet margin:
(a) is always greatest after a step climb
(b) decreases during a constant Mach number descent
(c) is always positive below M_{MO}
(d) increases during a constant IAS descent

Q13.13 By what percentage does V_A alter when an aeroplane's mass decreases by 19%?
(a) 10% lower
(b) 4.36% lower
(c) It remains the same
(d) 19% lower

Q13.14 In straight and level flight C_L is 0.4. An increase of 1° of the angle of attack increases C_L by 0.09. A vertical gust changes the angle of attack by 5°. The revised instantaneous load factor is:
(a) 2.125
(b) 1.09
(c) 2.00
(d) 3.18

Q13.15 V_A is:
(a) the speed that should not be exceeded in a climb
(b) the maximum speed at which maximum up-elevator deflection is permitted
(c) the maximum speed at which rolls are permitted
(d) the turbulence speed for transport aeroplanes

Q13.16 V_{MO}:
(a) should be chosen between V_C and V_D
(b) is equal to the design speed for maximum gust intensity
(c) is the calibrated airspeed at which M_{MO} is reached at FL350
(d) should not be greater than V_C

Q13.17 The three vertical speeds in fps for the clean configuration that determine the shape of the gust load factor diagram are:
(a) 25; 55; 75
(b) 35; 55; 66
(c) 28; 56; 66
(d) 15; 56; 65

Q13.18 The turn indicator shows a right turn and the slip indicator is left of the neutral position. The action required to coordinate the turn is:
   (a) decrease the right bank
   (b) increase the rate of turn
   (c) increase the right bank
   (d) increase the right rudder

Q13.19 The approximate total lift of an aeroplane of 50 000 N gross mass in a 45° banked turn is:
   (a) 50 000 N
   (b) 80 000 N
   (c) 70 000 N
   (d) 60 000 N

Q13.20 In a level, coordinated turn the load factor n (i) ....and the stalling speed Vs (ii) ....
   are:
   (a) (i) less than 1; (ii) lower than in straight and level flight
   (b) (i) more than 1; (ii) lower than in straight and level flight
   (c) (i) less than 1; (ii) higher than in straight and level flight
   (d) (i) more than 1; (ii) higher than in straight and level flight

Q13.21 Exceeding V_A by a small amount may result in:
   (a) permanent structural damage if the elevator is deflected fully up
   (b) structural failure if the elevator is deflected fully up
   (c) permanent structural damage because of excessive dynamic pressure
   (d) structural failure if a turn is executed

Q13.22 At high altitude an aeroplane flies at a Mach number that ensures a 0.3g buffet margin. So that the buffet margin is increased to 0.4g at the same speed the pilot must:
   (a) increase the angle of attack
   (b) decrease the altitude
   (c) extend the flaps to the take-off setting
   (d) increase the altitude

Q13.23 If there is only one fixed value of V_A it is:
   (a) the speed at which the aeroplane will stall at the manoeuvring load factor at the MTOM
   (b) the maximum speed in smooth air
   (c) the speed that ensures that the maximum manoeuvring load factor is not exceeded with unrestricted use of the elevators
   (d) the speed to be used in rough air

Q13.24 The bank angle to be used in a rate one turn depends on the:
   (a) TAS
   (b) mass
   (c) load factor
   (d) wind

Q13.25 Regarding the gust load of an aeroplane for every other factor remains constant, which of the following statements is correct:
   (i) the gust load increases when the mass decreases.
   (ii) the gust load increases when the altitude increases.
      (a) (i) incorrect; (ii) incorrect
      (b) (i) incorrect; (ii) correct
      (c) (i) correct; (ii) incorrect
      (d) (i) correct; (ii) correct

Q13.26 If all other factors remain constant, which of the following has the effect of increasing the load factor?
(a) increased aeroplane mass
(b) decreased air density
(c) aft CG
(d) vertical gust

Q13.27 Regarding the gust load of an aeroplane if every other factor remains constant, which of the following statements is correct:
(i) the gust load increases with increased aspect ratio of the wing.
(ii) the gust load increases when the speed increases.
   (a) (i) correct; (ii) correct
   (b) (i) incorrect; (ii) correct
   (c) (i) incorrect; (ii) incorrect
   (d) (i) correct; (ii) incorrect

Q13.28 Is it permissible for a transport aeroplane to exceed the 'buffet-onset Mach number?
(a) Yes, during the approach.
(b) Yes, to fly faster at extremely high altitudes.
(c) No.
(d) Yes.

Q13.29 The speed limiting the right side of the gust manoeuvre envelope is:
(a) V_{MO}
(b) V_D
(c) V_C
(d) Vflutter

Q13.30 The formula to determine wing loading is:
(a) 1/Bank angle
(b) Mass/Lift
(c) Lift/Mass
(d) Mass/Wing Area

Q13.31 The following device is used to ensure the correct response when a jet aeroplane enters or leaves a turn:
(a) a yaw damper
(b) vortex generators
(c) a dorsal fin
(d) differential ailerons

Q13.32 The following device is used to ensure the correct response when a jet aeroplane enters or leaves a turn:
(a) a yaw damper
(b) vortex generators
(c) a dorsal fin
(d) roll-control spoilers

Q13.33 To maintain altitude and airspeed during a turn the pilot must:
(a) increase the angle of attack
(b) increase the angle of attack and increase the thrust
(c) increase the thrust
(d) decrease the radius of the turn

Q13.34 The percentage increase of lift, over that required in straight and level flight, required to maintain altitude during a 45° banked turn is:
(a) 10%
(b) 20%
(c) 41%
(d) 50%

Q13.35 A heavy aeroplane and a light aeroplane enter a turn at the same TAS and with the same bank angle. The radius of turn of the heavy aeroplane is . . . . that of the light aeroplane.
   (a) the same as
   (b) greater than
   (c) less than
   (d) double

Q13.36 The radius of turn attained at a TAS of 300 kt with 45° of bank if g is assumed to be 10 m/s^2.
   (a) 3376 ft
   (b) 7829 ft
   (c) 9450 ft
   (d) 2398 ft

Q13.37 The following device is used to ensure the correct response when a jet aeroplane enters or leaves a turn:
   (a) a yaw damper
   (b) vortex generators
   (c) a dorsal fin
   (d) aileron-rudder coupling

Q13.38 The effect that mass has on the radius of turn for the same bank angle is that:
   (a) it increases proportionately as mass increases
   (b) it decreases proportionately as mass increases
   (c) mass does not affect the radius of turn
   (d) it increases proportionately as mass decreases

Q13.39 To maintain a constant Mach number in a level turn the pilot has to increase:
   (a) the angle of attack and maintain the thrust setting
   (b) the angle of attack and increase the thrust
   (c) the thrust and maintain the same angle of attack
   (d) the thrust and decrease the angle of attack

Q13.40 In a right turn the slip indicator is left of neutral. One of the ways to coordinate the turn is to apply:
   (a) a greater rate of turn
   (b) more left rudder
   (c) more right rudder
   (d) less right bank

Q13.41 The manner in which a pilot can increase the rate of turn and decrease the radius of turn simultaneously is to:
   (a) increase the bank angle and increase the airspeed
   (b) decrease the bank angle and increase the airspeed
   (c) decrease the bank angle and decrease the airspeed
   (d) increase the bank angle and decrease the airspeed

Q13.42 In a minimum radius turn if the flap is lowered:
   (a) the aeroplane will stall
   (b) increase the turn radius provided that there is sufficient thrust to prevent the aeroplane from stalling and that V_{LE} is not a factor
   (c) decrease the turn radius provided that there is sufficient excess thrust to maintain a safe speed and the V_{FE} is not a factor
   (d) increase the rate of turn but not affect the turn radius

Q13.43 The correct design gust value at MSL from the following is:
   (a) At V_C 56 fps
   (b) At V_D 55 fps
   (c) At V_B 26 fps
   (d) At V_D 66 fps

Q13.44 The relationship of V_A to V_S is:
  (a) $V_{S1g} = V_A\sqrt{n}$
  (b) $V_A = V_{S1g}\sqrt{n}$
  (c) $V_A = V_{S1g} \times n^2$
  (d) $V_A = V_{S1g} \times n$

Q13.45 Which of the following statements regarding V_A is correct?
  (a) At speeds below V_A stalling with full control deflection will damage the airframe.
  (b) In normal operations V_A should never be exceeded.
  (c) Full control deflection is possible without exceeding the limiting load factor up to V_A.
  (d) Manoeuvring the aeroplane at speeds in excess of V_A will permanently damage the airframe.

Q13.46 The positive limiting load factor for large jet transport aeroplanes at V_B is:
  (a) 1.0
  (b) 1.5
  (c) 1.75
  (d) 2.5

Q13.47 V_A is the speed to be used as the maximum:
  (a) turbulence speed for transport aeroplanes
  (b) climbing speed for transport aeroplanes
  (c) speed at which maximum nose-up-elevator deflection is permitted
  (d) descending speed for transport aeroplanes

Q13.48 The effect on V_A of a mass decrease of 19% is a decrease of:
  (a) 10%
  (b) 19%
  (c) 5%
  (d) 4.4%

Q13.49 The limiting load factor for a light utility aeroplane is:
  (a) 4.4
  (b) 5.0
  (c) 2.5
  (d) 2.0

Q13.50 If the maximum cruise altitude limited by the minimum load factor is exceeded:
  (a) the load factor may be exceeded if turbulence is encountered
  (b) the load factor may be exceeded if a sudden turn is made
  (c) Mach buffet will be immediately experienced
  (d) Mach buffet may be induced by turbulence

Q13.51 The positive limiting load factor for large jet transport aeroplanes with flaps extended is:
  (a) 1.5
  (b) 2.0
  (c) 2.5
  (d) 4.4

Q13.52 V_A is the speed at which:
  (a) a gust factor of 56 fps is determined
  (b) an aeroplane should fly in turbulence
  (c) the maximum cruise ceiling is determined
  (d) full elevator deflection is possible without exceeding the load limits

Q13.53 The following values can be determined from the buffet-onset boundary chart:
  (a) M_{CRIT} at various altitudes and masses
  (b) the stalling speed for various altitudes and masses
  (c) the low-speed and the high-speed buffet at various altitudes and masses
  (d) all of (a), (b) and (c) above

Q13.54 The manoeuvre diagram right limitation is the speed:
   (a) V_D
   (b) V_{MO}
   (c) V_C
   (d) V_A

Q13.55 Which of the following actions would increase the buffet margin?
   (a) Increase the Mach number.
   (b) Decrease the Mach number.
   (c) Increase the altitude with the same Mach number.
   (d) Decrease the altitude with the same Mach number.

Q13.56 Mach buffet margin:
   (a) decreases with decreased altitude at constant Mach number
   (b) is constant at M_{MO}
   (c) improves using the cruise climb technique
   (d) increases with decreased altitude at a constant IAS

Q13.57 The load factor in straight and level flight is altered by:
   (a) rearward movement of the CG
   (b) a vertical gust
   (c) increased mass
   (d) increased air density

Q13.58 V_{MO} is:
   (a) greater than V_C
   (b) between V_C and V_D
   (c) less than V_C
   (d) less than V_D

Q13.59 The maximum acceptable cruising altitude is limited by the minimum acceptable load factor because exceeding that altitude:
   (a) turbulence may induce Mach buffet
   (b) turbulence may exceed the load factor limit
   (c) a sudden bank may exceed the load factor limit
   (d) Mach buffet will occur immediately

Q13.60 To produce a coordinated turn when the turn and slip needle shows right and the ball is to the left it is necessary to:
   (a) increase the bank angle to the right
   (b) increase the rudder input to the right
   (c) decrease the bank angle to the right
   (d) increase the radius of turn

Q13.61 Which of the following statements is true?
   (a) Extending flaps in severe turbulence moves the CP aft, which increases the margins to the structural limits.
   (b) In severe turbulence, the limiting factors are the stall and the margin to the structural limitations.
   (c) Extending flaps in severe turbulence reduces the speed and increase the margins to the structural limits.
   (d) Extending flaps in severe turbulence reduces the stalling speed and reduces the risk of exceeding the structural limitations.

Q13.62 The C_L of an aeroplane in straight and level flight is 0.42. A $1°$ increase of the angle of attack increase the C_L by 0.1. A vertical gust increases the angle of attack by $3°$. The instantaneous load factor is:

    (a) 1.49

    (b) 2.49

    (c) 1.71

    (d) 0.74

Q13.63 An aeroplane continues straight and level flight with wings level after suffering a port engine failure, which of the following statements is correct?

    (a) turn indicator neutral; slip indicator left of neutral

    (b) turn indicator left of neutral; slip indicator left of neutral

    (c) turn indicator left of neutral; slip indicator neutral

    (d) turn indicator neutral; slip indicator neutral

14 Climb and Descent Aerodynamics

14.1 Climbing Flight

The gradient of climb is the ratio of a height gain to the horizontal ground distance travelled, expressed as a percentage. It is also the ratio of the ROC in feet per minute to the horizontal speed in feet per minute. Whereas, the angle of climb is the inclination of the climb path above the horizontal measured in degrees.

Obstacle avoidance is of primary importance during the climb, especially just after take-off. Therefore, the minimum safe climb gradient is a predominant requirement during the take-off climb. Of secondary importance is the rate of climb, which governs the time taken to reach a given altitude.

14.2 The Forces in a Climb

The thrust required to climb at a given IAS, is greater than that required to maintain level flight at the same IAS. This is because the thrust has to counteract the effects of drag **plus** the component of mass that acts in the same direction as drag. This component is the product of the mass and the sine of the climb angle. The lift requirement in a climb is less than that required for level flight because it only has to counteract the component of mass at right angles to the airflow (i.e. mass × cosine climb angle). This is shown in Figure 14.1. However, it is still considered sufficiently correct to assume that total lift equals total aeroplane mass up to a climb angle of approximately 15° because the cosine of 15° is 0.9659 and the error due to this assumption is less than 4%.

Thrust Required = Drag + (Mass × sine climb angle)

Lift Required = Mass × cosine climb angle

The ability of an aeroplane to climb is determined by the excess power available above that required and the excess thrust available over the total drag. Most climbs are established using a constant power/thrust setting, normally maximum continuous power/thrust.

The Principles of Flight for Pilots   P. J. Swatton
© 2011 John Wiley & Sons, Ltd

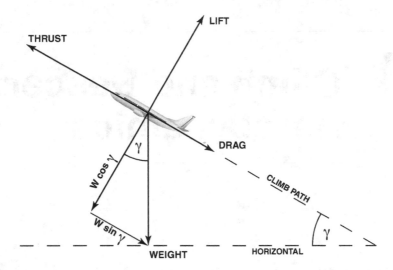

Figure 14.1   The Forces Acting in a Climb.

The performance of an aeroplane in a climb is adversely affected by increased altitude, temperature, flap setting or mass either individually or in any combination. It is also adversely affected if a turn is executed during the climb or if an attempt is made to accelerate whilst climbing.

14.3   The Effect of the Variables on the Climb

14.3.1   Altitude

As with take-off, low air density causes the engines to produce a low thrust output. Thus, the climb gradient and rate of climb will be lower at high altitude and/or high air temperature than would be the case for the same mass and flap setting in a denser atmosphere. However, total drag does not change with increased altitude provided the other conditions remain constant.

14.3.2   Mass

A heavy mass adversely affects the performance of an aeroplane because it has to fly at a faster speed to develop sufficient lift to sustain the mass. To increase the speed attained, for the same atmospheric conditions, thrust setting and flap setting, the gradient of climb and rate of climb must be decreased. Therefore, the climb performance of an aeroplane is directly proportional to the mass of the aeroplane, i.e. heavy aeroplanes have lower gradients and rates of climb when compared with those of light aeroplanes.

14.3.3   Flap Setting

Climbing with flaps extended increases the lift generated. However, this bonus is offset by the greatly increased drag, consequently any flap setting, no matter how small, reduces the climb performance and increases the thrust requirement. *A climb with flaps extended will produce a lower gradient and rate of climb than would be the case with flaps retracted for the same mass and atmospheric conditions.* The gradient of climb and the rate of climb attained are inversely proportional to the amount of flap extended; i.e. larger flap settings produce lower climb gradients and rates of climb.

If the requirement is to maintain a particular climb gradient then the use of flap will cause the aeroplane to adopt a flatter attitude during the climb and the stalling speed to be lower than would be the case with flaps retracted.

If the flaps are retracted whilst climbing, it will cause an immediate loss of lift and the aeroplane will lose height or sink unless the angle of attack is increased to counter this effect. To avoid the pilot having to make large attitude changes during the climb when retracting flap, the manufacturers of large aeroplanes recommend that flap retraction be made in stages, particularly if the aeroplane is at a high gross mass.

Before retracting flap some large aeroplanes have to fly level to accelerate to the appropriate flap-retraction speed, because they have insufficient excess of thrust and power available over that required to continue climbing **and** accelerate.

14.3.4 Wind Component

A *wind component will affect the distance travelled in a climb to a given altitude but has no affect on the rate of climb*, because it is the rate at which the height changes per unit of time. A headwind will cause the groundspeed to be less than the TAS, thus the distance travelled to reach a particular height will be less than it would have been in still air. Consequently, the climb gradient and climb angle will be increased in a headwind and decreased in a tailwind.

14.4 Climb Gradient

Climb gradient is defined as the ratio of the height change to the horizontal distance travelled during the climb. It should not be confused with the climb angle, which is the angle of inclination of the climb path to the horizontal plane. *The tangent of the climb angle is equal to the climb gradient.* For example, a climb angle of 3° is equivalent to a climb gradient of 5.2%. The climb gradient in **still air** is the ratio of the height change to the horizontal **air-distance** travelled. The **wind-effective** climb gradient is that which applies when there is any wind velocity and is the ratio of the height change to the horizontal **ground-distance** travelled.

The excess thrust available over the total of that required determines the climb gradient, i.e. (drag + [mass × sine climb angle]). The magnitude of the climb gradient is directly affected by the values of thrust, drag, mass, TAS, wind component, altitude, air temperature and flap setting. Increasing the mass increases the value of (drag + [mass × sine climb angle]) and therefore diminishes climb performance.

For a constant IAS climb, the angle of attack remains constant throughout the climb, but the gradient of climb and pitch angle gradually decrease as the climb progresses. This is because the reduced air density with increased altitude decreases the thrust available; as a consequence the excess thrust available for climbing is diminished. However, the climbing speed to attain a given gradient is increased for an increased mass and decreased for an increased flap setting.

A further effect when climbing at a constant IAS is that the reduced air density causes both the TAS and the groundspeed to increase with increased altitude. Thus, the distance travelled to attain a given height gain will increase and the climb gradient will decrease progressively at a greater rate throughout the climb. Great care must be exercised during a constant IAS climb to ensure that M_{MO} is not inadvertently exceeded.

The *climb gradient is decreased by a tailwind* because a greater distance is travelled over the ground during the climb than would be the case in still air. Other factors that adversely affect the climb gradient either individually or in any combination are: *increased altitude, increased air temperature, increased mass, increased IAS, increased flap setting and any acceleration.*

14.5   Climb-Gradient Calculations

14.5.1   Method 1

Although in practice the climb gradient is equal to the tangent of the climb angle, *for climb gradients up to 15° it is safe to assume that lift equals mass and that the climb gradient equals the sine of the climb angle*. This is because the values of the sine and tangent of 15° and below are almost equal. Sine 15° = 0.2588 and tangent 15° = 0.2679. The error caused by this assumption is less than 4%. The climb gradient can be calculated from the formula:

$$\textbf{(1) Climb Gradient} = \frac{\textbf{(Total thrust} - \textbf{Total drag) kg}}{\textbf{Mass kg}} \times \textbf{100}$$

This formula can also be used to find the maximum mass to achieve a given climb gradient:

$$\textbf{(2) Maximum Mass kg} = \frac{\textbf{(Total thrust} - \textbf{Total Drag) kg}}{\textbf{Climb Gradient}} \times \textbf{100}$$

The values of thrust and drag can be calculated as follows:

$$\textbf{(3) Total Thrust} = \text{Number of operative engines} \times \frac{\text{Newtons per engine}}{g \text{ m/s}^2} \text{ kg}$$

(4) Total Drag = Lift ÷ Lift/Drag ratio kg

Therefore, because lift is assumed to equal mass in these problems, then:

(5) Total Drag = Mass ÷ Lift/Drag ratio kg

The component of drag derived from the Mass = M sine climb angle.
    Then, the rearward element of mass, M sine climb angle = (Total Thrust − Drag) kg.

Example 14.1

Given: The thrust of a twin-engined turbojet aeroplane is 40 000 Newtons per engine. The minimum permissible climb gradient is 2.4% in still air. Assume gravity (g) = 10 m/s². Drag = 29 000 Newtons. Calculate the maximum permissible mass (in kg) of the aeroplane with one engine inoperative that will enable it to attain the minimum permissible gradient of climb.

Solution 14.1

$$\text{Gradient} = \frac{\text{(Total Thrust} - \text{Total Drag)}}{\text{Total Mass N}} \text{ N} \times 100$$

$$\text{Mass N} = \frac{\text{(Total Thrust} - \text{Total Drag)}}{\text{Gradient}} \text{ N} \times 100$$

$$\text{Mass} = \frac{(40\,000 - 29\,000)}{2.4} \times \frac{100 \text{ kg}}{10}$$

$$= 45\,833 \text{ kg}$$

Example 14.2

Given: A twin turbojet aeroplane that weighs 75 000 kg is climbing with both engines operating. The lift/drag ratio is 15:1. The thrust of each engine is 38 000 Newtons. Assume gravity (g) = 10 m/s². Calculate the gradient of climb.

Solution 14.2
Total Thrust = 2 × 38 000 ÷ 10 = 7 600 kg
Total Drag = 75 000 ÷ 15 = 5 000 kg

$$\text{Gradient} = \frac{(\text{Thrust} - \text{Drag})}{\text{Mass kg}} \text{ kg} \times 100$$

$$\text{Gradient} = \frac{(7600 - 5000)}{75\,000} \times 100 = \textbf{3.47\% in still air}$$

Example 14.3

Given: The total thrust available for a four engined turbojet aeroplane weighing 85 000 kg is 240 000 Newtons. Calculate the rate of climb of this aeroplane with one engine inoperative if the thrust required is 110 000 Newtons and the climbing speed is 190 kt TAS. Assume gravity (g) = 10 m/s/s.

Solution 14.3
Thrust available = 240 000 N ÷ 4 × 3 = 180 000 N = 18 000 kg
Thrust required = 110 000 N = 11 000 kg

$$\text{Gradient} = \frac{(\text{Thrust} - \text{Drag})}{\text{Mass}} \times 100 = \frac{(18\,000 - 11\,000)}{85\,000} \times 100 = \frac{7000 \times 100}{85\,000} = 8.24\%$$

Rate of climb in still air = gradient of climb × TAS = 8.24 × 190 = **1 565 fpm**

Example 14.4

Given: An aeroplane has a thrust to mass ratio of 1:5 at take-off. If the lift to drag ratio in the climb is 20:1, determine the climb gradient.

Solution 14.4
Because the lift and mass are assumed to be equal for all climbs having a climb gradient of 15° or less, then the lift can be expressed in terms of mass. Therefore, the thrust/mass ratio 1:5 becomes 4:20. The gradient of climb is calculated from the formula:

$$\text{Gradient} = \frac{(\text{Total thrust} - \text{Total drag})}{\text{Total mass}} \times 100 = \frac{(4-1)}{20} \times 100 = 15\%$$

Example 14.5

Given: An aeroplane of mass 100 000 kg achieves a climb gradient of 6%. If the conditions remain the same, what is the maximum mass at which it can attain a 5% climb gradient?

Solution 14.5
Maximum mass = 100 000 × 6 ÷ 5 = 120 000 kg

14.5.2 Method 2

The gradient of climb can be found from two basic formulae:

(1) $\text{Gradient } (\%) = \dfrac{\text{Height Difference ft}}{\text{Distance Travelled ft}} \times 100 \qquad (\text{nm} = \text{ft} \div 6080)$

If the still-air gradient is given, then the still-air distance travelled can be found by transposing the formula as follows:

(2) $\text{Distance Travelled (ft)} = \dfrac{\text{Height Difference ft}}{\text{Still-Air Gradient}} \times 100$

If the calculation needs to take into account of the wind effect, the formula becomes:

(3) $\text{Ground Distance Travelled (ft)} = \dfrac{\text{Height Difference ft}}{\text{Still-Air Gradient}} \times 100 \times \dfrac{\text{Groundspeed kt}}{\text{True Airspeed kt}}$

The ground distance travelled in the climb can also be found by using the rate of climb (ROC) as follows:

(4) $\text{Distance Travelled in Climb nm} = \dfrac{\text{Height Difference in feet}}{\text{Rate of Climb in fpm}} \times \dfrac{\text{Groundspeed nm}}{60}$

The still-air gradient can also be found using the ROC as follows:

(5) $\text{Still-air gradient of Climb } \% = \dfrac{\text{ROC fpm}}{\text{TAS kt}} \times \dfrac{6000}{6080}$

This formula, when transposed, can be used to determine the rate of climb (ROC):

(6) $\text{ROC fpm} = \text{Still-air climb gradient} \times \text{TAS kt} \times \dfrac{6080}{6000}$

To convert a still-air gradient to become a wind effective gradient then use the following formula:

(7) $\text{Wind Effective Gradient} = \text{Still-Air Gradient} \times \dfrac{\text{TAS kt}}{\text{G/S kt}}$

To account for the wind component in formulae (5) and (6) use the following:

(8) $\text{Wind effective gradient of Climb } \% = \dfrac{\text{ROC fpm}}{\text{G/S kt}} \times \dfrac{6000}{6080}$

This formula, when transposed, can be used to determine the rate of climb (ROC):

(9) $\text{ROC fpm} = \text{wind effective climb gradient} \times \text{G/S kt} \times \dfrac{6080}{6000}$

Example 14.6

Calculate the load factor for an aeroplane climbing at a climb gradient of 15%.

Solution 14.6
Gradient = Height gain ÷ Distance travelled = Tangent climb angle = Climb gradient
Climb gradient as a decimal = $15 \div 100 = 0.15$
Shift tangent climb gradient = climb angle
Shift tangent $0.15 = 8.53°$

Cosine climb angle = Load factor

Cosine $8.53° =$ **Load factor 0.989**

14.6 Rate of Climb

The rate of climb is defined as the height change over a given time period, usually expressed in feet per minute or metres per second. *Wind component has no effect on the rate of climb.* The magnitude of the rate of climb is determined by the excess power available over that required. The value of the rate of climb is diminished by high altitude, high temperature, high mass and high flap setting either individually or in any combination. *Any acceleration or a turn made whilst climbing also reduces the rate of climb.* The effect of the variables on climb performance is summarised at Table 14.1.

Table 14.1   The Effect Summary of the Variables on Climb Performance.

Variable Factor	Change	Effect on	
		Gradient	**ROC**
Mass	Increase	Decrease	Decrease
Flap Setting	Increase	Decrease	Decrease
Altitude	Increase	Decrease	Decrease
Temperature	Increase	Decrease	Decrease
Wind Component	Headwind	Increase	None
Speed	Increase	Decrease	Decrease

14.7 Rate-of-Climb Calculations

The rate of climb is the vertical speed of an aeroplane, which is usually specified in feet per minute (fpm) and is the vertical component of the forward speed of the aeroplane along the climb path. These speeds can be shown diagrammatically by a triangle. The forces acting on the aeroplane in the climb can also be represented by the same triangle. Both triangles are depicted in Figure 14.2. Remember that wind has no effect on the rate of climb.

It can be seen that triangles ABC and DEF are similar. The angle of climb is ϕ. ROC is vertically upwards, whereas mass is vertically downward.

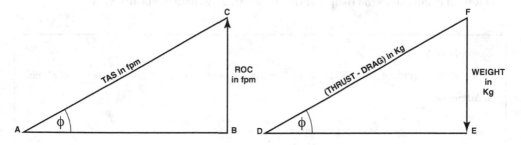

Figure 14.2   The Similar Climb Triangles.

In triangle ABC: $\sin \phi = \dfrac{\text{ROC in fpm}}{\text{TAS in fpm}}$

Whereas in triangle DEF: Thrust Required = Drag + Mass $\sin \phi$ (See Figure 14.2).

Therefore, $\sin \phi = \dfrac{(\text{Thrust} - \text{Drag}) \text{ in kg}}{\text{Mass in kg}}$ (by Transposition).

By combining both triangles it is evident that:

$$\frac{\text{ROC in fpm}}{\text{TAS in fpm}} = \frac{(\text{Thrust} - \text{Drag}) \text{ in kg}}{\text{Mass in kg}}$$

Therefore, **ROC in fpm in still air** $= \dfrac{(\textbf{Thrust} - \textbf{Drag}) \textbf{ in kg}}{\textbf{Mass in kg}} \times \textbf{TAS in fpm}$

Example 14.7

Given: A twin turbojet aeroplane weighing 75 000 kg is climbing with both engines operating. The lift/drag ratio is 15:1. The thrust of each engine is 38 000 Newtons. $g = 10$ m/s². TAS = 300 kt. Calculate the rate of climb.

Solution 14.7
Total thrust $= 2 \times 38\,000 \div 10 = 7600$ kg.
Drag $= 75\,000 \div 15 = 5000$ kg.
TAS $= 300 \times 6080 \div 60 = 30\,400$ fpm.

$\text{Rate of climb} = \dfrac{(\text{Thrust} - \text{Drag}) \text{ kg}}{\text{Mass kg}} \times \text{TAS fpm}.$

$\text{Rate of climb} = \dfrac{(7600 - 5000)}{75\,000} \times 30\,400 = \textbf{1053.9 fpm}.$

Although the rate of climb is unaffected by the wind component, because it is a change of height in a given time period, when it is derived from a climb gradient the wind component must be accounted because it directly affects the value of the climb gradient. The approximate rate of climb can be calculated from the formula:

(10) Rate of climb (in still air) = Still-air gradient of climb × TAS

(11) Rate of climb (with wind component) = Wind effective climb gradient × G/S

Example 14.8

Given: Still-air gradient 2.8%; TAS 180 kt; Wind component 20 kt tail. Calculate the rate of climb.

Solution 14.8

$$\text{Still-air gradient} = \frac{\text{ROC}}{\text{TAS}} \times \frac{6000}{6080}$$

$$\text{Rate of climb} = \text{Still-air gradient} \times \text{TAS} \times \frac{6080}{6000}$$

$$= \textbf{511 fpm}$$

Example 14.9

Given: TAS 120 kt; Glide Slope 4°. Calculate the rate of descent in feet per minute.

Solution 14.9

$$\text{TAS } V = 120 \text{ kt} = \frac{120 \times 6080}{60} = 12\,160 \text{ fpm.}$$

Vertical velocity $= V \sin \emptyset = 12\,160 \sin 4° = 848.2$ fpm.

14.8   Vx and Vy

Vx is the IAS at which to fly to attain the greatest height gain in the shortest horizontal distance, that is the maximum gradient of climb. Vy is the IAS at which to fly to obtain the maximum height gain in the minimum time, that is the maximum rate of climb. Vx is always less than Vy, except at the absolute ceiling where the speeds are equal. The mathematical proof of this fact is in Chapter 19. *Neither speed is affected by wind component.*

The speeds Vx and Vy were originally introduced for the guidance of pilots of small underpowered propeller-driven aeroplanes that had difficulty in clearing obstacles after take-off or in climbing to a new altitude quickly when directed by Air Traffic Control. It was later introduced for low-powered jet aeroplanes for the same reasons. Subsequently, it has become usual for manufacturers to specify these speeds for modern jet aircraft even though it is unnecessary, because these aircraft have a large surplus of power/thrust available over that required.

Both speeds increase in value with increased mass but the speeds quoted in the Aeroplane Flight Manual are usually for the maximum authorised take-off mass and normally there is no provision made for other masses. Increased flap setting decreases both speeds. If the speed is changed from the value of Vx or Vy appropriate to the mass of the aeroplane and flap setting then it is off-optimum and will result in a reduced gradient or rate of climb.

14.9   Vx

Vx is defined as the speed at which the maximum gradient of climb is attained. It can be seen from Figure 14.1 that the sine of the climb angle is equal to the total thrust minus the total drag divided by the mass. The maximum angle of climb is attained at the IAS at which the excess of total thrust available over total thrust required is greatest. The total thrust required and total-drag curves are identical, because of this, Vx is also the IAS at which the excess thrust available over total drag is at a maximum. For small angles up to 15° it has long been accepted that because the sine of the angle is almost equal to the tangent of the same angle then the same formula can be used to calculate the climb gradient.

i.e. Climb Gradient = (Total thrust available − Total drag) ÷ Total mass × 100.

If compressibility is ignored, the Drag (thrust required) v EAS curve does not change with increased altitude. *Therefore, at a constant mass, in terms of IAS, the stalling speed Vs, Vx and VIMD remain the same at all altitudes.* However, the thrust available decreases because of the decreased air density and as a result the curve depicting the thrust available, whilst retaining its original MSL shape, adopts a lower position on the graph as altitude increases (See Figure 14.3).

Therefore, at a constant mass, the IAS value of *Vx does not alter with changes of altitude or temperature.* The value of Vx is directly proportional to the mass of the aeroplane and is the speed at which the ratio of the ROC and the forward speed is maximised.

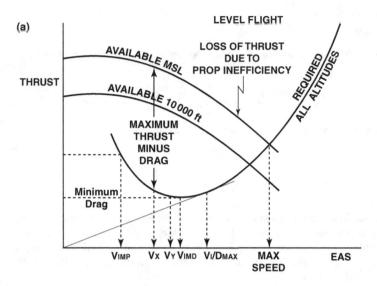

Figure 14.3   (a) Vx for a Piston/Propeller Aeroplane.

Vx can be shown graphically on the thrust available/drag curves as depicted in Figures 14.3(a) and 14.3(b) as the speed at which the largest vertical separation of the total thrust-available curve and the total thrust-required curve is attained. It can be seen from Figure 14.3(a) that, for a propeller-driven aeroplane, Vx occurs at a lower speed than VIMD but greater than VIMP. However, for a jet-engined aeroplane, VIMD and Vx are the same speed. (See Figure 14.3(b)).

The relationship of Vx to VIMD for both types of aircraft is constant at all altitudes, for all masses. Although VIMD produces the best lift/drag ratio in level flight, it does not do so in the climb or descent because in both cases lift is less than mass. Furthermore, because thrust available varies little with speed for a jet-engined aeroplane, the best speed to climb is *Vx, which occurs at the speed of best lift/drag ratio, it is not exactly at, but close to the speed of minimum drag, VIMD.*

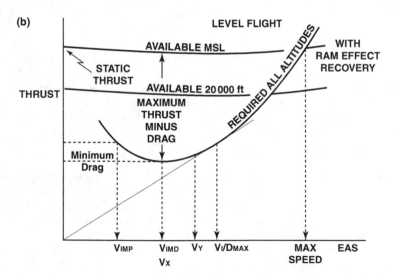

Figure 14.3   (b) Vx – Jet Aircraft.

If a climb is established at any speed other than Vx, it will result in a lower climb angle and climb gradient being attained. *Great care must be exercised when climbing at an IAS of Vx, because it is possible to inadvertently exceed the maximum permitted operating Mach number, M*ᴍᴏ*. Climbing at Vx will ensure the greatest vertical separation from any obstacle encountered during the climb. The still-air climb gradient attained equates approximately to the ratio of the ROC to the TAS in feet per minute. The holding speed for most jet aeroplane is V*ɪᴍᴅ *which is equal to Vx.*

14.10   Vʏ

Because power is the **rate** of doing work then when dealing with a **rate** of climb the power curves must be used. The actual rate of climb at any particular speed is determined by the excess power available over that required at that speed. To attain the maximum rate of climb the aircraft must be flown at the speed attained, at the point on the graph, at which the excess power available over that required is greatest. This speed is referred to as Vʏ, the value of which is directly proportional to aeroplane mass. F*or piston/propeller aeroplanes at low altitude the value of V*ʏ *is approximately equal to the holding speed.*

The speed attained at the tangent to the power-required curve from the origin is Vᴍᴅ. Figure 14.4(a) shows that for propeller-driven aeroplane Vʏ is a lower speed than Vᴍᴅ but greater than Vx and Vᴍᴘ. However, Figure 14.4(b) shows that for a jet aeroplane Vʏ is a higher speed than Vᴍᴅ, but less than the maximum TAS. For both types of aeroplane, *the value of V*ʏ *is always greater than the value of Vx up to the absolute ceiling at which altitude they are equal.*

If a climb is established at any speed other than Vʏ, for a piston/propeller aeroplane the rate of climb will be less than the maximum. For a jet aircraft the rate of climb will be unaffected by not climbing at Vʏ, provided the speed is within 10% of Vʏ this difference occurs because of the shape of the power-available curves of each type of aeroplane.

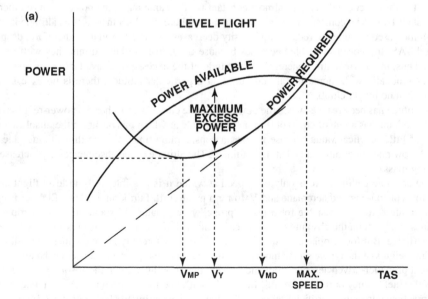

Figure 14.4   (a) Vʏ for a Piston/Propeller Aeroplane.

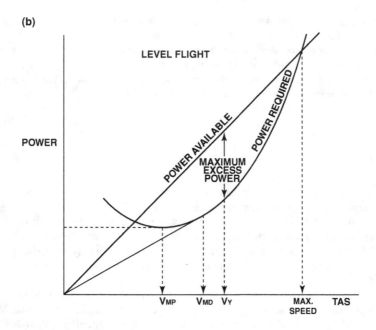

Figure 14.4 (b) V$_Y$ for a Jet Aircraft.

14.11 Aircraft Ceiling

Power is the rate of doing work and is the product of the thrust and TAS. The rate of climb is determined by the amount of excess power available over that required. Although with increased altitude the total drag, i.e. the thrust required, remains almost constant for the same mass, the power required increases. At a constant IAS and a constant mass in a normal atmosphere, the TAS increases as altitude increases. But with increased altitude the reduced air density decreases the thrust available; therefore, despite the increased TAS, the power available decreases because the thrust available diminishes with increased altitude. Thus, in a normal atmosphere, the magnitude of the excess power available over that required decreases gradually with increasing altitude until, at a particular altitude, there is no excess and the aeroplane can no longer climb.

This altitude has been reached when the power-available curve just touches the power-required curve; and the aeroplane has a rate of climb of 0 fpm. The altitude at which this occurs is the **absolute ceiling**, but it is of little practical value because it takes a considerable time to reach this altitude, due to the extremely low rates of climb attained at high altitude. The altitude of the absolute ceiling increases with decreasing mass.

At the absolute ceiling there is only one speed at which it is possible to fly in level flight and that is V$_{MP}$ for a piston-engined aeroplane and V$_{MD}$ for a jet aircraft. Furthermore, level-flight cruise speed at the absolute ceiling is unstable for a piston/propeller aeroplane and stable for a jet aeroplane. Jet aeroplanes rarely attain the absolute ceiling because other limitations, such as pressurization, are often more restrictive. But for aeroplanes of the same type having different types of engines, the ceilings and stabilising heights of the type with the most powerful engines are higher than those for the less powerful type. For a propeller-driven aeroplane it is more likely that it will not attain the absolute ceiling because of propeller inefficiency or it is above 'full-throttle height', and the required power cannot be attained.

For practical purposes an artificial ceiling called the **service ceiling** has been introduced and is defined as that altitude at which, with all engines operating, the maximum rate of climb that can be attained is

500 fpm (2.5 m/s) for jet aircraft and 100 fpm (0.5 m/s) for piston/propeller aeroplanes. In scheduled performance, this is referred to as the **gross** ceiling. A further artificial ceiling is introduced for safety reasons known as the **net** ceiling at which a maximum rate of climb that can be attained is 750 fpm for a jet and 150 fpm piston-engined aeroplane, respectively. The **aerodynamic** ceiling is that altitude at which the low-speed buffet and the high-speed buffet are equal.

The same requirements exist in scheduled performance for the one-engine-inoperative configuration and produce similar artificial ceilings. However, these ceilings in this configuration are known as **stabilising altitudes**. They are considerably lower than the ceilings for the all-engines-operating configuration. A similar set of stabilising altitudes exists for aeroplanes having three or more engines, for the two-engines-inoperative configuration. These again are lower than the one-engine-inoperative stabilising altitudes.

At high masses both the ceilings and the stabilising heights will be lower than those for low masses in the same conditions.

14.12   Vy at the Absolute Ceiling

In Figures 14.5(a) and 14.5(b), which are Power v TAS graphs, the absolute ceiling is shown as the point at which the power-required and the power-available curves just touch. Figure 14.5(a) shows that for a propeller aeroplane, because the power-available curve is concave to the horizontal axis, the point of contact with the power-required curve occurs at a speed lower than that at the point of tangency, i.e. less than VMD. Thus, the speed Vy for this type of aeroplane is less than VMD but greater than Vx and VMP. At the absolute ceiling, the only speed at which level flight can be maintained is Vy for both types of aeroplane.

The absolute ceiling for a jet aircraft is depicted by Figure 14.5(b) and is the altitude at which a straight line from the origin of the graph, which is the power available, is tangential to the power-required curve. The speed at the point of tangency is Vy. This confirms that at the absolute ceiling Vx, Vy and VMD are the same TAS for a jet aeroplane and is shown to be greater than VMP. (See Figure 14.6(b)).

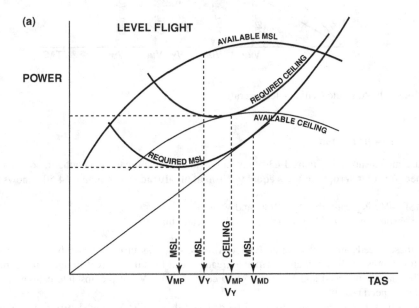

Figure 14.5   (a) Absolute Ceiling - Piston/Propeller Aircraft.

14.12.1 Piston/Propeller Aeroplanes

The power curves depicted in Figure 14.5(a) show that in a normal atmosphere with constant conditions:

a. At MSL – V_Y is less than V_{MD} and greater than V_{MP} and V_X.
b. At the absolute ceiling – V_Y is less than V_{MD} but equal to V_{MP} and V_X.

When all of these speeds are plotted for a piston/propeller aeroplane on an EAS v Altitude graph, as in Figure 14.6(a), then V_X, V_{IMP} and V_{IMD} are plotted as vertical straight lines; V_Y and the maximum EAS converge on V_X at the absolute ceiling. The maximum altitude at which V_{IMD} can be attained is at the intersection of the maximum speed line with that of V_{IMD}. There is only one speed at which it is possible to maintain level flight at the absolute ceiling, and that is V_X, which is unstable. As shown in Figure 14.6(a) the sequence of speeds from lowest to highest for a propeller aeroplane is V_{IMP}, V_X, V_Y and V_{IMD}.

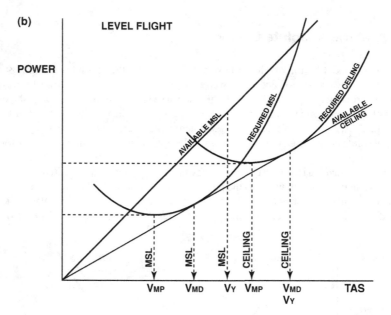

Figure 14.5   (b) Absolute Ceiling - Jet Aircraft.

14.12.2 Jet Aeroplanes

The thrust curves shown in Figure 14.3(b) reveal that at a constant mass and configuration in a normal atmosphere **for a jet aeroplane V_X is equal to V_{IMD} at all altitudes**. But Figure 14.5(b) shows that:

a. At MSL – V_Y is greater than both V_{MD} and V_{MP}.
b. At the absolute ceiling - V_Y is equal to V_{MD} but greater than V_{MP}.

If all of these speeds are plotted on an EAS v Altitude graph as in Figure 14.6(b), then one vertical straight line represents both V_X and V_{IMD} at a greater speed than the vertical straight line representing V_{IMP}. The maximum speed and V_Y converge to coincide with V_X at the absolute ceiling, and is the lowest stable speed possible.

From Figures 14.6(a) and 14.6(b) it can be deduced that whilst V_X and V_{IMD}, in terms of EAS, remain constant with increased altitude, in terms of EAS V_Y decreases in value to become equal to V_X

at the absolute ceiling for both aeroplane types. Theoretically, the speed change is 1 kt per 2000 ft for a piston/propeller aeroplane and 1 kt per 4000 ft for a jet aeroplane.

The relationship of Vx to VIMD is constant irrespective of altitude or aeroplane type. VIMP also has a fixed relationship to Vx but a lower value. However, in terms of TAS both Vx and Vy increase in value with increased altitude. As shown in Figure 14.6(b) the sequence of speeds from lowest to highest for a jet aeroplane is VIMP, Vx and VIMD, and Vy.

14.13   The Effect of the Variables on Vx and Vy

14.13.1   Mass

If the aeroplane mass changes then the value of Vx and Vy will change in direct proportion to the mass change. An increased mass requires that the speeds of Vx and Vy be increased.

14.13.2   Flap

The selection of flap will also alter the value of these speeds in inverse proportion to the angle of flap selected. An increased flap angle setting requires the speeds Vx and Vy be reduced.

14.13.3   Altitude

Figures 14.6(a) and 14.6(b) show that in a normal atmosphere, for a constant mass and configuration, in terms of EAS, the value of the speeds for both Vx and VIMD do not change with increased altitude because they were derived from the thrust curves. For a piston/propeller aeroplane Vx is less than VIMD, but for a jet aeroplane, Vx is equal to VIMD. Table 14.2 summarises the effect of the variables on the values of Vx and Vy.

14.13.4   Temperature

Because Vx was derived from the thrust curves then temperature will not alter its value, however, Vy was derived from the power curves and is affected by ambient temperature. A high temperature decreases the air density, which for a given value of Vy TAS will decrease its value as an EAS.

14.13.5   Wind Component

Wind component has no affect on the values of Vx and Vy.

Table 14.2   The Effect Summary of the Variables on Vx and Vy.

Variable Factor	Change	Effect on EAS	
		Vx	Vy
Mass	Increase	Increase	Increase
Flap Setting	Increase	Decrease	Decrease
Altitude	Increase	None	Decrease
Temperature	Increase	None	Decrease
Wind Component	Increase	None	None

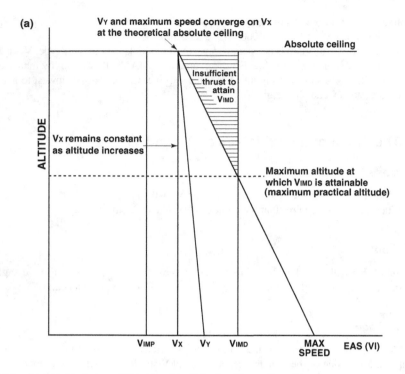

Figure 14.6   (a) The Variation of Speeds with Altitude - Piston/Propeller.

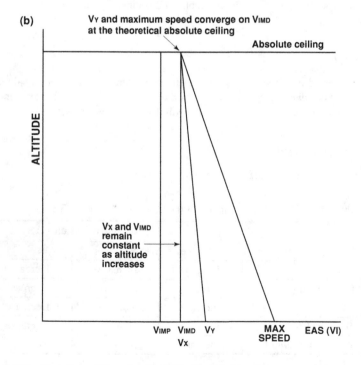

Figure 14.6   (b) The Variation of Speeds with Altitude - Jet Aeroplane.

14.14   The Effect of Climbing-Speed Variations

The speed to attain the maximum gradient of climb is Vx and the maximum rate of climb is Vy. For both piston-engined and jet-engined aeroplanes Vx is always a lower speed than Vy except at the absolute ceiling. If the power is constant but the speed changes during the climb it will affect both the gradient of climb and the rate of climb attained. Climbing with flap extended decreases Vx and Vy and also the maximum gradient and the maximum rate of climb; so both curves in Figure 14.7 will move down to the left.

If the aeroplane has a climbing speed equal to that at point A in Figure 14.7 and it accelerates, then the gradient of climb will increase and the rate of climb will increase because the speed is getting closer to Vx and Vy. However, if the speed of an aeroplane is at point B and it accelerates, then the gradient of climb will decrease but the rate of climb will increase because the speed is getting further away from Vx, but at the same time closer to Vy.

At point C, if the aeroplane accelerates the gradient of climb will decrease and the ROC will increase. At point D the effect of accelerating would be to decrease both the climb gradient and the rate of climb. If the change of speed in the preceding paragraphs were decelerations then the effect would be the reverse.

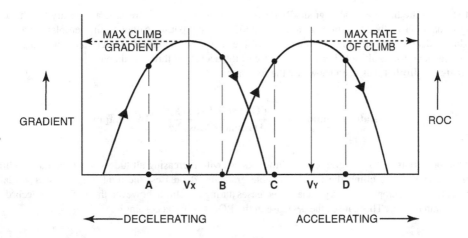

Figure 14.7   Speed Changes v Gradient and Rate of Climb.

The simple way to solve such a problem is to write Vx to the left and Vy to the right and an arrow pointing to the right to indicate increasing speed. Then insert the given speed at the appropriate place on the diagram and an arrow to show either acceleration or deceleration as the case may be. If this arrow is pointing toward the speed of Vx then the gradient of climb is increasing or away from it is decreasing. Similar reasoning applies to Vy.

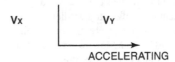

Figure 14.8   Speed between Vx and Vy Aircraft Accelerating.

For example, if the speed is between Vx and Vy and the aeroplane is accelerating then the gradient of climb is decreasing, because the speed is getting further away from Vx; the rate of climb is increasing, because the speed is approaching Vy, as shown in Figure 14.8.

14.15   Factors Affecting the Climb

Angle of Attack. The relationship of IAS, angle of attack and C_L is shown in Figure 5.9 for level, unaccelerated flight. The relationship in the climb is similar and shows that the value of the IAS determines the angle of attack and the C_L.

Climb Gradient. The climb gradient in still air is dependent on the relative values of the thrust, total drag and the aeroplane mass. The relationship is shown by the formula:

$$\textbf{Climb Gradient } \% = \frac{\textbf{(Thrust − Drag) kg}}{\textbf{Aeroplane Mass kg}} \times \textbf{100}$$

In the climb the thrust available gradually decreases because of the decreasing air density but the total drag remains virtually constant at all altitudes up to the tropopause. Although the aeroplane mass will gradually decrease during the climb because of the fuel used the climb gradient and pitch angle will gradually decrease as altitude increases because of the decreased thrust available.

Rate of Climb. It has been shown by the formula:

$$\textbf{Rate of Climb fpm } = \frac{\textbf{(Thrust − Drag) kg}}{\textbf{Aeroplane Mass kg}} \times \textbf{TAS in fpm}$$

For the same reason that the gradient of climb reduces with increasing altitude so does the rate of climb decrease. During the climb the thrust available gradually decreases, the total drag remains virtually constant and the aeroplane mass gradually decreases during the climb, however, the TAS varies according to the climb regime. This causes the decrease in the ROC to vary accordingly.

14.16   The Glide Descent

A descent may be with or without power. If it is with power, it will be at a very low setting to maintain a particular descent angle at a constant speed. A descent without power is a glide descent and has no thrust vector. The remaining force vectors, lift, mass and drag, however, must remain balanced.

The potential energy derived from the mass of the aeroplane, at altitude and descending at an angle enables the mass vector, which acts vertically downward, to be resolved into two component vectors. The component that acts in the direction of flight, parallel to the longitudinal axis, directly opposes and balances the drag vector. The component vector of the mass that acts downward at 90° to the longitudinal axis opposes and balances lift. Figure 14.9 depicts the forces in a descent. Only the aeroplane configuration and angle of attack affect the descent path angle and gradient.

In Figure 14.9 the descent angle is λ. Therefore, the mass can be divided into two components; W sinλ, acting in a forward direction, and W cosλ, downward at right angles to the descent path. A summary of the relationship between the forces in a glide descent is:

Forward vector = Mass × sinλ; Downward vector = Mass × cosλ;
Lift = W cosλ. So Thrust = W sinλ. Therefore, the Drag = W sinλ.

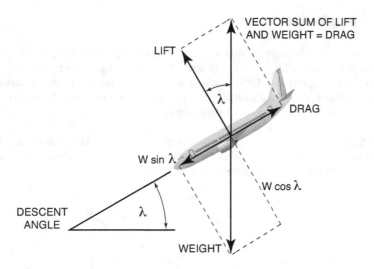

Figure 14.9   The Forces Acting in a Glide Descent.

This shows that the *lift required to descend is less than the mass* of the aeroplane and the *forward vector (equivalent to thrust) is equal to the drag*. This means that the lift required in a descent is less than that required in level flight for an aeroplane of the same mass. For glide angles of less than 15° it is safe to assume that total lift equals total mass because the cosine of the descent angle almost equals 1 and level-flight performance curves can be used. From these details it is possible to calculate the still-air gradient of descent and, if the TAS is known, the rate of descent using the following formulae:

$$\text{(1) Still-air descent gradient \%} = \frac{(\textbf{Drag kg} - \textbf{Thrust kg})}{\textbf{Mass kg}} \times 100$$

$$\text{(2) Rate of descent in fpm} = \frac{(\textbf{Drag kg} - \textbf{Thrust kg})}{\textbf{Mass kg}} \times \textbf{TAS fpm}$$

$$\text{(3) Still-air descent gradient \%} = \frac{\textbf{Rate of descent fpm}}{\textbf{TAS kt}} \times \frac{\textbf{6000}}{\textbf{6080}}$$

$$\text{(4) Wind effective descent gradient \%} = \frac{\textbf{Rate of descent fpm}}{\textbf{G/S kt}} \times \frac{\textbf{6000}}{\textbf{6080}}$$

14.16.1   The Glide Variables

There are three variables in a glide descent, the IAS, the configuration and the glide angle that can be manipulated to achieve either of two goals, maximum range or maximum endurance. When descending at a constant Mach number the IAS increases and the C_L decreases in a standard atmosphere; great care must be taken to ensure that V_{MO} is not exceeded. If the goal is to take as long as possible to descend, then the aeroplane must descend at the lowest rate of descent, which is gliding for endurance.

When the aim is to cover as much ground distance as possible during the descent this is gliding for range. The attainment of either goal is dependent on the speed flown during the descent. The optimum gliding speed for maximum endurance is less than the optimum speed for maximum range. Achievement of the goal is only possible by using the optimum nose-down pitch angle during the descent.

14.17   Gliding for Maximum Range

Figure 14.10 shows an aeroplane at position A and the forces acting on that aeroplane in a glide descent. The vector AB represents the lift, BC the drag and CA the mass. Also shown is the height as AD and the horizontal ground distance as DE. The ratio of the horizontal ground distance to the height of the aeroplane is the Glide Ratio, which to attain the maximum glide range must be as large as possible. In other words, the higher the aeroplane at the commencement of the descent the greater is the range covered during the descent.

Triangles ABC and ADE are both right-angle triangles and both contain an angle equal to the glide angle. Therefore, the triangles are similar and the sides of them are proportional. This means that the Glide Ratio = Ground Distance ÷ Height = Lift ÷ Drag = C_L ÷ C_D.

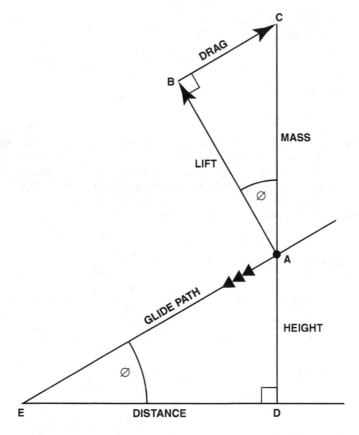

Figure 14.10   The Similar Descent Triangles.

To cover the maximum ground distance during a glide descent the aeroplane must be flown at the speed that will produce the maximum lift to drag ratio. The drag must be at the lowest possible level and the angle of attack set to give the best lift/drag ratio, for which the speed is close to, but not, V_{IMD}. This produces a specific value of C_L and its associated angle of attack at a shallow glide angle. The Lift/Drag ratio changes with angle of attack and configuration at a constant mass. If the descent is made at a constant Mach number then the IAS will increase and consequently the C_L will decrease as the descent progresses.

The TAS at which this occurs derives from the power/TAS curves and is the speed vertically beneath the point of tangency of a straight line from the origin to the power-required curve as shown in Figure 14.11.

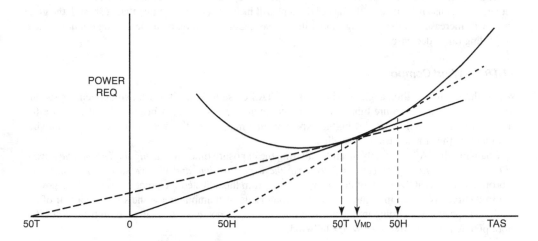

Figure 14.11 The Gliding Speed for Maximum Range.

To ensure the drag remains at its lowest level, the *EAS must remain constant at the speed that achieves the best lift/drag ratio*, the magnitude of which will decrease as mass reduces. This speed is not V_{IMD} because lift is less than that required in level flight but it is close to that speed. For glide angles of less than 15° it is safe to use the level-flight thrust/power curves. If the glide angle exceeds 15° then the lift required is less and the gliding speed should be reduced by a factor of $\sqrt{\cos \lambda}$, where λ = the gliding angle.

To calculate the distance travelled during a glide descent having been given the lift/drag ratio and the height of the aeroplane then the lift = the distance travelled and the drag = the height of the aeroplane in feet.

Example 14.9

Given: Lift/Drag Ratio 15:1; Aeroplane Height 20 000 ft. Calculate the distance travelled in a glide descent in nm.

Solution 14.9.
Distance travelled = 20 000 × 15 ÷ 6080 = 49.34 nm.

14.18 The Effect of the Variables on a Glide Descent

14.18.1 Speed

In a normal atmosphere, gliding for maximum range will cause the TAS and the rate of descent to decrease continuously because the best lift/drag ratio must be maintained. Flying at the speed that produces the maximum lift/drag ratio ensures the maximum vertical separation from any obstacle encountered during the descent. The still-air distance travelled during the descent is equal to the product of the lift/drag ratio and the height difference (approximately).

Max. Still-Air Glide Descent Distance = Max. Lift/Drag Ratio × Height Difference

Because a constant EAS must be maintained throughout the descent its conversion to a TAS gradually increases during the descent. This means that the rate of descent and the groundspeed gradually increase with decreased altitude.

If, as the mass decreases the EAS is not reduced appropriate to the aeroplane mass, as it should be, but kept at a constant value then the rate of descent will increase at a greater rate than it should, the glide angle will increase and the lift/drag ratio will decrease. Consequently, both the gliding endurance and the gliding range decrease.

14.18.2   Wind Component

Whilst the minimum glide angle in still air is at a TAS close to V_{MD}, if a wind component exists due consideration to the exposure time to the wind component is essential. A headwind will decrease the ground distance travelled. (Reducing the exposure time maximizes the distance travelled.) A tailwind will increase the ground distance travelled.

Changing the TAS is the only means of altering the exposure time. *Increasing the TAS in a headwind will increase the rate of descent and decrease the exposure time and vice versa in a tailwind.* The appropriate TAS is at the point of tangency of a line from the adjusted the point of origin to the power-required curve, as shown in Figure 14.11. As a rough rule of thumb increase the speed by a half of the headwind speed when gliding into wind. The same rule does not apply to a tailwind; in most cases it is more practical not to slow down with a tailwind.

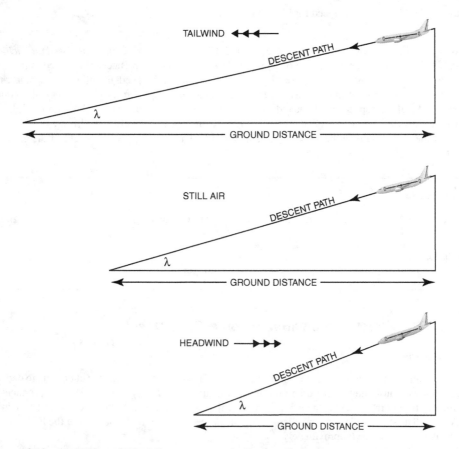

Figure 14.12   The Effect of Wind Component on the Descent Path.

To obtain the required TAS it is necessary to adjust the EAS and is, therefore, a compromise between the speed for the best lift/drag ratio and the speed required because of the wind component. Figure 14.12 shows the effect the wind component has on the descent path.

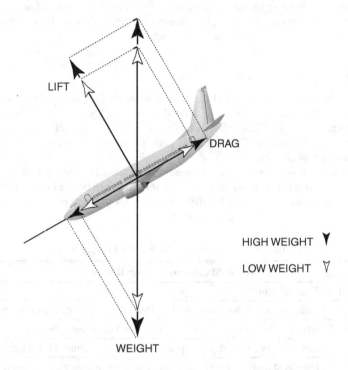

Figure 14.13 The Effect of Increased Mass on the Descent.

14.18.3 *Mass*

Figure 14.13 shows that if the mass is increased the lift and drag increase by the same proportion therefore the Lift/Drag ratio and the glide angle remain unchanged. Changes in mass do not affect the gliding angle or gliding maximum range, provided the speed and the angle of attack are appropriate for the revised mass. The EAS is higher for a heavy mass and lower for a light mass.

Thus, the rate of descent and groundspeed for a heavy aeroplane are greater than those for a lighter aeroplane. Consequently, the exposure time to the wind component is decreased. Hence, it is more beneficial when gliding for range into wind, for the aeroplane to be heavy (i.e. do not jettison fuel) because the higher TAS reduces exposure time to the headwind. To increase gliding endurance reduce the aeroplane mass by jettisoning fuel or for a glider by dumping the water ballast, which will require the aeroplane to be flown at a lower speed resulting in a decreased rate of descent on the same glide path.

Figure 14.13 shows the effect that increased mass has on the magnitude of the four forces during the descent. For an increased mass, the vectors for the thrust, lift and drag all increase proportionately. They, therefore, remain balanced but the forward speed is increased, which results in a higher groundspeed and increased rate of descent. The glide angle is unaffected by mass provided the speed to obtain the best lift/drag ratio for the mass is maintained. If, for instance, the mass decreases but the speed is not decreased then the angle of attack remains unaltered, consequently the descent gradient and rate of descent will increase because of the steeper descent path resulting in a decreased gliding range but the lift/drag ratio will decrease.

14.18.4   Angle of Attack

Maximum range in a glide descent is attained at the pitch angle that produces the optimum angle of attack, i.e. that which gives the best lift/drag ratio. Increasing the angle of attack above the optimum will decrease the range because of the increased drag. Decreasing the angle of attack below the optimum will increase the forward speed, thus decreasing the time taken to descend and therefore the distance travelled.

14.18.5   Flap

The use of flap is restricted to the period of the descent when the IAS has fallen below the maximum speed for lowering flap (V$_{LO}$). It is, therefore, limited to the final approach path. Using flap increases parasite drag, decreases the Lift/Drag ratio, increases the effective angle of attack and decreases the maximum glide range. To maintain the same glide path it is necessary to lower the nose, which will increase the nose-down pitch angle.

The forward speed is less, because of the increased drag generated by the flaps, and the stalling speed is lower because of the increased lift, which enables the touchdown to be at a lower forward and vertical speed. It also affords a steeper approach path giving a better view of the runway. Table 14.3 summarises the effect of the variables on maximum gliding range.

Table 14.3   The Effect of the Variables on Maximum Gliding Range.

Effect on Maximum Gliding Range	
Variable	**Value/Setting**
Speed	Adjust to attain the maximum Lift/Drag Ratio appropriate to the mass
AoA	Above or below the optimum for the speed decreases the gliding range
Mass	Has no effect, provided the speed and AoA are adjusted for the mass
Flap	Increased flap decreases range because of the increased pitch angle
Wind	A headwind decreases the maximum range
Altitude	The higher the start altitude the greater the maximum range

14.19   Gliding for Maximum Endurance

The speed used during the descent is the rate of movement of the aeroplane along the descent path, V, and can be divided into two components: horizontal and vertical. The horizontal component is the aeroplane's forward movement across the ground, its groundspeed, and the vertical component is its rate of descent.

It is essential to reduce the vertical component to a minimum to produce the maximum endurance. This means that the descent speed of the aeroplane and the rate of descent must be as low as possible throughout the descent. The distance travelled during the descent is of no consequence. Therefore, the speed to fly is less than V$_{MD}$ at V$_{MP}$. In Figure 14.14, V is the descent TAS in feet per minute and the rate of descent in fpm is V $\times$ sinø, where ø is the glide angle. Therefore the glide angle must be the minimum possible.

The power required is the product of the TAS and the drag. Therefore, the lowest rate of descent is attained at the speed for which the least power is required, which is close to C$_{Lmax}$. For any given mass this speed occurs at the lowest point on the power-required curve that may be determined by the point at which a horizontal line is tangential to the curve. Figure 9.7 illustrates this fact.

Mass directly affects gliding endurance. *For heavy masses, both the rate of descent and the EAS increase, consequently this results in decreasing the gliding endurance.* The wind component does not affect the gliding endurance. Table 14.4 summarises the effect of the variables on gliding endurance.

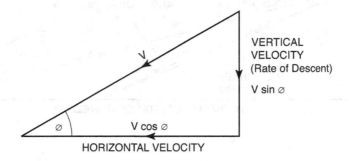

Figure 14.14   The Descent Speed Components.

Table 14.4   Effect of Variables on Maximum Gliding Endurance.

Effect on Maximum Gliding Endurance	
Variable	**Value/Setting**
Speed	V$_{MP}$ appropriate to the mass
AoA	Optimum for speed to attain the minimum glide angle
Mass	Increased mass increases the ROD and decreases the endurance
Flap	Increased flap decreases endurance because of the increased drag
Wind	Has no effect on endurance
Altitude	The higher the start altitude the greater the maximum endurance

14.20   Climbing and Descending Turns

In a climbing turn the rate of climb is the same for both wings but the outer wing travels faster than the inner wing. It therefore travels further than the inner wing and because of the climbing attitude of the aeroplane it has a higher effective angle of attack. Consequently, the outer wing will stall before the inner wing. See Figure 14.15(a).

During a descending turn the rate of descent is the same for both wings and again the outer wing travels faster and further than the inner wing. Because of the descending attitude of the aeroplane the inner wing has the larger effective angle of attack. As a result, the inner wing will stall before the outer wing and the aeroplane is likely to roll further into the turn that could develop into a spin. See Figure 14.15(b).

During a climbing or descending turn should a stall occur, because of a wing drop, the height loss experienced during recovery will be greater than that suffered from a similar occurrence in unbanked flight. This is because the wings have to be levelled before the recovery can commence during which time there is an additional height loss.

(a) CLIMBING TURN

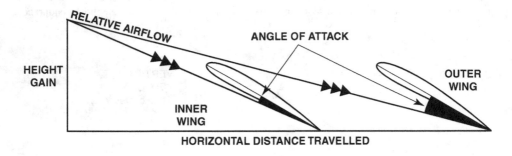

(b) DESCENDING TURN

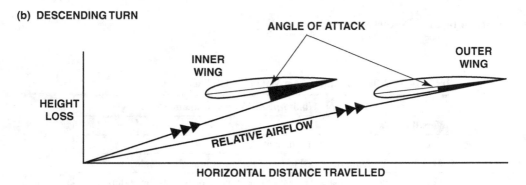

Figure 14.15   Climbing and Descending Turns.

Self-Assessed Exercise 14

Q14.1 If the TAS is 100 kt on a glide slope of 3°, the rate of descent is:
(a) 530 fpm
(b) 300 fpm
(c) 250 fpm
(d) 1000 fpm

Q14.2 In a glide descent at the optimum speed, the effect that increasing the nose-down pitch angle has on the glide distance is that:
(a) it increases
(b) it decreases
(c) it has no effect
(d) it increases but only in a headwind

Q14.3 The effect that increased mass has on a glide descent is:
(a) it increases the vertical speed
(b) it increases the vertical and the horizontal speeds
(c) it increases the distance travelled
(d) it decreases the distance travelled

Q14.4 Two identical aeroplanes are descending at idle thrust and the optimum angle of attack but at different masses. Which of the following statements is true?
(a) The heavier aeroplane will glide further.
(b) The lighter aeroplane will glide further.
(c) The vertical and forward speeds of the heavier aeroplane will be greater.
(d) There is no difference in their glide characteristics.

Q14.5 Which of the following combinations of characteristics will affect the glide angle?
(a) configuration and angle of attack
(b) mass and altitude
(c) altitude and configuration
(d) configuration and mass

Q14.6 In a glide descent at the optimum gliding speed, if the pilot increases the pitch attitude the glide distance will:
(a) remain the same
(b) may increase or decrease depending on aeroplane type
(c) decrease
(d) increase

Q14.7 In a glide, at minimum glide angle speed, at a lower mass at the same speed (i) the rate of descent will .... (ii) the glide angle will .... (iii) the C_L/C_D ratio will ....
(a) (i) increase; (ii) increase; (iii) decrease
(b) (i) decrease; (ii) remain constant; (iii) decrease
(c) (i) increase; (ii) increase; (iii) remain constant
(d) (i) increase; (ii) remain constant; (iii) increase

Q14.8 Which of the following will increase the glide distance?
(a) increased aeroplane mass
(b) decreased aeroplane mass
(c) a headwind
(d) a tailwind

Q14.9 Which of the following factors will increase the glide time?
(a) increased aeroplane mass
(b) a headwind
(c) a tailwind
(d) decreased aeroplane mass

Q14.10 In a glide descent, to fly for the maximum endurance the speed should be that for:
(a) maximum lift
(b) minimum drag
(c) critical Mach number
(d) the minimum angle of descent

Q14.11 What effect does increased aeroplane mass have on a glide descent?
(a) None.
(b) Glide angle decreases.
(c) The lift/drag ratio decreases.
(d) The speed for the best angle of descent increases.

Q14.12 Which of the following statements for a glide descent at the maximum lift to drag ratio speed is correct?
(a) A tailwind component increases the fuel and the time to descend.
(b) A tailwind component decreases the ground distance travelled.
(c) A tailwind component increases the ground distance travelled.
(d) A headwind component increases the ground distance travelled.

Q14.13 Which of the following statements for a glide descent at the maximum lift to drag ratio speed is correct?
(a) The mass of the aeroplane does not have any effect on the speed of descent.
(b) The higher the gross mass the greater is the speed for descent.
(c) The higher the gross mass the lower is the speed of descent.
(d) The higher the average temperature (OAT) the lower is the speed of descent.

Q14.14 The lift coefficient decreases during a glide descent at a constant Mach number because the:
(a) TAS decreases
(b) glide angle decreases
(c) IAS increases
(d) aircraft mass decreases

Q14.15 In a steady descent (descent angle GAMMA) the equilibrium of the forces is given by the formula:
(a) $T + D = -M \sin GAMMA$
(b) $T + M \sin GAMMA = D$
(c) $T - M \sin GAMMA = D$
(d) $T - D = M \sin GAMMA$

Q14.16 Two identical aeroplanes at different masses are descending at idle thrust. Which of the following statements correctly describes their descent characteristics?
(a) At a given angle of attack, both the vertical and the forward speeds are greater for the heavier aeroplane.
(b) There is no difference between the descent characteristics of the two aeroplanes.
(c) At a given angle of attack the heavier aeroplane will always glide further than the lighter aeroplane.
(d) At a given angle of attack, the lighter aeroplane will always glide further than the heavier aeroplane.

Q14.17 An aircraft has a thrust: mass ratio of 1:4 at take-off. If the lift: drag ratio in the climb is 12:1 the climb gradient will be:
(a) 3.0%
(b) 8.3%
(c) 33.0%
(d) 16.7%

Q14.18 What is the effect of extending flap on the value of Vx and Vy?
(a) Vx increases and Vy increases
(b) Vx increases and Vy decreases
(c) Vx decreases and Vy increases
(d) Vx decreases and Vy decreases

Q14.19 What is the effect of an increased flap setting on the gradient and rate of climb?
    (a) the gradient of climb increases and the rate of climb increases
    (b) the gradient of climb increases and the rate of climb decreases
    (c) the gradient of climb decreases and the rate of climb increases
    (d) the gradient of climb decreases and the rate of climb decreases

Q14.20 At a constant mass, the effect that an increase of altitude has on the value of V_X and V_Y is:
    (a) V_X increases and V_Y increases
    (b) V_X constant and V_Y decreases
    (c) V_X decreases and V_Y increases
    (d) V_X constant and V_Y constant

Q14.21 For a given power setting in the climb, if the speed is increased:
    (a) the rate of climb will increase if the speed is below V_Y
    (b) the gradient of climb will increase if the speed is above V_X
    (c) the rate of climb will decrease if the speed is below V_Y
    (d) the gradient of climb will decrease if the speed is below V_X

Q14.22 The wind effective climb gradient is approximately equal to:
    (a) ROC in fpm ÷ TAS in kt
    (b) ROC in fpm ÷ ground distance in nm
    (c) TAS in kt. ÷ horizontal distance in nm
    (d) ROC in fpm. ÷ Groundspeed in kt

Q14.23 The effect an increase of mass has on the values of the ROC and the climb speeds is:
    (a) ROC decreases and the climb speeds remain the same
    (b) ROC is unchanged and the climb speeds are unchanged
    (c) ROC increases and the climb speeds increase
    (d) ROC decreases and the climb speeds increase

Q14.24 The effect an increase of mass has on the value of V_X and V_Y is:
    (a) V_X increases and V_Y increases
    (b) V_X increases and V_Y decreases
    (c) V_X decreases and V_Y increases
    (d) V_X decreases and V_Y decreases

Q14.25 The speed of V_Y is flown to obtain:
    (a) maximum range for a piston-engined aeroplane
    (b) maximum endurance for a piston-engined aeroplane
    (c) maximum gradient of climb
    (d) maximum rate of climb

Q14.26 The effect a headwind will have on a climb to a specified altitude, compared with still air, is to:
    (a) increase the climb time
    (b) decrease the climb time
    (c) decrease the ground distance travelled
    (d) decrease the fuel used

Q14.27 An aircraft is climbing at a constant power setting and a speed of V_X. If the speed is reduced and the power setting maintained the climb gradient will (i). . . . . In addition, the rate of climb will (ii). . . . . :
    (a) (i) decrease; (ii) increase
    (b) (i) increase; (ii) increase
    (c) (i) increase; (ii) decrease
    (d) (i) decrease; (ii) decrease

Q14.28 Given: Still-air climb gradient 4%; TAS 400 kt; Wind Component 50 kt tail; The ground distance travelled in a climb from 8000 ft to 32 000 ft is:
    (a) 99 nm
    (b) 88 nm
    (c) 111 nm
    (d) 123 nm

Q14.29 Given: A four-engined jet aeroplane weighing 37 500 kg has a lift: drag ratio of 14:1 and a total thrust of 75 000 Newtons. g = 10 m/s. The climb gradient is:
   (a) 1.286%
   (b) 27%
   (c) 7.86%
   (d) 12.86%

Q14.30 Given: TAS = 194 kt; ROC = 1000fpm. The climb path is:
   (a) 3°
   (b) 3%
   (c) 5°
   (d) 8%

Q14.31 Given: Climb Gradient = 2.8%; Climb Mass = 110 000 kg; Required Climb Gradient = 2.6%. The maximum climb mass to achieve the required gradient is:
   (a) 121 300 kg
   (b) 106 425 kg
   (c) 118 462 kg
   (d) 102 150 kg

Q14.32 Given: A three-engined jet aeroplane has 50 000 Newtons of thrust per engine and a total drag of 72 069 Newtons. The gross gradient required with one engine inoperative is 2.75%. The maximum take-off mass is:
   (a) 101 567 kg
   (b) 286 781 kg
   (c) 174 064 kg
   (d) 209 064 kg

Q14.33 Given: TAS = 194 kt; ROC = 1000 fpm. The climb gradient is:
   (a) 3°
   (b) 3%
   (c) 5°
   (d) 5%

Q14.34 The maximum angle of climb speed for a jet-engined aeroplane is approximately:
   (a) 1.1Vs
   (b) highest C_L/C_D ratio speed
   (c) highest C_L/C_D^2 ratio speed
   (d) 1.6Vs

Q14.35 How does the value of Vx vary with wind component?
   (a) It increases in a headwind.
   (b) It decreases in a tailwind.
   (c) It is not affected by the wind component.
   (d) It decreases in a headwind.

Q14.36 Increased mass will cause the climb performance to:
   (a) improve
   (b) be unchanged
   (c) be unchanged if the short-field technique is used
   (d) degrade

Q14.37 Given: Climb gradient = 3.3%; TAS 100 kt; Still air. The approximate ROC is:
   (a) 33.0 m/s
   (b) 330 fpm
   (c) 3300 fpm
   (d) 3.30 m/s

Q14.38 The speed to fly to obtain the maximum obstacle clearance in the climb is:
    (a) $1.2V_S$
    (b) V_Y
    (c) V_2
    (d) V_X

Q14.39 Whilst in a positive climb:
    (a) V_X is sometimes greater and sometimes less than V_Y
    (b) V_X is always greater than V_Y
    (c) V_Y is always greater than V_{MO}
    (d) V_X is practically always less than V_Y

Q14.40 The greatest rate of climb at a constant mass:
    (a) Decreases with increasing altitude because the thrust available decreases due to the decreased air density.
    (b) Increases with increasing altitude because drag decreases due to the decreased air density.
    (c) Increases with increasing altitude due to the higher TAS.
    (d) Is independent of altitude.

Q14.41 At the same altitude a higher gross mass decreases the gradient and rate of climb, whereas:
    (a) V_Y and V_X are unaffected
    (b) V_Y and V_X are decreased
    (c) V_Y and V_X are increased
    (d) V_X is decreased and V_Y is decreased

Q14.42 A higher ambient temperature:
    (a) increases the angle of climb but decreases the rate of climb
    (b) does not affect climb performance
    (c) decreases the angle of climb and the rate of climb
    (d) decreases the angle of climb but increases the rate of climb

Q14.43 Which of the following speeds will give the maximum obstacle clearance during a climb?
    (a) The speed at which the ratio of the rate of climb and the forward speed is maximum
    (b) $V_2 + 10$ kt
    (c) V_Y
    (d) V_2

Q14.44 Apart from lift, the forces that determine the angle of climb are:
    (a) mass and drag only
    (b) thrust and drag only
    (c) mass and thrust only
    (d) mass, drag and thrust

Q14.45 How do (i) the maximum angle of climb and (ii) the maximum rate of climb vary with increasing altitude?
    (a) (i) increase; (ii) decrease
    (b) (i) decrease; (ii) increase
    (c) (i) decrease; (ii) decrease
    (d) (i) increase; (ii) increase

Q14.46 On a twin-engined piston aeroplane with variable-pitch propellers, for a given mass and altitude, V_{IMD} is 125 kt and the holding speed (minimum fuel flow) is 95 kt. The maximum rate of climb speed is:
    (a) less than 95 kt
    (b) between 95 kt and 125 kt
    (c) at 125 kt
    (d) at 95 kt

Q14.47 In an unaccelerated climb:
   (a) thrust equals drag plus the uphill component of the gross mass in the flight-path direction
   (b) thrust equals drag plus the downhill component of the gross mass in the flight path direction
   (c) lift is greater than the gross mass
   (d) lift equals mass plus the vertical component of the drag

Q14.48 Which of the following provides the maximum obstacle clearance during a climb?
   (a) $1.2V_S$
   (b) The speed for the maximum rate of climb
   (c) The speed at which the flaps may be selected UP one notch
   (d) The speed for the maximum climb angle V_X

Q14.49 During a constant Mach number descent the speed that may be inadvertently exceeded is:
   (a) V_{NE}
   (b) V_D
   (c) M_{MO}
   (d) V_{MO}

Q14.50 Which of the following equations expresses approximately the unaccelerated percentage climb gradient for small climb angles? Climb Gradient =
   (a) $[(\text{Thrust} - \text{Drag})/\text{Mass}) \times 100]$
   (b) $[(\text{Thrust} + \text{Drag})/\text{Lift}) \times 100]$
   (c) $[(\text{Thrust} - \text{Mass})/\text{Lift}) \times 100]$
   (d) $[(\text{Lift}/\text{Mass}) \times 100]$

Q14.51 A constant headwind component:
   (a) increases the angle of the climb path
   (b) increases the maximum rate of climb
   (c) decreases the angle of the climb path
   (d) increases the maximum endurance

Q14.52 V_X and V_Y with take-off flaps will be:
   (a) higher than that for the clean configuration
   (b) lower than that for the clean configuration
   (c) same as that for the clean configuration
   (d) changed so that V_X increases and V_Y decreases compared with the clean configuration

Q14.53 Acceleration in a climb with a constant power setting:
   (a) improves the climb gradient if the airspeed is below V_X
   (b) improves the rate of climb if the airspeed is below V_Y
   (c) decreases the rate of climb and increases the angle of climb
   (d) decreases the rate of climb and the angle of climb

Q14.54 What factors determine the distance travelled over the ground by an aeroplane in a glide descent?
   (a) the wind component and the aeroplane mass
   (b) the wind component and C_{Lmax}
   (c) the wind component, aeroplane mass and power loading
   (d) the wind component and the optimum L/D ratio

Q14.55 The factors that will increase the ground distance travelled in a glide descent are:
   (a) decreased aeroplane mass
   (b) a tailwind
   (c) a headwind
   (d) increased aeroplane mass

Q14.56 Regarding the best lift/drag ratio speed, which of the following statements is correct?
   (a) Induced drag is greater than parasite drag
   (b) It is the speed used to obtain the maximum endurance for propeller-driven aeroplanes
   (c) The glide angle is minimum
   (d) Parasite drag is greater than induced drag

Q14.57 The speed to obtain the minimum glide angle occurs at:
  (a) C_L/C_{Dmax}
  (b) C_{Lmax}
  (c) C_L/C_D^2max
  (d) C_L^3/C_D^2max

Q14.58 The relationship of the TAS for the minimum sink rate ($V_R/Dmin$) to the speed for the best glide angle (V_{Best} glide) is:
  (a) $V_R/Dmin < V_{Best}$ glide
  (b) $V_R/Dmin = V_{Best}$ glide
  (c) $V_R/Dmin > V_{Best}$ glide
  (d) $V_R/Dmin > V_{Best}$ glide or $V_R/Dmin < V_{Best}$ glide depending on aeroplane type

Q14.59 An aeroplane descends at 160 kt and 1000 fpm. In this condition:
  (a) lift equals mass
  (b) lift is less than drag
  (c) drag is less than the combined forces that move the aeroplane forward
  (d) mass is greater than lift

Q14.60 The lift of an aeroplane in a climb is approximately equal to:
  (a) Mass × cosine climb angle
  (b) Mass × (1 – sine climb angle)
  (c) Mass × (1 – tan climb angle)
  (d) Mass ÷ cosine climb angle

Q14.61 The angle of climb is dependent on the amount by which the (i).........exceeds the (ii).........:
  (a) (i) lift; (ii) mass
  (b) (i) thrust; (ii) drag
  (c) (i) thrust; (ii) mass
  (d) (i) power; (ii) drag

Q14.62 Given: A 50 000 kg twin-engined aeroplane in a climb has a lift/drag ratio of 12:1 and the thrust per engine of 60 000 N per engine. Assuming that $g = 10$ m/s² the climb gradient is:
  (a) 24.0%
  (b) 12.0%
  (c) 15.7%
  (d) 3.7%

Q14.63 The effect that increased mass has on the gliding range is that it:
  (a) increases range
  (b) decreases range
  (c) has no effect on range
  (d) does not affect the range but it increases the gliding angle

Q14.64 In a descent which of the following statements is true?
  (a) lift is less than mass
  (b) lift is less than drag
  (c) drag is less than thrust
  (d) lift equals mass

Q14.65 In a descent:
  (a) mass and lift are equal
  (b) mass is greater than lift
  (c) mass is less than lift
  (d) mass and drag are equal

Q14.66 Which of the following formulae is correct for the lift in a steady climb?
  (a) M cos ø
  (b) M/ cos ø
  (c) M sin ø
  (d) M tan ø

Q14.67 The maximum gliding range of an aeroplane depends on the wind and:
  (a) C_{Lmax}
  (b) Minimum L/D ratio
  (c) V_{IMP}
  (d) Maximum L/D ratio

Q14.68 Other than wind what factors affect the gliding range?
  (a) Mass
  (b) L/D ratio
  (c) C_{Lmax}
  (d) Mass and power required

Q14.69 Which of the following will enable the maximum ground distance to be covered in a glide descent?
  (a) headwind
  (b) increased mass
  (c) tailwind
  (d) decreased mass

Q14.70 The coefficient to obtain the minimum glide angle is:
  (a) C_L/C_D min
  (b) C_L^2/C_D^2
  (c) C_L/C_{Dmax}
  (d) C_D^2/C_L^2

Q14.71 In a constant IAS descent which of the following statements is true?
  (a) lift is less than drag
  (b) lift is equal to mass
  (c) drag is less than thrust
  (d) lift is less than mass

Q14.72 The minimum glide angle speed occurs at an angle of attack that corresponds to:
  (a) C_L/C_{Dmax}
  (b) $C_L^3/C_D^2{}_{max}$
  (c) $C_L/C_D^2{}_{max}$
  (d) C_L max

Q14.73 The relationship of the TAS to obtain the shallowest glide angle (V_{MGA}) and the TAS to obtain the lowest rate of descent (V_{MDR}) is:
  (a) V_{MGA} is least
  (b) V_{MDR} is least
  (c) they are equal
  (d) dependent on aeroplane type

Q14.74 The speed to obtain the minimum sink rate compared with V_{MD} is:
  (a) less than V_{MD}
  (b) more than V_{MD}
  (c) equal to V_{MD}
  (d) cannot be compared with V_{MD} because they come from different graphs

Q14.75 In a constant IAS climb, the speed that may not be exceeded is:
  (a) M_D
  (b) M_{MO}
  (c) V_{NO}
  (d) V_A

Part 6
Other Aerodynamic Considerations

15 High-Speed Flight

15.0.1 General Introduction

High-speed flight can be subdivided into subsonic, transonic, supersonic and hypersonic regimes. In the transonic speed range both subsonic and supersonic speeds exist in the airflow around the aeroplane but it is the Mach number that determines an aeroplane's handling characteristics. The flight regimes are subdivided by speed as shown in Table 15.1.

Table 15.1   Speed Regimes.

Regime.	Lowest speed	Highest speed
Subsonic	Less than Mach 0.8	Critical Mach Number
Transonic	Critical Mach Number	Mach 1.3
Supersonic	Mach 1.3	Mach 5.0
Hypersonic	Mach 5.0	No Limit

An aeroplane when moving through the air transmits a disturbance pressure wave in all directions. That which travels ahead of the aeroplane is particularly important because it 'warns' the air in front of the aeroplane of the approach of the aircraft and enables it to divide to allow its passage with the least amount of disturbance.

Because air is compressible the changing air pressure created by the movement of the aeroplane is accompanied by a change of air temperature and a change of air density. Owing to this, the speed of movement generated by the pressure wave is equal to the speed of sound. Sound waves are audible pressure waves travelling at the speed of sound.

As the forward speed of the aeroplane increases the distance the air ahead of the aeroplane that is influenced by the approaching pressure wave decreases; there is also a change to the airflow and pressure patterns surrounding the aeroplane. Eventually, this results in changes to the manoeuvrability, stability and control characteristics of the aeroplane.

The Principles of Flight for Pilots   P. J. Swatton
© 2011 John Wiley & Sons, Ltd

15.1 High-Speed Definitions

The terms used throughout this chapter are defined as follows:

a. **Free-Stream Mach number (MFS).** The Mach number of the air at a point, unaffected by the passage of the aeroplane but measured relative to the speed of the aeroplane, is the Free Stream Mach number. (MFS). If the speed of the aeroplane is above the critical Mach number (MCRIT) a shockwave may well form, even if the MFS is below Mach 1.

b. **Local Mach number (ML).** This is the ratio of the speed of the airflow at a point on the aeroplane to the speed of sound at the same point.

c. **Critical Mach number (MCRIT).** As MFS increases so do the local Mach numbers. MCRIT is that MFS at which any ML has reached unity. It is the lowest speed in the transonic range.

d. **Critical Drag Rise Mach number (MCDR).** This is the MFS at which because of shockwaves, the CD for a given angle of attack increases significantly.

e. **Detachment Mach number (MDET).** This is the speed at which the bow shockwave of an accelerating aeroplane attaches to the leading edge of the wing or detaches if the aeroplane is decelerating.

f. **Indicated Mach number.** TAS is the difference between pitot pressure and static pressure. LSS is a function of static pressure and air density. Because air density is common to both TAS and LSS, both can be expressed as pressure ratios; this is what the Machmeter measures. Indicated Mach number is therefore the ratio of the dynamic pressure to the static pressure.

g. **Shock Stall.** The airflow over an aeroplane's wings is disturbed when flying at or near the critical Mach number. This causes the separation of the boundary layer from the upper surface of the wing behind the shockwave and is the shock stall, which is described in greater detail later in this chapter.

h. **The Speed of Sound (a).** The speed of propagation of a very small pressure disturbance in a fluid in specified conditions. The local speed of sound through the air is equal to 38.94 multiplied by the square root of the absolute temperature (A). Therefore, as air temperature decreases so also does the local speed of sound (LSS).

$$LSS = 38.94\sqrt{A} \text{ in kt}$$

i. **True Mach number (M).** The ratio of the speed of an object to the value of the speed of sound in the same environmental conditions is the Mach number; it has no units of measurement because it is a ratio.

$$\text{Mach number} = TAS \div LSS \text{ in kt}$$
$$\text{Therefore, Mach number} = TAS \div 38.94\sqrt{A}$$

15.2 High-Speed Calculations

The temperature at absolute zero is equal to $-273A$. Therefore, $0\,°C = +273A$
The absolute temperature at msl in a standard atmosphere $= +273A + 15 = +288A$.
Therefore, the LSS in a standard atmosphere $= 38.94\sqrt{A} = 38.94\sqrt{288} = 661$ kt at msl.
Then, in a standard atmosphere the speed of sound, Mach 1, at mean sea level = the local speed of sound $= 661$ kt.

Example 15.1

Given: An aeroplane flying at M0.75 at FL330 where the ambient temperature is –45 °C. Its TAS is:

Solution 15.1
Absolute air temperature = +273 – 45 = +228A

Mach Number = TAS ÷ LSS. Therefore, TAS = Mach number × LSS
LSS = $38.94\sqrt{A}$ = $38.94\sqrt{228}$ kt = 588 kt.
TAS = 0.75 × 588 kt = **441 kt**

Example 15.2

Given: An aeroplane flying at M0.81 at FL310 where the ambient temperature is –48 °C. Its TAS is:

Solution 15.2
Absolute air temperature = +273 – 48 = +225A

Mach Number = TAS ÷ LSS. Therefore, TAS = Mach number × LSS
LSS = $38.94\sqrt{A}$ = $38.94\sqrt{225}$ kt = 584 kt.
TAS = 0.81 × 584 kt = **473 kt**

15.3 The Shockwave

15.3.1 Compressibility

An aeroplane flying at low, subsonic, speeds causes air-pressure changes as a result of its movement but they are relatively small. It is convenient, therefore, to treat the air as though it could not be compressed. However, the air-pressure changes caused by the movement of an aeroplane at high speed, close to the speed of sound, are considerable and cannot be ignored. The air is compressible and its effect on the aeroplane must be accounted for when it is travelling at high speed.

An aeroplane moving through the air creates a pressure wave around it, which is propagated away from it in all directions simultaneously at the speed of sound. For an aeroplane flying at a relatively low speed, less than the speed of sound, the air-pressure wave ahead of the aeroplane can move away from the aeroplane. Consequently, air-pressure, air-density, air-temperature and air-velocity changes take place ahead of the leading edge of the wings and are a gradual process.

If, however, the aeroplane is travelling at the speed of sound then the air-pressure wave ahead of the aeroplane is unable to move away from the aeroplane and a wave of compressed air builds up at the leading edge of the wing. This means that the rise of static air pressure, air density, air temperature and the local speed of sound together with the decrease of the Mach number and total pressure take place rapidly almost instantaneously across the depth of the compression wave. This phenomenon is often referred to as the normal shockwave or the shock front, which is positioned at ninety degrees to the direction of travel of the aeroplane. The air behind the shockwave travels at subsonic speeds. The compression of a normal shockwave is greater than that of an oblique shockwave at the same Mach number.

15.3.2 Shockwave Formation

To an observer on the ground there is no perceptible noise ahead of the aircraft but as the shockwave passes overhead, at the aeroplane's groundspeed, there is an intense 'bang' followed by the normal aircraft noise. A second 'bang' may then be heard that originates from the shockwave formed around the wings and fuselage and is explained later in this chapter.

An aeroplane does not have to be flying at the local speed of sound for the airflow over the wings to be supersonic. A normal shockwave can develop at any time the aeroplane is in the transonic speed range. The function of the wing is to accelerate the air passing over its upper surface. The airflow may be subsonic at the leading edge of the wing but can be accelerated to become supersonic by the trailing edge of the wing, despite the fact that the aeroplane is flying at a speed of less than Mach 1.0. This is transonic flight, which usually occurs between Mach 0.75 and Mach 1.2.

At speeds less than Mach 0.75 all of the airflow over the wing is subsonic and as the aeroplane accelerates the first shockwave will form on the upper surface at the wing root. The onset of the formation of the shockwave is delayed if the wings have a low thickness/chord ratio. Such a wing also diminishes the large transonic variations of C_L and C_D. A normal shockwave can occur at several different points at the same time on an aeroplane in transonic flight, whereas at speeds in excess of Mach 1.2 all of the airflow around the aeroplane is supersonic.

When the air over the upper surface of the wing becomes sonic, a shockwave forms because the pressure waves over the rear of wing at the wing root are attempting to move forward and meet the air moving aft from the leading edge at exactly the same speed but in the opposite direction. The meeting point is usually just aft of the point of maximum camber because this is where the air over the upper surface of the wing is accelerating at the greatest rate.

The meeting point of the two air-pressure waves is the shockwave and is where the airflow changes from being sonic or supersonic to become subsonic. *Air passing through the shockwave experiences an increase of pressure, temperature and density.* If the aeroplane speed continues to increase the area of supersonic air on top of the wing spreads backward towards the trailing edge, resulting in the shockwave moving aft. See Figure 15.1(a).

The curvature of the underside of a wing is usually less than that of the upper surface, therefore the acceleration of the airflow is less, consequently the airflow does not become sonic until the aeroplane is travelling at a higher airspeed. Therefore, at lower airspeeds the only shockwave occurs on the upper surface and first appears at the wing root. At speeds above this, a shockwave forms beneath the lower surface of the wing ahead of the upper surface shockwave. The meeting point of the two pressure waves occurs nearer the trailing edge. See Figure 15.1(b).

If the aeroplane continues to accelerate then both the upper surface and the lower surface shockwaves will move aft. When the aeroplane's speed reaches Mach 1 both shockwaves reach the trailing edge of the wing together. See Figure 15.1(c). The profile of the pressure distribution over the upper surface of the wing during acceleration is irregular but when all of the airflow is supersonic the profile becomes rectangular.

As the speed increases to approximately Mach 1.05 a second shockwave forms ahead of the leading edge of the wing and is maintained in this position by the higher stagnation pressure. This shockwave is known as the ***detached bow wave*** and remains just ahead of the leading edge until further acceleration, to approximately Mach 1.3, causes it to attach itself to the leading edge to become the ***attached bow wave***. Both the leading-edge and the trailing-edge shockwaves remain in their positions irrespective of any further speed increase. If the speed does increase further then both shockwave profiles will just become more oblique. See Figure 15.1(d).

15.4 Air-Pressure-Wave Patterns

The changing pressure-wave pattern with the increasing speed of an aeroplane is three-dimensional and is shown in Figure 15.2. The speed of propagation in the diagram is 'a' and is equal to the local speed of sound. The vector 'v' shows the speed of the aeroplane. There are three speeds depicted; subsonic, sonic and supersonic.

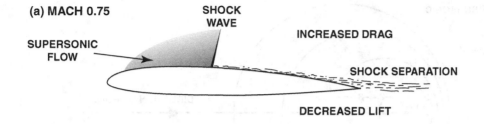

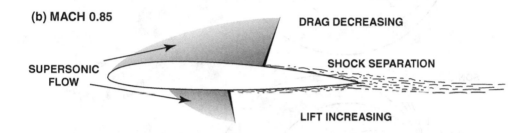

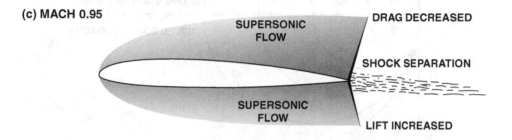

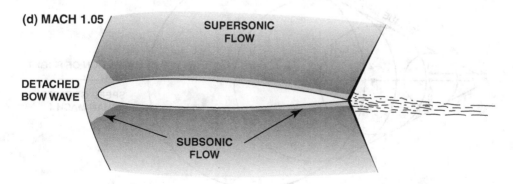

Figure 15.1 Shockwave Formation.

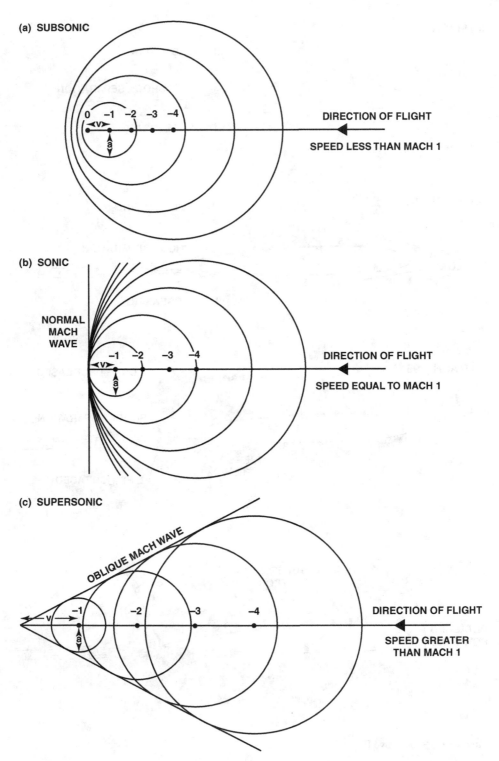

Figure 15.2 Air-Pressure Waves.

15.4.1 Subsonic

The length of the vector 'v' in Figure 15.2(a) is less than the length of 'a', therefore, although the pressure waves are compressed ahead of the aeroplane, they maintain their separation.

15.4.2 Sonic

In Figure 15.2(b) the length of vector 'v' is equal to the length of 'a.' This shows that the air-pressure waves do not move ahead of the aeroplane but compress to form a straight-line Mach wave at right angles to the direction of movement of the aeroplane and is referred to as a 'normal shockwave.' It is a plane of discontinuity normal to the local airflow direction at which the compression of the air and the loss of total pressure are greater than those of an oblique shockwave.

The *loss of total pressure in the shockwave is due to the conversion of the flow kinetic energy into heat energy*. This effect can be minimized if the airflow just ahead of the shockwave is only just supersonic. The total temperature of a normal shockwave is higher than that of the oblique shockwave.

The air behind the shockwave is subsonic and, although there are large variations to the lift and drag, the direction of the airflow is unchanged. This can cause variations to the pitching moment and to the trim of the aeroplane; it may also affect the operation of the controls. Shock-induced separation can occur behind a strong normal shockwave independent of the angle of attack.

15.4.3 Supersonic

The length of 'v' in Figure 15.2(c) is longer than the radius 'a', indicating that the aeroplane is travelling faster than the speed of sound. This produces a boundary beyond which no pressure wave passes; this is the 'oblique Mach wave.' It is a three-dimensional boundary and is therefore conic in shape and within which all disturbances caused by the aeroplane's movement are contained. The aeroplane centre of pressure is further aft than at subsonic speeds and the speed of the airflow after passing through an oblique shockwave is greater than that of the speed of sound. The air temperature increase is less than a normal shockwave.

15.5   The Shockwave Deflection Angle

The angle that the oblique shockwave makes, relative to the direction of flight, is determined by the forward speed of the aeroplane. The sine of the semi-apex angle of the cone is equal to the speed of sound (which is 1) divided by the actual Mach number. Therefore, the faster the aeroplane travels the smaller the shockwave apex angle becomes. For example, for an aeroplane travelling at Mach 2 then the sine of the semi-apex angle would be equal to $1/2$. This would produce a pressure wave cone with a semi-apex angle equal to 30° to the direction of flight. See Figure 15.3. Therefore, the whole apex angle of the cone is 60°. Examples of this are shown in Table 15.2.

The airflow speed behind the shockwave has a reduced Mach number, but if the wave whole apex angle is less than 70° the airflow speed will remain supersonic. The direction of flow behind the shockwave is turned through an angle equal to the deflection angle of the obstruction causing the shockwave.

Table 15.2   Shockwave Deflection Angle.

Actual Mach Number	Semi-Apex Deflection Angle
1.00	90°
1.15	60°
1.41	45°
2.00	30°

FLIGHT AT MACH 2

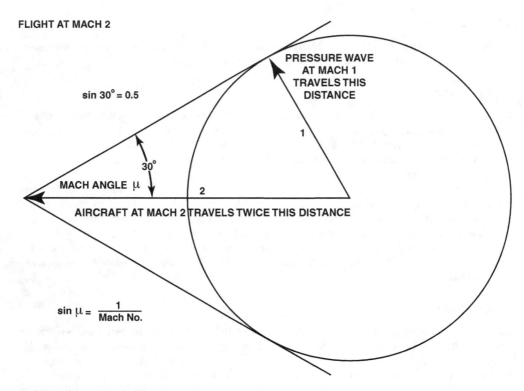

Figure 15.3 Deflection Semi-Apex Angle at Mach 2.

15.6 The High-Speed CP

The position of the centre of pressure (CP) during transonic flight is determined by the angle of attack and therefore is speed dependent. At Mach 0.75 the CP is located at approximately 20% of the length of the chordline from the leading edge of the aerofoil. As the speed increases the CP moves progressively aft, increasing the longitudinal static stability and requiring an increasing amount of nose-up pitch input of the stabiliser, at Mach 1.0 the CP is at approximately 45% of the MAC from the leading edge, until at Mach 1.4 the CP positioned at approximately 50% of the length of the chordline from the leading edge of the aerofoil.

For swept-wing aeroplanes this rearward movement of the CP restricts their maximum speed. Transferring fuel between tanks to move the CG can generate a correcting moment that counteracts the rearward movement of the CP. However, this method is limited because as the flight progresses and the fuel is used, although the aeroplane mass decreases enabling it to travel faster, there is less fuel available for transference to move the CG and counteract the effect of the movement of the CP.

15.7 Critical Mach Number (MCRIT)

The critical Mach number (MCRIT) is the lower limit of a speed band, known as the transonic range, in which the local Mach number may be either subsonic or supersonic. Its value varies with the angle of attack and is the free-stream Mach number at which sonic speed is reached over the upper surface of a wing. In other words, it is the speed at which the local Mach number is equal to the free-airstream Mach number. It is also the highest speed possible without supersonic flow over the wing. For example, if the air passing over a wing at some point reaches Mach 1, but the wing is only moving at Mach 0.8 then the critical Mach number is 0.8.

Generally, a thicker wing with a large amount of camber has a lower critical Mach number than a thin wing with little camber because the airflow over its upper surface accelerates to a higher speed than it would over a thin wing. At the design stage of an aeroplane the inclusion of the following characteristics either individually or in any combination will increase the value of the critical Mach number but the largest increase will result from a combination of **a** and **b**:

a. A swept wing having the same wing area and loading.
b. A thin aerofoil cross-section (low thickness to chord ratio).
c. Area ruling. (Described later in this chapter).

The critical Mach number is inversely proportional to the angle of attack, i.e. MCRIT decreases with increasing angle of attack and vice versa. This is because, as the angle of attack increases the local peak velocity of the airflow over the wing also increases. It therefore reaches Mach 1 sooner than it would have done at a lower angle of attack, thus the critical Mach number is less.

As a flight progresses and fuel is burnt the aeroplane mass decreases and it is able to accelerate to a higher maximum speed. If the pilot maintains level flight and does not touch the throttles the aeroplane will continually accelerate and the angle of attack must be continually decreased to allow for the increased IAS. In such a case, the airflow peak velocity will not be attained until the aeroplane is at a higher maximum speed. Thus, the critical Mach number increases continuously throughout a flight assuming the aeroplane is permitted to accelerate in level flight and the angle of attack is adjusted accordingly.

If, in level flight, the pilot maintains a constant IAS as the aeroplane mass decreases then the angle of attack decreases for that speed but not as much as it would have done had the aeroplane been allowed to accelerate. As a result, as the flight progresses, the angle of attack diverges more from that appropriate to the maximum speed for its mass, and although the critical Mach number will increase it is still less than it could be.

At speeds in excess of the critical Mach number, a swept-wing aeroplane loses lift in the area of the wing roots, which decreases stick-force stability and produces a sudden increase to the drag coefficient. As a result, the aeroplane will experience buffeting, the first evidence of the formation of a shockwave on the upper surface of the wing at the wing root, and a tendency to pitch nose-down.

During acceleration the pressure-distribution pattern on the upper surface of the wing becomes irregular as the airflow becomes supersonic at MCRIT. However, with continued acceleration, when the whole airflow becomes supersonic the pressure pattern steadies and becomes a geometrical parallelogram, some refer to as a rectangular pattern.

15.8   The Effect of a Shockwave

15.8.1   Wave Drag

Shockwave formation on the upper surface of a wing causes the airflow to separate the boundary layer from the surface of the wing behind the shockwave. This creates an adverse pressure gradient towards the trailing edge of the wing that not only diminishes lift but is also the origin of a powerful type of drag known as 'wave drag.' Such drag absorbs a considerable amount of the thrust available, increases the fuel flow, decreases the specific air range and reduces the safe endurance. Flying at subsonic speeds can minimize wave drag.

The speed at which this phenomenon occurs is the critical drag Mach number, which is approximately 1.1 to 1.15 times the critical Mach number. Wave drag cannot happen at subsonic speeds, but is particularly important at MCRIT because it results in a marked decrease of the lift/drag ratio in the transonic range. See Figure 15.1(a).

As the aeroplane accelerates the shockwave moves aft, the drag begins to decrease and the lift starts to increase. See Figure 15.1(b). At a speed of Mach 1 the drag has decreased further, a bow wave forms and the lift has increased further. However, the lift generated at Mach 1 is less than it would be for the same angle of attack at subsonic speeds because the bow wave decreases the value of C_L at supersonic speeds. See Figure 15.1(c). In supersonic flight, as shown in Figure 15.1(d), the pressure-distribution pattern around the aerofoil is rectangular.

The adverse effects of the shockwave are minimized if the waves are kept as small as possible. This is achieved by designing the wings such that they are thin with little camber, to reduce the pressure gradient, and have a sharp leading edge, to decrease the bow wave. The use of vortex generators mounted well aft of the leading edge on the upper surface of the wing decreases shockwave-induced boundary-layer separation and improves the response of trailing-edge controls behind the shockwave. The disadvantages experienced because of the shockwave are diminished if the 'area rule' (described later in this chapter) is used in the aeroplane design. The effects of a shockwave are summarized in Table 15.3.

Table 15.3   The Effects of a Normal or Oblique Shockwave.

Consideration	Factor Value After Passing through a Normal or Oblique Shockwave
Air Temperature	Increased (due to compression)
Static Pressure	Increased (due to compression)
Total Pressure	Decreased (Normal shockwave suffers greater loss than an oblique shockwave)
Air Density	Increased (Oblique shockwave has less compression)
LSS	Increased (due to the temperature increase)
Speed of Airflow	Decreased to less than Mach 1 (normal shockwave) Decreased but still above Mach 1.0 (oblique shockwave).

If an aeroplane not designed to operate at transonic or supersonic speeds did fly at these speeds, then such an occurrence could adversely affect the stability of the aeroplane to such an extent that control may be extremely difficult or even impossible.

15.8.2   Drag Divergence Mach Number

It can be seen in Figure 15.4 that wave drag increases the total drag up to the passage of the shockwave after which the total drag rapidly decreases as the speed is increased further. The Mach number at which the aerodynamic drag of an aerofoil or airframe commences to rapidly develop with increased speed is the drag divergence Mach number. Wave drag can cause the coefficient of drag to increase to ten times its low-speed value.

Because the drag divergence Mach number is always greater than, but still close to, the critical Mach number it is a transonic effect. In Figure 15.4 it is shown at 0.78 Mach but it may be higher for aeroplanes having wings with a large degree of sweepback. However, a wing specifically shaped to make the drag divergence Mach number as high as possible and at the same time decrease the total drag at high subsonic and low transonic speeds is the 'supercritical wing,' which is described later in this chapter. Total drag at speeds in excess of Mach 1.0 comprises induced drag, parasite drag and wave drag.

15.9   The Flying Controls

The deflection of a conventional hinged-flap type of control creates a force ahead of the control surface that is normally greater than that which the control surface itself produces. When there is a shockwave

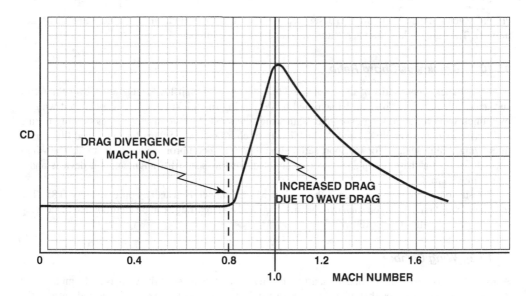

Figure 15.4   Drag Divergence Mach Number.

established on the lifting surface ahead of the control surface this is no longer the case because pressure variations cannot be transmitted forward through the shockwave.

The shockwave moves slightly aft when an aileron is deflected downwards. This effectively diminishes the result of deflecting the control surface because only the airflow passing over the control surface is modified. This effect is exacerbated when the shockwave reaches the trailing edge of the aerofoil and the control effectiveness is decreased even further. In transonic flight the ailerons are less effective than in subsonic flight because aileron deflection only partially affects the pressure distribution around the wing.

These characteristics have led to the substitution of the tailplane and elevator combination in high-speed aeroplanes with a fully flying horizontal stabiliser. On these aeroplanes the conventional wing-tip ailerons are replaced with differential spoilers and/or inboard ailerons that are for use when the aeroplane is flying at transonic speeds. They are fitted partially to combat the adverse effects just described but primarily to avoid the adverse effects of wing twist that can negate or even reverse the effect of conventional ailerons when deflected at high speed. The complete reversal of the aileron controls occurs at high speeds generally outside of the flight envelope for most aeroplanes. However, the effect of aeroelastic-distortion, the combination of decreased wing rigidity and high airspeed, could manifest itself as a reduced rate of roll. See Chapter 3.

The movement of the shockwave rearward to the trailing edge of the aerofoil during transition to supersonic speeds means that the shockwave must transit the control surface. Any deflection of the control surface during this transition will move the shockwave, which results in significant changes in the location and magnitude of the aerodynamic forces acting on the control. Consequently, the control surface will experience a high frequency vibration referred to as 'buzz.'

Because of the increased input forces required from the pilot during flight at transonic speeds and the large variation of the stick forces encountered during the transition of the shockwave over the control surface, most high-speed aeroplanes are fitted with fully power-operated controls. (See page 223.)

15.10   The Effect of the Aerofoil Profile

The thickness of an aerofoil has a significant effect on its performance at transonic speeds. To minimize the adverse effects of the shockwave during transonic flight it is essential for the aeroplane to have a high critical Mach number. There are two design features that if incorporated, will increase the critical Mach number. They are the low thickness/chord ratio and low wing camber.

15.10.1 Thickness/Chord Ratio

A low thickness/chord ratio of the aerofoil section decreases the airflow acceleration over the upper surface of the aerofoil. This can be achieved by decreasing the thickness of a short-chorded wing or by a thick wing having a long chordline. Thin wings are subject to the disadvantages of wing flexure and twisting, low maximum C_L and the inability to fit chord-extending flaps for low-speed flight. But, thick wings are unsuitable for transport aeroplanes; therefore a compromise has to be used. The wings normally used for these aeroplanes incorporate adequate low-speed lift augmentation devices and a higher maximum C_L.

For subsonic transport aeroplanes the critical Mach number is usually approximately 0.85. The thickness/chord ratio has to be such that the fuel, flaps and undercarriage can be incorporated in the wing design and is usually 10% to 15%.

15.10.2 Wing Camber

A low camber is most beneficial, particularly if the maximum thickness of the wing is at the midpoint of the chordline. This ensures that when the shockwave forms it is less intense and is located towards the trailing edge. It also reduces the surface area behind the wave that is affected by the turbulence.

15.11 Swept Wings

The primary reason that the swept-wing design is used for most jet transport aeroplanes is to increase the value of the critical Mach number for that type of aeroplane. It is the lowest speed of the free airflow that when passing over some part of the aeroplane becomes supersonic. Usually, it is the upper surface of the wing, over which the airflow accelerates to a speed of Mach 1.

Therefore, a swept wing delays the onset of the effects of compressibility and delays the airflow from becoming supersonic. It is best employed for aeroplanes that operate in the transonic regime of flight because the sweepback necessary to delay the drag rise at extremely high Mach numbers or continuous flight in the supersonic regime is too great to be practical.

15.12 The Effect of Sweepback

The critical Mach number of a wing of a given thickness/chord ratio and aspect ratio can be increased by including a high-angled sweepback in the design. The angle of sweepback of such wings is limited by the practicality of their construction. Although it is assumed that any sweepback is better than none, to be of any significant value the sweepback should be at least 30°. Nevertheless, the inclusion of a relatively small angle of sweepback in the design of any wing increases the critical Mach number.

15.12.1 The Advantages of Sweepback

The advantages of an aeroplane having swept-back wings are:

a. M_{CRIT} is increased in direct proportion to the sweep angle.
b. C_D is decreased in direct proportion to the angle of sweep.
c. Drag divergence is delayed to a higher speed.
d. Static directional stability is improved. (See Chapter 11.)
e. Static lateral stability is improved in a similar way to dihedral. (See Chapter 11.)
f. For a given aspect ratio and wing loading, the aeroplane is less sensitive to gusts than a straight wing.

15.12.1.1   Increased M_{CRIT}

It is the airflow at right angles to the leading edge of a wing that determines the magnitude of the pressure distribution around the wing and thus the amount of lift developed. Consequently, the critical Mach number for the wing is determined by this component of the airflow speed over the upper surface of the wing. The component normal to the leading edge of the wing is equal to the true airspeed of the free airflow multiplied by the cosine of the angle subtended between the direction of the free airflow and the normal to the wing leading edge.

An example of this feature is shown in Figure 15.5(a); if the free-flowing airflow speed is Mach 0.75 and the aeroplane has a straight leading edge then the acceleration over the upper surface of the wing would produce a speed of approximately Mach 0.80. If the same airspeeds were experienced by an aeroplane with a swept wing with 30° of sweep, see Figure 15.5(b), then the flow perpendicular to the leading edge of the wing would be Mach 0.8 × cos 30° = Mach 0.69. This reduction of the true airspeed (M0.80 – M0.69) is equivalent to 11% of the LSS.

In a standard atmosphere at 30 000 ft the temperature is –45 °C and the local speed of sound is 589 kt. A reduction of 11% at this altitude is equal to 65 kt. This means that an aeroplane with 30° of sweep can fly 65 kt faster before the critical Mach number is reached than a straight-wing aeroplane having the same wing area and wing loading.

However, the swept-wing aeroplane will generate less lift than a straight-wing aeroplane having the same wing area; this loss can be partially regained by increasing the angle of attack. Theoretically, in this example, M_{CRIT} for the swept-wing aeroplane should be equal to the M_{CRIT} of the straight-wing aeroplane multiplied by 1/cos 30°. In other words, theoretically M_{CRIT} for the swept-wing aeroplane is 15.5% higher than the M_{CRIT} for the straight-winged aeroplane, but in practice the increase actually achieved is closer to 8%.

In Figure 15.5(b) the total airflow over the upper surface of the wing can be divided into the following components:

a. Perpendicular to the leading edge of the wing = Upper surface airflow speed × cosine of the angle subtended between the longitudinal axis and the perpendicular to the wing leading edge. This component determines the value of the critical Mach number.
b. Parallel to the leading edge of the wing = the speed of movement of the airflow towards the wing tip, i.e. the spanwise flow = Upper surface airflow speed × sine of (the angle subtended between the longitudinal axis and the perpendicular to the wing leading edge). This component determines the rate at which the boundary layer will build up at the wing tip.

15.12.1.2   Aerodynamic Effects

Further effects that a swept wing has on the performance of an aeroplane are summarised in Table 15.4.

a. Drag divergence Mach number and the peak drag rise Mach number increase because the speed component affecting the pressure distribution is less than that of the free-stream velocity. The peak drag rise is delayed to approximately that speed normal to the leading edge that produces sonic flow.
b. Any change to C_L, C_D or C_M is decreased in magnitude due to the effect of compressibility.

15.12.2   The Disadvantages of Sweepback

Despite their advantages, swept wings have the following disadvantages:

a. Trailing-edge controls, such as flaps and wing-tip ailerons, are less effective because they are not at right angles to the airflow. Some flap systems only produce an increase of lift of 50% of that which would have been produced by the same flap on a straight-winged aeroplane.
b. A swept wing of the same wing area and aspect ratio as that of a straight-winged aeroplane has a greater wing span, which increases its mass. This causes greater bending and stress towards the

wing tip. It is also subject to the twisting effect of the wing in high-speed flight that diminishes the effectiveness of wing-tip ailerons.

c. When combined with taper there is a strong tendency for the wing to tip stall first. This is because, although taper produces a strong local lift coefficient towards the wing tip similar to sweepback, there is a strong spanwise flow of the boundary layer towards the wing tip, particularly at high angles of attack, that results in a low-energy pool at the wing tip which easily separates from the wing surface.

Table 15.4 The Theoretical Approximate Effects of Sweepback. (Assuming a wing of moderate aspect ratio in transonic flight.)

Sweep Angle (A)	Theoretical Approximate Percentage Change			
	Increase to M_{CRIT}	Increase to Peak Drag Mach Number	Decrease to Drag Rise	Decrease to Reduction of C_{Lmax}
0°	0%	0%	0%	0%
15°	3.5%	5%	5%	3.5%
30°	15.5%	15.5%	15%	15%
45°	41.0%	41%	35%	30%
60°	100%	100%	60%	50%

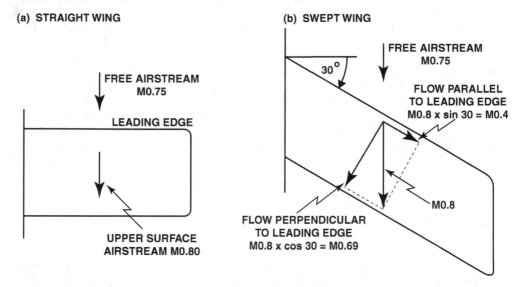

Figure 15.5 The Effect of Wing Sweep.

15.13 Remedial Design Features

To overcome some of these disadvantages some jet transport aeroplanes having a swept-wing design are fitted with two sets of ailerons, outboard are the low-speed ailerons and inboard are the high-speed ailerons. See Figure 15.6.

15.13.1 Low-Speed Ailerons

The **low-speed ailerons**, mounted just inboard of the wing tip, are provided for use at the speeds experienced during take-off, take-off climb and landing. They are effective for such use because of the large aerodynamic force couple they generate, due to their long moment arm, produces the required rate of roll at these low speeds. These ailerons are de-activated when the slats and flaps are retracted or the speed is above a predetermined value during high-speed flight.

15.13.2 High-Speed Ailerons

The **high-speed ailerons** are used for the cruise portion of a flight because, being mounted inboard towards the wing root where the wing is thicker they do not cause twisting of the wing. They also have the advantage of generating a smaller aerodynamic force couple, which is essential to produce a lower roll rate. High rates of roll are not required at high speeds. Inboard ailerons are less effective at transonic speeds than at subsonic speeds because their deflection only partially affects the pressure distribution around the wing. These ailerons are locked during low-speed flight.

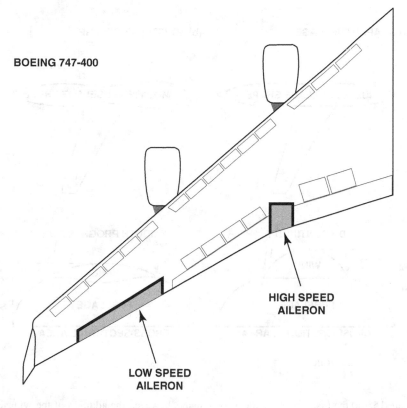

BOEING 747-400

HIGH SPEED
AILERON

LOW SPEED
AILERON

Figure 15.6   Swept-Wing Ailerons.

15.14   Area Rule

At the design stage of an aeroplane the name given to the principle adopted when matching parts of an aeroplane together so that they produce the minimum total drag is the 'area rule.' The area referred to is the cross-sectional area of the fuselage, wings and tailplane. The shape of each component of the aeroplane is not as critical in the creation of drag as is the rate of change of the shape.

It was found from wind-tunnel tests that to reduce the number and magnitude of the shockwaves, the perfect aerodynamic cross-sectional shape is a smooth cigar shape that is pointed at both ends. The area rule states that provided the cross-sectional area of the aeroplane is similar to that of the ideal shape, the wave drag of the aeroplane will be similar to that of the ideal shape, irrespective of its actual shape. Wave drag is related to the curvature of the distribution of the volume of an aeroplane. The smoother the curvature the less is the magnitude of the wave drag.

Large volumes of the airframe, such as the wings, are usually positioned at the widest part of the fuselage and other necessary features such as the cockpit, the tailplane and engine intakes have to be spread out along the fuselage to maintain the cross-sectional area distribution as near perfect as possible.

For transonic aeroplanes it was discovered that indenting the fuselage to produce a waist and adding more volume to the rear of the aeroplane minimized transonic wave drag. Thus, the volume of the fuselage has to be decreased wherever a projection such as the wings and tailplane are positioned so that the cross-sectional area distribution changes smoothly with no discontinuities. This is sometimes referred to as the 'coke bottle' effect.

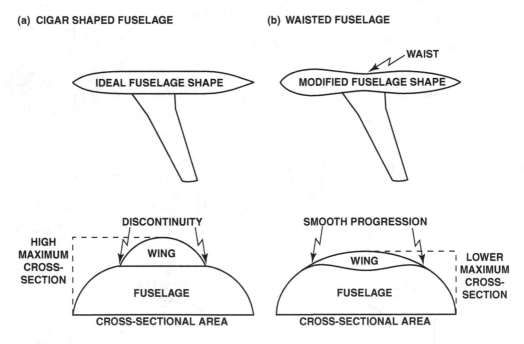

Figure 15.7 The Area Rule.

In Figure 15.7(a) the fuselage is the ideal cigar shape, however, the addition of the wings creates a cross-sectional area that has a high maximum point and two discontinuities that will create an enormous amount of wave drag. By modifying the shape of the fuselage to include a waist to account for the additional volume of the wings, as shown in Figure 15.7(b), the area cross-section has a lower maximum and a smooth progression of area reducing the wave drag considerably. But this shape causes production difficulties.

As an alternative to waisting the fuselage at the wing root, extra large bulging nacelles can be added to the trailing edge of the wing, this assists in the smooth transition of the cross-sectional area distribution along the fuselage; this characteristic is particularly evident on the TU 95. Airbus has also introduced a similar design feature by housing the flap tracks in large protruding fairings at the trailing edge of the wings, thus bringing it closer to the ideal cross-sectional area distribution. Not only do the fairings achieve this purpose but they also reduce the strength of the shockwave by almost eliminating the shock separation of the airflow over the rearward half of the upper surface of the wing.

Boeing also utilised the area-rule principle when designing the 747. The hump, which encompasses the cockpit and upper passenger deck, was introduced to increase the forward fuselage cross-sectional area and so smooth the volume distribution over the whole length of the aeroplane. This feature delayed the onset of the transonic wave drag and improved its cruise performance to such an extent that it normally operates at a higher speed than most other transport aeroplanes despite its size.

The ideal shape for supersonic aeroplanes is determined by the angle of the Mach cone and is biased rearward. Thus, the wings are located back further along the fuselage so that the shape of the aeroplane remains within the Mach cone.

15.15   High-Speed-Flight Characteristics

15.15.1   High-Speed Buffet

High-speed buffet is the rapid movement of the control surfaces caused by turbulent airflow passing over them. It can only occur when the Mach number is above the critical Mach number and is likely to be accompanied by a tendency for the aeroplane to pitch nose-down. The maximum operating altitude of an aeroplane can be restricted by the load factor. Above that altitude any turbulence may induce Mach buffet, which could cause the aeroplane to exceed the maximum permitted load factor.

The airflow behind a shockwave is turbulent and can cause a rapid vibrating movement of the horizontal stabiliser surfaces and is often referred to as the 'shock stall' and is the primary cause of high-speed buffet. A secondary cause of the rapid movement of the control surfaces, especially with supercritical aerofoil sections, is when the wing reaches a speed of MCRIT the relatively weak shockwave that forms moves rapidly backward and forward over the surface of the wing producing similar indications to those of the shock stall.

15.15.2   Tuck Under

The marked nose-down pitching tendency often experienced by aeroplanes accelerating from high subsonic to transonic speeds is known as 'tuck under' and usually occurs at a relatively high Mach number just above MCRIT. It can only happen when the speed is greater than the critical Mach number, and is caused by any one of the following three reasons:

a. The position of the shockwaves on the upper and lower surfaces of the wing do not coincide when flying at a speed between MCRIT and Mach 1. The lower shockwave is forward of the shockwave on the upper surface, which causes the CP to move aft. This creates a nose-down pitching moment. See Figure 15.1(b).
b. For swept-wing aeroplanes the shockwave forms on the thick inboard wing roots first, thus decreasing the lift on this part of the wing. This causes the CP to move aft because of the sweepback and produces a nose-down pitching moment.
c. Due to the shockwave, the loss of the airflow downwash over the horizontal stabiliser decreases its downward balancing force, which results in a nose-down pitching moment.

15.15.3   The Shock Stall

In the transonic range, the 'shock stall' is the separation of the boundary layer from the upper surface of the wing behind a shockwave. This causes the total lift generated to decrease. Unlike the normal high-incidence angle stall, the 'shock stall' occurs at low angles of attack and at an unexpectedly high IAS. For swept-wing aeroplanes because the wing-tips stall first the aeroplane has a nose-up pitch tendency.

The angle of attack of the high-speed shock stall decreases as the Mach number increases and C_{Lmax} *decreases in a similar manner.* Because the turbulent airflow behind a shockwave results in an increased loss of lift, any large rapid increase of speed can induce a premature stall.

C_{Lmax} at high subsonic speeds is less than that theoretically possible with an incompressible flow and occurs at a lower speed. The maximum velocity of the airflow over the upper surface of the wing increases with increased angle of attack; which decreases the value of MCRIT. This fact has great significance when considering manoeuvring at high speed because a shockwave is induced at a lower Mach number than in level flight.

15.15.4 The Buffet Boundary

The normal high-incidence angle stalling speed for an aeroplane is specified as an IAS. When operating at high altitude, in the extremely low ambient temperatures present at those altitudes, the TAS at which a high-incidence angle stall, for a normal, clean aeroplane will occur is very high. By the same reasoning the local speed of sound, Mach 1, in the temperatures experienced at high altitudes in terms of TAS is low; as a result the value of the MCRIT TAS is also low.

The cruising TAS of most jet transport aeroplanes operating at high altitude is just above MCRIT. The difference between the normal high-incidence angle stalling speed and MCRIT can be as little as 20 kt. The 'buffet boundary' is the speed range between the high-speed buffet and the low-speed buffet, which decreases with increased altitude. The separation of the cruising speed from the buffet-onset speed is the 'buffet margin.' There is a low-speed margin and a high-speed margin. See Figure 13.4.

15.15.5 Coffin Corner

For any given mass and 'g' loading there is one altitude, which is temperature dependent, at which the high-incidence stalling speed and the critical Mach number are equal, this is colloquially known as 'coffin corner' or 'Q corner.' It refers to the apex of a triangular shape at the top of the flight envelope chart where the stalling speed and critical Mach number lines converge to a point.

The colloquial name alludes to the fact that at this altitude the pilot is flying on a knife-edge that demands great skill. If the speed is reduced the aeroplane will stall and lose altitude and if the pilot increases the speed of the aeroplane it will lose lift, because of the shockwave and flow separation that will cause the aeroplane to violently pitch nose-down and lose altitude. In other words, 'tuck under' occurs.

When flying at the 'Q' corner altitude, if an aeroplane turns, the inner wing slows down and could fall below the low-speed stalling speed, whilst the outer wing speeds up and could exceed MCRIT. If such is the case then the aeroplane will have exceeded both limits at the same time. Alternatively, turbulence at high altitude could result in the aeroplane exceeding the 'g' limitations.

15.16 Speed Instability

The centre of pressure moves aft initially because of the aft movement of the shockwave and creates a nose-down tendency; this usually occurs from M0.80 to M0.98. The nose-down pitch causes the aeroplane to accelerate and this creates a further nose-down pitch, which further exacerbates the situation and continues the acceleration. Thus, the speed is unstable.

The rearward movement of the CP in the transonic range continues as the aeroplane accelerates, causing a marked increase to the longitudinal stability. If corrective action is not taken the CP continues to move rearward beyond the transonic envelope making the situation extremely dangerous. Manual recovery from this situation is extremely difficult.

The effect a nose-down pitch angle with increased speed has on the handling characteristics of an aeroplane is such that at a particular Mach number the aeroplane's speed becomes unstable. The large amount of up deflection of the elevator that is required to maintain straight and level flight decreases the manoeuvring capability of the aeroplane.

To prevent such a situation arising, every aeroplane designed to operate in the high subsonic or in the transonic speed range is fitted with a Mach trimming device. This system automatically moves the elevators to counteract the pitching moments expected at any given Mach number by increasing the nose-up pitch. It is therefore very sensitive to the Mach number. Such a system maintains the aeroplane's longitudinal stability and is known as the Mach trimmer.

15.16.1 The Mach Trimmer

The function of a Mach trimmer is to minimize the adverse effects of rearward movement of the CP at high Mach speeds. It compensates for the nose-down pitching moment experienced by aeroplanes in the transonic speed range. To do this it corrects the change to stick-force stability of swept-wing aeroplanes when flying above a specific Mach number by decreasing the elevator (stabilizer) incidence by an amount greater than that required for the longitudinal trim change.

In so doing the Mach trimmer ensures positive stability and corrects insufficient stick-force stability at high Mach numbers. It adjusts the elevator according to the Mach number and prevents 'tuck under.' In other words, it maintains the stick-force gradient required to sustain level flight as the speed increases, which for aeroplanes fitted with power-operated controls is approximately 1 pound for every 6 kt change of speed in level flight.

A Mach trimmer should be fitted to aeroplanes that demonstrate unconventional elevator stick-force characteristics at transonic Mach numbers. In the event of this equipment failing, the speed is limited to specified Mach number.

15.16.2 Lateral Instability

The formation of a shockwave does **not** occur at precisely the same location on both wings or even at the same Mach number. Consequently, when the aeroplane attains M$_{CRIT}$, because of the asymmetry of the shockwave formation one wing will drop because it generates less lift than the other. This phenomenon causes lateral instability and will occur between M$_{CRIT}$ and the local speed of sound.

Lateral design features incorporated to increase the lift of the downgoing wing in a sideslip, such as dihedral and/or sweepback, cause a greater local acceleration of the airflow that increases the Mach number of the air passing over that wing. This results in the drag on that wing increasing and creates a shockwave or intensifies the shockwave if one already exists and exacerbates the situation.

15.17 The Supercritical Wing

A supercritical wing is one that has been designed with specific streamwise segments to delay the onset of the adverse effects of the shockwave, i.e. wave drag and its severity and decreased lateral stability, when the aeroplane is flying in the transonic speed range or just above M$_{CRIT}$. It increases the cruise Mach number, the range and the endurance of the aeroplane.

The main features of a supercritical wing are:

a. The upper surface is flattened but the lower surface is reflex cambered towards the trailing edge. Thus, the airflow over the upper surface is not accelerated as much as conventional aerofoils, which delays the formation of the shockwave to a higher speed. See Figure 15.8(a).

b. Compared with a conventional aerofoil section it has a larger leading-edge radius, a flatter upper surface and both positive and negative camber and enables a relatively thick wing to be used for approximately the same cruise Mach number. See Figure 15.8(b).

(a) CONVENTIONAL AEROFOIL

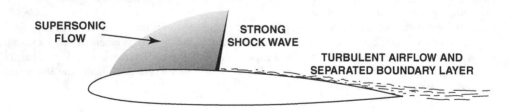

(b) SUPERCRITICAL AEROFOIL

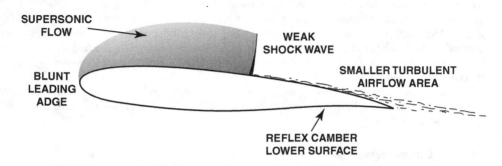

Figure 15.8   The Supercritical Wing.

c. The curved lower surface increases the lift generated by the aft end of the aerofoil and replaces some of the lift lost from the flat upper surface. Because of the rear-loaded lift generation it is often called an aft-loaded wing. See Figure 15.8(b).

d. The leading edge of the supercritical wing has a greater radius than traditional aerofoils that suffer from excessive wave drag and the loss of longitudinal stability (tuck under). By ensuring that its profile is carefully designed any expansion waves will be reflected from the subsonic boundary back to the shockwave, thereby weakening its intensity. See Figure 15.8(b).

15.18   Supersonic Airflow

15.18.1   The Convex Corner Mach Wave (Expansion Wave)

The expansion wave, sometimes called the Mach wave or a Mach line, is a pressure wave travelling at the speed of sound that has its smooth flow upset by a small pressure disturbance that creates a weak shockwave. It is a line radiating from a point, along which the *expansion results in decreased pressure, decreased density, decreased temperature, increased velocity and an increased Mach number behind the expansion wave*. This is expansive flow.

When a supersonic airflow encounters a sharp convex corner it allows the flow lines to spread out and expand, causing a small decrease of pressure and creates a Mach line. An expansion fan consisting of an infinite number of diverging expansion waves or Mach lines is formed and is centred on the sharp corner. If the corner is a smooth curve the expansion fan is centred at the centre of curvature.

The angle at which the Mach line intercepts the flow lines is the Mach angle. Each wave in the expansion fan turns the airflow in small steps until it flows parallel to the new surface. The final Mach line of the fan is at a more acute angle to the new direction of flow than the initial Mach line was to the original flow direction. In other words, the Mach angle is more acute after the change of flow direction. See Figure 15.9.

Across the expansion fan the airflow gradually accelerates and the Mach number increases, whilst the static pressure, ambient temperature and air density progressively decrease, but the change is infinitely small (i.e. the air pressure and density are greater ahead of the wave). Therefore, the process is one that does not involve a change to its capacity to undergo spontaneous change, it is an isentropic process. The stagnation properties of pressure, temperature and density remain constant. The effects caused by an expansion shockwave are summarised in Table 15.5.

Table 15.5 The Effects of an Expansion Shockwave.

Consideration	Factor Value Beyond Expansion Wave
Static Temperature	Decreased (due to expansion)
Static Pressure	Decreased (due to expansion)
Total Pressure	Increased
Air Density	Decreased (due to expansion)
LSS	Decreased (due to the temperature decrease)
Speed of Airflow (the Mach Number)	Increased

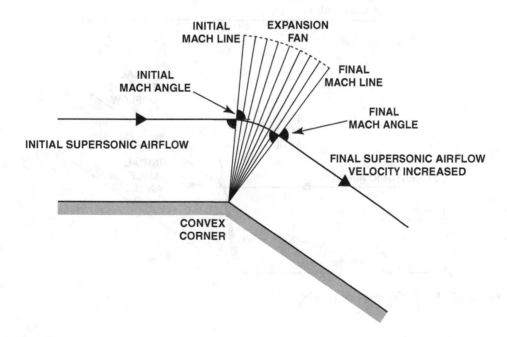

Figure 15.9 The Convex-Corner Expansion Wave.

The supersonic flow direction can be changed through a large angle by employing a series of expansion fans created by a succession of convex corners. A curved-surfaced corner achieves the same effect, which makes the changes even more gradual than at a single corner. Including convex curves in the design of an aeroplane not only reduces the number and magnitude of the shockwaves but also is beneficial in subsonic flight.

15.18.2 The Concave-Corner Shockwave

The concave-corner shockwave is caused by the convergence of an infinite number of Mach lines when a supersonic airflow encounters a concave corner and has the same properties as an oblique shockwave. At the shockwave the convergence results in a suddenly decreased flow speed and suddenly increased temperature, increased pressure and increased density. This is compressive flow. The shockwave produced is strong and makes a more acute angle with the final airflow direction than it does with the initial airflow direction. See Figure 15.10. The effects of a compressive shockwave are summarised in Table 15.6.

Table 15.6   The Effects of a Compressive Shockwave.

Consideration	Factor Value Beyond Compressive Shockwave
Static Temperature	Increased
Static Pressure	Increased
Total Pressure	Decreased
Air Density	Increased
LSS	Increased
Speed of Airflow (the Mach number)	Increased

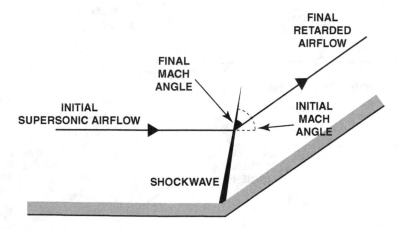

Figure 15.10   The Concave-Corner Shockwave.

Self-Assessment Exercise 15

Q15.1 The critical Mach number of an aeroplane is increased by:
  (a) swept wings
  (b) vortex generators
  (c) control deflection
  (d) wing dihedral

Q15.2 High-speed buffet is induced by:
  (a) boundary-layer separation due to shockwaves
  (b) boundary-layer control
  (c) expansion waves on the upper surface of the wing
  (d) movement of the CG

Q15.3 In transonic flight the ailerons are less effective than in subsonic flight because:
  (a) the air pressure is less behind the shockwave
  (b) aileron deflection downward moves the shockwave forward
  (c) aileron deflection only affects the air ahead of the shockwave
  (d) aileron deflection only partially affects the pressure distribution around the wing

Q15.4 At the same Mach number, the normal shockwave compared with the oblique shockwave has a:
  (a) greater expansion
  (b) smaller compression
  (c) smaller expansion
  (d) greater compression

Q15.5 When flying at a constant IAS as the mass decreases the M$_{CRIT}$:
  (a) is independent of the angle of attack
  (b) decreases
  (c) increases
  (d) remains constant

Q15.6 Given Mach number 0.8 and TAS 400 kt. The speed of sound is:
  (a) 500 kt
  (b) 320 kt
  (c) 480 kt
  (d) 600 kt

Q15.7 The function of the Mach trimming device is to minimize the adverse effects of:
  (a) compressibility on the stabiliser
  (b) movement of the CP when at high speed
  (c) increased drag due to shockwave formation
  (d) uncontrolled changes in the stabiliser setting

Q15.8 The flight regime between M$_{CRIT}$ and Mach 1.3 is the:
  (a) transonic range
  (b) supersonic range
  (c) hypersonic range
  (d) subsonic range

Q15.9 The increase of drag above the critical Mach number is due to:
  (a) wave drag
  (b) increased angle of attack
  (c) increased interference drag
  (d) increased skin friction

Q15.10 MCRIT is the free-stream Mach number that produces the first evidence of:
    (a) local sonic flow
    (b) buffet
    (c) shockwave
    (d) supersonic flow

Q15.11 After air has passed through a normal shockwave. Which of the following statements is correct?
    (a) The temperature increases.
    (b) The total pressure decreases.
    (c) The temperature decreases.
    (d) The velocity increases.

Q15.12 In the case of a supersonic flow being retarded by a normal shockwave, the loss of total pressure can be minimized if the Mach number in front of the shockwave is:
    (a) supersonic
    (b) less than 1
    (c) equal to 1
    (d) small but still supersonic

Q15.13 Which of the following flight phenomena can happen at Mach numbers below MCRIT?
    (a) Mach buffet
    (b) shock stall
    (c) Dutch roll
    (d) tuck under

Q15.14 When air has passed through a normal shockwave the Mach number is:
    (a) less than before but still greater than 1
    (b) equal to 1
    (c) higher than before
    (d) less than 1

Q15.15 The speed at which 'tuck under' will occur is:
    (a) only at the critical Mach number
    (b) only below the critical Mach number
    (c) above or below the critical Mach number depending on the angle of attack
    (d) only above the critical Mach number

Q15.16 The type of flow separation that occurs at the lowest angle of attack is the:
    (a) shockstall
    (b) high-speed stall
    (c) low-speed stall
    (d) deep stall

Q15.17 Compared with an oblique shockwave at the same Mach number, a normal shockwave has a:
    (a) lower static temperature
    (b) greater loss in total pressure
    (c) greater total pressure
    (d) greater total temperature

Q15.18 The two methods of increasing the critical Mach number are:
    (a) thin aerofoils and wing dihedral
    (b) positive cambering of the aerofoil and wing sweep back
    (c) thick aerofoils and wing dihedral
    (d) thin aerofoils and wing sweep back

Q15.19 'Tuck under' is caused by (i) ....movement of the CP (ii) ....change of the downwash angle at the stabiliser.
    (a) (i) aft; (ii) decreasing
    (b) (i) forward; (ii) decreasing
    (c) (i) aft; (ii) increasing
    (d) (i) forward; (ii) increasing

Q15.20 If the Mach trimmer fails:
  (a) the aeroplane mass must be reduced
  (b) the maximum Mach number is limited
  (c) the CG must be moved aft
  (d) the speed must remain constant

Q15.21 'Tuck under' may occur at:
  (a) high Mach numbers
  (b) low Mach numbers
  (c) all Mach numbers
  (d) only at low altitude

Q15.22 Critical Mach number is the:
  (a) speed at which there is supersonic airflow over the whole aeroplane
  (b) the highest speed for which the aeroplane is certificated
  (c) the highest speed without supersonic flow over any part of the aeroplane
  (d) the speed at which there is subsonic airflow over all parts of the aeroplane

Q15.23 When comparing the characteristics of a straight wing with those of a swept wing that have the same wing area and loading, the swept wing has the advantage of:
  (a) increased longitudinal stability
  (b) lower stalling speed
  (c) higher critical Mach number
  (d) greater strength

Q15.24 The formula for the Mach number is:
  (a) $M = IAS/LSS$
  (b) $M = TAS/LSS$
  (c) $M = LSS/TAS$
  (d) $M = TAS \times LSS$

Q15.25 Which of the following (i) aerofoils and (ii) angle of attack will produce the lowest MCRIT values?
  (a) (i) thin; (ii) large
  (b) (i) thin; (ii) small
  (c) (i) thick; (ii) large
  (d) (i) thick; (ii) small

Q15.26 In the transonic range lift will decrease at the shock stall due to the:
  (a) first appearance of a shockwave on the wing upper surface
  (b) first appearance of the bow wave
  (c) separation of the boundary layer at the shockwaves
  (d) attachment of the shockwave on the wing trailing edge

Q15.27 When air is passing through a shockwave the density will:
  (a) decrease
  (b) remain constant
  (c) decrease and then increase
  (d) increase

Q15.28 In supersonic flight, all disturbances produced by an aeroplane are:
  (a) in front of the aeroplane
  (b) very weak and negligible
  (c) within a conical area, dependent on the Mach number
  (d) outside a conical area, dependent on the Mach number

Q15.29 The application of the area rule decreases:
  (a) form drag
  (b) wave drag
  (c) skin friction
  (d) induced drag

Q15.30 At what speed does the front of a shockwave move over the earth's surface?
(a) the speed of sound at ground level
(b) the speed of sound at cruising level
(c) the TAS of the aeroplane
(d) the groundspeed of the aeroplane

Q15.31 If the Mach number of an aeroplane in supersonic flight is increased, the shockwave angles will:
(a) decrease
(b) increase
(c) remain constant
(d) decrease up to a specified Mach number and then a slow increase

Q15.32 When air is passing through a shockwave the static air temperature will:
(a) increase
(b) decrease
(c) remain constant
(d) decrease up to a specified Mach number and then slowly increase

Q15.33 Some aeroplane fuselages have a waist, the reason for this is:
(a) the application of the area rule
(b) to increase the strength of the wing root
(c) to make the engine intakes more efficient
(d) to improve the low-speed characteristics

Q15.34 Which of the following statements regarding the normal shockwave is correct?
(a) The airflow changes from supersonic to subsonic.
(b) The airflow changes direction.
(c) The airflow changes from subsonic to supersonic.
(d) The airflow expands when passing the aerofoil.

Q15.35 To increase M_{CRIT} a conventional aerofoil should:
(a) have a low thickness chord ratio
(b) have a large camber
(c) be used with a high angle of attack
(d) have a large leading-edge radius

Q15.36 The Mach trim system will:
(a) maintain a constant Mach number automatically
(b) transfer fuel automatically, dependent on the Mach number
(c) adjust the elevator trim tab, dependent on the Mach number
(d) adjust the stabiliser, dependent on the Mach number

Q15.37 A normal shockwave:
(a) is a discontinuity plane of an airflow, in which the pressure drops suddenly
(b) is a discontinuity plane of an airflow, which is always normal to the surface
(c) can occur at different points on an aeroplane in transonic flight
(d) is a discontinuity plane in an airflow, in which the temperature suddenly falls

Q15.38 Regarding the speed of sound, which of the following statements is correct?
(a) it varies with the square root of the absolute temperature
(b) It always increases if the air density decreases
(c) It is independent of altitude
(d) It doubles if the temperature increases from 10 °C to 40 °C

Q15.39 Which statement regarding the expansion wave is correct?
(i) The density in front of the expansion wave is higher than behind.
(ii) The static pressure in front of an expansion wave is higher than behind.
(a) (i) correct; (ii) incorrect
(b) (i) incorrect; (ii) correct
(c) (i) incorrect; (ii) incorrect
(d) (i) correct; (ii) correct

Q15.40 When are outboard ailerons deactivated?
(a) when the undercarriage is retracted
(b) when the undercarriage is extended
(c) when flaps and slats are retracted or the speed is above a predetermined value
(d) when flaps and slats are extended or the speed is below a predetermined value

Q15.41 The speed of sound is affected by the air:
(a) temperature
(b) density
(c) pressure
(d) humidity

Q15.42 A Mach trimmer:
(a) is necessary to compensate for the autopilot at high Mach numbers
(b) does not affect the shape of the elevator position v IAS curve for a fully hydraulically controlled aeroplane
(c) corrects for insufficient stick force at high Mach numbers
(d) increases the stick force per g at high Mach number

Q15.43 Which of the following statements is correct regarding the Mach trimmer?
(a) A Mach trimmer corrects the change in stick-force stability of a swept-wing aeroplane above a specific Mach number.
(b) A straight-wing aeroplane always needs a Mach trimmer to fly Mach number close to M$_{MO}$.
(c) A Mach trimmer reduces the stick-force stability of a straight-wing aeroplane to zero at high Mach numbers.
(d) The Mach trimmer corrects the natural tendency of a swept-wing aeroplane to pitch-up.

Q15.44 How does decreased aeroplane mass affect M$_{CRIT}$ at a constant IAS?
(a) M$_{CRIT}$ decreases as a result of flying at a higher angle of attack
(b) M$_{CRIT}$ increases as a result of flying at a lower angle of attack
(c) M$_{CRIT}$ increases as a result of compressibility effects
(d) M$_{CRIT}$ decreases

Q15.45 Which of the following phenomena can only occur at speeds above M$_{CRIT}$?
(a) elevator stall
(b) Mach buffet
(c) Dutch roll
(d) speed instability

Q15.46 An aeroplane is flying in the transonic range with an increasing Mach number the shockwave on the upper surface of the wing:
(a) remains unchanged
(b) disappears
(c) moves aft
(d) moves forward

Q15.47 If a symmetrical aerofoil is accelerated from subsonic to supersonic speed the CP will move:
(a) aft to the trailing edge
(b) forward to the leading edge
(c) forward to the midchord
(d) aft to the midchord

Q15.48 For a swept-wing aeroplane in straight and level flight the consequences of exceeding M$_{CRIT}$ are:
(a) buffeting of the aeroplane and a tendency to pitch-down
(b) an increase of speed and a tendency to pitch-up
(c) buffeting of the aeroplane and engine imbalance
(d) buffeting of the aeroplane and a tendency to pitch-up

Q15.49 Mach number is the ratio between the:
    (a) TAS and the local speed of sound
    (b) TAS and the speed of sound at sea level
    (c) IAS and the local speed of sound
    (d) IAS and the speed of sound at sea level

Q15.50 When air has passed through an expansion wave, the static pressure is:
    (a) decreased
    (b) decreased or increased depending on the Mach number
    (c) increased
    (d) unchanged

Q15.51 Which statement regarding the expansion wave is correct?
    (i) The temperature in front of the expansion wave is higher than behind.
    (ii) The speed in front of an expansion wave is higher than behind.
    (a) (i) incorrect; (ii) incorrect
    (b) (i) correct; (ii) incorrect
    (c) (i) correct; (ii) correct
    (d) (i) incorrect; (ii) correct

Q15.52 When air has passed through a shockwave the speed of sound is:
    (a) increased
    (b) not affected
    (c) decreased
    (d) decreased up to a specified Mach number and then increased

Q15.53 The loss of total pressure in a shockwave is due to:
    (a) the boundary-layer surface friction is higher
    (b) the kinetic energy in the flow is changed into heat energy
    (c) the speed reduction is too great
    (d) the static pressure decrease is comparatively high

Q15.54 When moving from a position ahead of the shockwave to a position behind the shockwave the density will (i) .... and the temperature will (ii) ....
    (a) (i) decrease; (ii) increase
    (b) (i) decrease; (ii) decrease
    (c) (i) increase; (ii) increase
    (d) (i) increase; (ii) decrease

Q15.55 'Tuck under' is:
    (a) the tendency of an aeroplane to nose-down when the control column is pulled back
    (b) the tendency of an aeroplane to nose-down when speed is increased into the transonic range
    (c) the tendency of an aeroplane to nose-up when the speed is increased into the transonic range
    (d) the vibration of the control column at high Mach numbers

Q15.56 Just above M$_{CRIT}$ the first evidence of a shockwave will appear:
    (a) on the upper surface of the wing
    (b) beneath the lower surface of the wing
    (c) on the leading edge of the wing
    (d) on the trailing edge of the wing

Q15.57 In supersonic flight the pressure distribution around an aerofoil is:
    (a) rectangular
    (b) triangular
    (c) irregular
    (d) the same as it is in subsonic flight

Q15.58 When air is passing through an expansion wave the local speed of sound will:
- (a) stay the same
- (b) decrease until a specific Mach number and then increase
- (c) decrease
- (d) increase

Q15.59 The movement of the CP when accelerating from subsonic to supersonic flight is:
- (a) forward
- (b) to a position near the leading edge
- (c) to a position near the trailing edge
- (d) to the midchord position

Q15.60 When air is passing through an expansion wave the Mach number will:
- (a) stay the same
- (b) decrease until a specific Mach number and then increase
- (c) increase
- (d) decrease

Q15.61 The bow wave will first appear at:
- (a) $M = Mcrit$
- (b) $M = 0.6$
- (c) $M = 1.3$
- (d) $M = 1.05$

Q15.62 A jet aeroplane equipped with inboard and outboard ailerons in level flight at its normal cruise Mach number. The inboard ailerons are (i) . . . . .and the outboard ailerons (ii) . . . .
- (a) (i) active; (ii) active
- (b) (i) inactive; (ii) inactive
- (c) (i) active; (ii) inactive
- (d) (i) inactive; (ii) active

Q15.63 MCRIT for an aerofoil equals the free-stream aerofoil Mach number at which:
- (a) Mach 1 is attained at a certain point on the upper surface of the wing
- (b) the maximum operating temperature is attained
- (c) a shockwave appears on the upper surface
- (d) a rectangular shape appears on the upper surface

Q15.64 In straight and level flight when the Mach number is slowly increased the first shockwave will occur:
- (a) on the underside of the wing
- (b) somewhere on the fin
- (c) somewhere on the horizontal tail
- (d) on the upper surface at the wing root

Q15.65 Shock stall is separation of the:
- (a) flow at high angles of attack and high Mach numbers
- (b) flow at the trailing edge of the wing at high Mach numbers
- (c) boundary layer behind the shockwave
- (d) flow behind the bow wave

Q15.66 The Mach trim system will prevent:
- (a) tuck under
- (b) Dutch roll
- (c) buffeting
- (d) shock stall

Q15.67 When air is passing through an expansion wave the static temperature will:
(a) stay the same
(b) decrease
(c) decrease until a specific Mach number and then increase
(d) increase

Q15.68 The critical Mach number can be increased by:
(a) increased aspect ratio
(b) swept wings
(c) positive wing dihedral
(d) a T-tail

Q15.69 In the transonic range the aeroplane characteristics are determined by the:
(a) IAS
(b) CAS
(c) Mach number
(d) TAS

Q15.70 MCRIT is the free-stream Mach number at which:
(a) CLmax is attained
(b) the critical angle of attack is reached
(c) somewhere on the airframe Mach 1 is reached locally
(d) Mach buffet occurs

Q15.71 When accelerating though the transonic range the wing CP will move aft and requires:
(a) a stability augmentation system
(b) much more thrust from the engines
(c) a higher IAS to compensate for the nose-down effect
(d) a pitch-up input of the stabiliser

Q15.72 The area of disturbance caused by the presence of a supersonic aeroplane is:
(a) in front of the Mach cone
(b) in front of the normal shockwave
(c) in front of the oblique shockwave
(d) within the Mach cone

Q15.73 The centre of lift of a symmetrical aerofoil on accelerating to supersonic speed moves:
(a) aft toward the centre of the chordline
(b) forward toward the 25% chordline
(c) aft of the 50% chordline
(d) forward of the 25% chordline

Q15.74 The static pressure (i) and the static temperature (ii) of supersonic airflow on passing through an oblique shockwave will:
(a) (i) increase, (ii) increase
(b) (i) increase, (ii) decrease
(c) (i) decrease, (ii) decrease
(d) (i) decrease, (ii) increase

Q15.75 The total pressure (i) and the density (ii) of supersonic airflow on passing through an oblique shockwave will:
(a) (i) increase, (ii) increase
(b) (i) decrease, (ii) increase
(c) (i) decrease, (ii) decrease
(d) (i) increase, (ii) increase

Q15.76 As air passes through an expansion wave the static pressure:
(a) increases
(b) does not change
(c) decreases
(d) decreases and then increases

Q15.77 The effect of Mach trim on stick forces of power-operated controls is to:
  (a) decrease the stick-force gradient to ensure manoeuvrability is maintained at high Mach numbers
  (b) maintain the required stick-force gradient
  (c) decrease the stick-force gradient to a high-speed stall
  (d) increase the stick-force gradient to improve the 'feel.'

Q15.78 The speed of the local airflow after passing through a normal shockwave is:
  (a) more than the speed of sound
  (b) equal to the speed of sound
  (c) initially more than the speed of sound then decreasing to less than the speed of sound
  (d) less than the speed of sound

Q15.79 The speed of the local airflow after passing through an oblique shockwave is:
  (a) equal to the speed of sound
  (b) initially more than the speed of sound then decreasing to less than the speed of sound
  (c) more than the speed of sound
  (d) less than the speed of sound

Q15.80 What does the Mach trim adjust?
  (a) stabilizer trim tab
  (b) fuel transfer
  (c) elevator trim
  (d) longitudinal trim

Q15.81 The Mach trim system operates at:
  (a) high Mach numbers only
  (b) low Mach numbers only
  (c) only at supersonic speeds
  (d) at all Mach numbers

Q15.82 The function of the Mach trim system is:
  (a) to prevent exceeding Mмo
  (b) to prevent high-speed tuck
  (c) to prevent Dutch roll
  (d) to prevent phugoidal oscillation

Q15.83 The pressure distribution on the upper surface of the wing as the airflow becomes supersonic during acceleration changes to:
  (a) rectangular
  (b) triangular
  (c) irregular
  (d) laminar

Q15.84 The pressure distribution on the upper surface of the wing when all of the airflow is supersonic during acceleration changes to:
  (a) rectangular
  (b) triangular
  (c) irregular
  (d) laminar

Q15.85 When operating in the transonic range, lift decreases when the shock stall occurs because:
  (a) of the appearance of the bow wave
  (b) of the shockwave attaching itself to the trailing edge of the wing
  (c) of the separation of the boundary layer at the shockwave
  (d) the speed has reached Mсrit

Q15.86 The large increase of drag experienced at high transonic speeds is due to the formation of the shockwave causing:
  (a) wave drag
  (b) increased parasite drag
  (c) increased induced drag
  (d) decreased thrust available

Q15.87 To minimize wave drag an aeroplane should be operated at:
  (a) the speed of sound
  (b) subsonic speeds
  (c) low supersonic speeds
  (d) high supersonic speeds

Q15.88 For a supersonic aeroplane the angle of the shockwave . . . . with increasing Mach number:
  (a) decreases
  (b) increases
  (c) remains the same
  (d) increases to Mach 1 and then decreases

Q15.89 The speed at which the shockwave moves over the ground is equal to:
  (a) the aeroplane's TAS
  (b) the aeroplane's groundspeed
  (c) the speed of sound at the aeroplane's cruising level
  (d) the speed of sound at ground level

Q15.90 When the laminar flow breaks down a shock stall occurs:
  (a) behind the leading edge of the wing
  (b) behind the trailing edge of the wing
  (c) at high angles of attack and high Mach number
  (d) behind the shockwave

Q15.91 A normal shockwave causes the airflow to:
  (a) increase speed from subsonic to supersonic
  (b) decrease speed from supersonic to subsonic
  (c) change direction
  (d) become laminar

Q15.92 The loss of total pressure of supersonic airflow caused by a normal shockwave can be minimized if the airflow speed ahead of the shockwave is:
  (a) the speed of sound
  (b) less than the speed of sound
  (c) just supersonic
  (d) much higher than the speed of sound

Q15.93 The semi-angle of the Mach cone:
  (a) decreases with increasing speed
  (b) increases with increasing speed
  (c) decreases with decreasing speed
  (d) remains constant irrespective of the speed

Q15.94 A normal shockwave:
  (a) is a plane of discontinuity at which the pressure decreases
  (b) is a plane of discontinuity at which the temperature decreases
  (c) can develop at any time the aeroplane is in the transonic speed range
  (d) is a plane of discontinuity that is at right angles to the aerofoil surface

Q15.95 The speed of sound is affected by:
    (a) air density
    (b) pressure altitude
    (c) air pressure
    (d) temperature

Q15.96 When flying at a constant IAS the effect of increased mass on MCRIT is that it:
    (a) decreases with increasing angle of attack
    (b) increases
    (c) decreases
    (d) increases with increasing angle of attack

Q15.97 On a swept-wing aeroplane increasing the speed above MCRIT causes:
    (a) buffeting and nose-up pitch
    (b) buffeting and nose-down pitch
    (c) structural failure
    (d) increased lift necessitating high-speed tuck under

Q15.98 The effect that a shock stall has on a swept-wing aeroplane is that it:
    (a) pitches nose-up
    (b) pitches nose-down
    (c) has no effect on longitudinal stability
    (d) increases longitudinal stability

Q15.99 Mach tuck occurs at:
    (a) low-pressure altitude
    (b) all Mach numbers
    (c) high Mach numbers
    (d) low Mach numbers

Q15.100 Mach number is equal to:
    (a) TAS multiplied by the local speed of sound
    (b) the local speed of sound divided by the TAS
    (c) TAS divided by the local speed of sound
    (d) the local speed of sound

Q15.101 As airflow passes through a shockwave the density (i) . . . . and the temperature (ii) . . . .
    (a) (i) increases; (ii) increases
    (b) (i) increases; (ii) decreases
    (c) (i) decreases; (ii) decreases
    (d) (i) decreases; (ii) increases

Q15.102 The handling characteristics in the transonic speed range are affected by the:
    (a) Mach number
    (b) IAS
    (c) TAS
    (d) CAS

Q15.103 As airflow passes through an expansion wave the Mach number:
    (a) decreases
    (b) increases
    (c) remains constant
    (d) is unaffected

Q15.104 The free-stream MCRIT is defined as:
    (a) the speed at which the local velocity reaches the speed of sound
    (b) the point at which the shock stall first occurs
    (c) equal to the speed of sound
    (d) the speed at which the shockwave reaches the trailing edge

Q15.105 Mach tuck is:
  (a) a nose-up pitch movement at transonic speed
  (b) a nose-down pitch movement at transonic speed
  (c) nose-up pitch caused by movement of the centre of pressure
  (d) caused by tip stalling of a swept-wing aeroplane

Q15.106 The value of MCRIT is increased most by:
  (a) a thin aerofoil with sweepback
  (b) sweepback with positive camber
  (c) a thin aerofoil with dihedral
  (d) sweepback with negative camber

Q15.107 The Mach trimmer:
  (a) compensates for a nose-up pitching moment at transonic speeds
  (b) increases the stick force per 'g' in the supersonic range
  (c) increases the nose-down pitching moment
  (d) compensates for a nose-down pitching moment at transonic speeds

Q15.108 The speed at which the bow wave forms is:
  (a) M1.3
  (b) M0.8
  (c) M1.05
  (d) M0.9

Q15.109 The lowest value of MCRIT results from (i) . . . . camber and a (ii) . . . . cross-section.
  (a) (i) large; (ii) thick
  (b) (i) small; (ii) thin
  (c) (i) small; (ii) thick
  (d) (i) large; (ii) thin

Q15.110 At low Mach numbers, which of the following can occur?
  (a) Mach tuck
  (b) Dutch roll
  (c) tuck under
  (d) shockwaves

Q15.111 In the event of the Mach trimmer becoming unserviceable:
  (a) fly at a constant speed
  (b) restrict the Mach number
  (c) reduce the speed to VMO
  (d) move the CG to the aft limit

Q15.112 Mach tuck occurs at:
  (a) low Mach numbers
  (b) all Mach numbers
  (c) high Mach numbers
  (d) low-pressure altitude

Q15.113 The reason that aeroplanes have a waisted fuselage is:
  (a) to increase the strength of the wing root
  (b) to improve stability at low speeds
  (c) to incorporate the engine nacelles
  (d) to comply with the area rule

Q15.114 In the transonic flight regime aft movement of the CP:
  (a) improves the directional stability
  (b) improves the static longitudinal stability
  (c) improves the static lateral stability
  (d) improves the sideslip stability

Q15.115 The normal shockwave when compared to the oblique shockwave has:
   (a) a lower static pressure
   (b) a higher total pressure
   (c) a higher total temperature
   (d) a lower total temperature

Q15.116 critical Mach number is the greatest speed at which:
   (a) the airflow over all parts of the aeroplane is supersonic
   (b) the airflow over some parts of the aeroplane is supersonic
   (c) the airflow is equal to the maximum operating Mach number
   (d) supersonic airflow has not been reached over any part of the aeroplane

Q15.117 As airflow passes through a shockwave the speed of sound:
   (a) decreases
   (b) remains unchanged
   (c) increases
   (d) is dependent on the type of shockwave

Q15.118 The Mach trimmer corrects for:
   (a) movement of the centre of pressure
   (b) movement of the centre of gravity
   (c) pitch-up
   (d) shockwave drag

Q15.119 The local speed of sound is:
   (a) dependent on the square root of the absolute temperature
   (b) is the same at all altitudes
   (c) increases with decreasing air density
   (d) is dependent on the mass of the aeroplane

Q15.120 Mach tuck is counteracted by:
   (a) increased thrust
   (b) decreased stabiliser incidence
   (c) increased stability
   (d) increased IAS

Q15.121 The aft movement of the CP when accelerating in the transonic speed range is compensated by:
   (a) stabiliser nose-up pitch
   (b) forward movement of the CG
   (c) stabiliser increased angle of incidence
   (d) spoilers

Q15.122 Increased drag in the transonic speed range is due to the:
   (a) skim friction
   (b) shockwave
   (c) angle of attack
   (d) parasite drag

Q15.123 Tuck under is caused by the (i) . . . . . movement of the CP and (ii) . . . . downwash striking the tailplane.
   (a) (i) forward; (ii) decreased
   (b) (i) forward; (ii) increased
   (c) (i) aft; (ii) increased
   (d) (i) aft; (ii) decreased

Q15.124 As the Mach number increases, Mach tuck is caused by the CP moving (i) .... and the
(ii) .... downwash striking the tailplane.
    (a) (i) forward; (ii) increased
    (b) (i) aft; (ii) decreased
    (c) (i) forward; (ii) decreased
    (d) (i) aft; (ii) increased

Q15.125 The speed range between the high-speed buffet and the low-speed buffet:
    (a) decreases during a constant IAS descent
    (b) increases during a constant IAS climb
    (c) increases during a descent at a constant IAS
    (d) remains constant irrespective of altitude

Q15.126 The static temperature of air passing through an expansion wave:
    (a) decreases
    (b) increases
    (c) remains constant
    (d) increases up to the tropopause and then decreases

Q15.127 Outboard ailerons are locked:
    (a) when flaps are retracted
    (b) when the undercarriage is extended
    (c) when flaps are extended
    (d) when the undercarriage is retracted

Q15.128 Shock-induced separation results in:
    (a) decreasing lift
    (b) constant lift
    (c) increasing lift
    (d) decreasing drag

16 Propellers

Essentially, a propeller is a fan, fitted with between two and six aerodynamic blades that is used to convert rotational motion into forward thrust for the propulsion of an aeroplane through the air. The blades of the propeller are small aerodynamic aerofoils mounted vertically on a horizontal shaft that is rotated by either a piston engine or a turbine-driven engine. The blades are in fact miniature wings that generate a pressure difference between the forward and rearward surface of the blade and is analogous to lift except that it acts horizontally and accelerates a mass of air rearward. Its motion is similar to the rotation of a screw through a solid material.

16.1 Propeller Definitions

The following definitions are used throughout the remainder of this chapter:
(See Figure 16.1 and Figure 16.2)

a. The **axis of rotation** is the shaft of the propeller.
b. The **blade angle**, sometimes called the pitch angle, is the angle subtended between the plane of rotation and the chordline of the propeller blade, which may be fixed or variable. When the aeroplane is moving it comprises two parts, the angle of attack and the helix angle, but when it is stationary the blade angle is equal to the angle of attack. It is usually referenced at 75% of the blade radius measured from the centre of the shaft.
c. The **blade angle of attack** is the angle subtended between the chordline of the propeller blade and the relative airflow.
d. The **blade back** is the rearward surface of a propeller blade equivalent to the lower surface of a wing.
e. The **blade face** is the forward surface of a propeller blade equivalent to the upper surface of a wing.
f. The **blade twist** is the effect on a propeller of the designed decreasing blade angle with increasing radius from the shaft.
g. The **effective pitch** is the distance that a propeller **actually** advances in one revolution. Effective Pitch = Geometric Pitch minus propeller slip.
h. The **forward velocity** of a rotating propeller is the sum of the TAS and the induced airflow, i.e. the air drawn through the disc by the propeller.
i. The **geometric pitch** is the distance that a propeller *should* theoretically advance in one revolution at zero angle of attack. It is greater in the cruise than it is in the climb. *Geometric Pitch = Effective Pitch plus Propeller Slip.*

The Principles of Flight for Pilots   P. J. Swatton
© 2011 John Wiley & Sons, Ltd

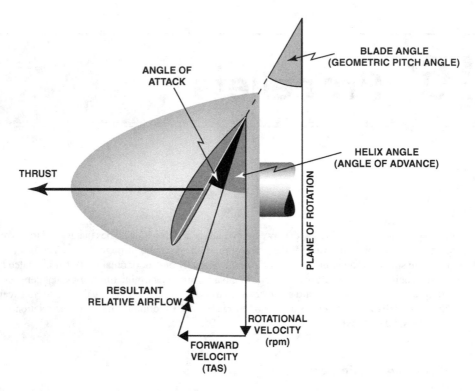

Figure 16.1   Propeller-Blade Definitions.

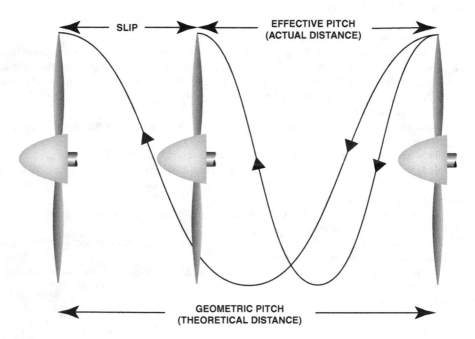

Figure 16.2   Propeller-Slip Definitions.

j. The **helix angle,** sometimes called the angle of advance, is the angle subtended between relative airflow and the plane of rotation of the propeller. It is equal to the blade angle minus the angle of attack.

k. The **leading edge** is the same as that of a wing and is the forward edge of the profile of a propeller blade.

l. The **propeller disc** is the area swept by a rotating propeller when viewed from the front.

m. The **propeller efficiency** is the ratio between thrust horsepower and shaft horsepower or the power available and the shaft power.

n. The **propeller slip** is the difference between the effective pitch and the geometric pitch and is the result of the air not being a perfect medium for propulsion. *Propeller Slip = Geometric Pitch minus Effective Pitch.*

o. The **reference section** of a propeller blade is at 75% of the blade radius measured from the propeller axis.

p. The **relative airflow (RAF)** is the airflow direction and speed resulting from the rotational velocity and the TAS + induced airflow acting together.

q. The **rotational velocity** of a propeller is equal to the circumference multiplied by the angular velocity (N). Rotational velocity = $2\pi rN$, where r is the radius from the shaft axis of rotation.

r. The **trailing edge** is the same as that of a wing and is the rearward edge of the propeller profile.

16.2 Basic Principles

The rotation of a propeller creates a total reactive force at right angles to the chordline of the propeller blade. This can be divided into two components, that which is perpendicular to the relative airflow, the lift, and that which is parallel to the relative airflow, the drag. If the same reactive force is divided into components relative to the flight path of the aeroplane then that which is at right angles to the flight path is the propeller torque and that which is parallel to the flight path is the thrust. See Figure 16.3.

As the maximum lift/drag ratio improves the total reactive force moves forward along the chordline. Consequently, more of the engine power is converted to become forward thrust. Propeller torque acts in the plane of rotation opposing the engine torque. This is the resistance to the propeller rotation that is totally undesirable because it is attempting to rotate the aeroplane in the opposite direction to the propeller.

The propeller blade angle of attack at any point is dependent on the ratio of the rotational speed to the TAS. Because the rotational velocity of the blade increases with distance from the propeller shaft and the fact that the aeroplane is moving forward, then the angle of attack of the blade varies at different distances from the shaft. The local angle of attack relies on the angular velocity. The most efficient angle of attack is 4° at the slowest moving part of the blade next to the shaft and 2° at the fastest moving part of the blade at the tip. The blade will stall if it attains an angle of attack of 15°. To ensure that the blade operates at the most efficient angle of attack at all points along its length it is designed with:

a. A geometric twist that changes the blade angle from high at the blade root to low at the blade tip.

b. A fineness ratio that is low at the blade root and high at the blade tip. See Figure 16.4.

The efficiency of any system is determined by the ratio of the output power to the input power. The force developed by the engine and delivered to the propeller via the shaft is the rotational force of the propeller, the propeller torque. The power delivered to the propeller is the product of the torque and the angular velocity of the propeller shaft (Torque × Rotational Speed).

Engine Power Delivered = Torque × 2 Angular Velocity (RPM)

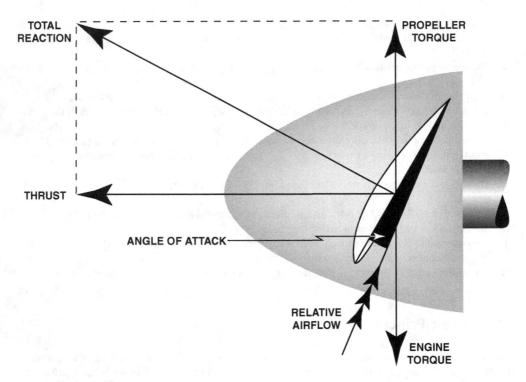

Figure 16.3   Propeller-Blade Forces.

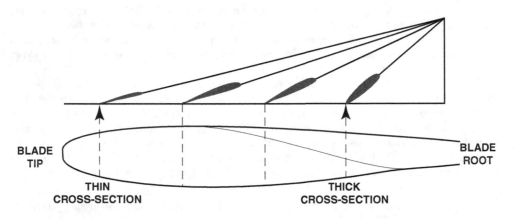

Figure 16.4   Propeller-Blade Twist.

The thrust developed by the propeller is the product of the mass of air moved and the acceleration given to it by the propeller. The power developed by the propeller is the product of the thrust and the TAS (Thrust × Axial Speed).

$$\textbf{Propeller Power Developed = Thrust} \times \textbf{TAS}$$

Propeller efficiency is the ratio of the propeller power developed to the engine power delivered. A well-designed propeller is typically 80% efficient when operating at its optimum blade angle of attack.

$$\text{Propeller Efficiency} = \frac{\textbf{(Thrust} \times \textbf{Axial Speed)}}{\textbf{(Torque} \times \textbf{Rotational Speed)}}$$

$$\text{Propeller Efficiency} = \frac{\textbf{(Thrust} \times \textbf{TAS)}}{\textbf{(Torque} \times \textbf{2 Angular Velocity)}}$$

16.3   Factors Affecting Propeller Efficiency

Propeller efficiency is the ratio of the thrust horsepower delivered by the propeller to the shaft (or brake) horsepower delivered by the engine. It is always less than 1 and is zero when the aeroplane is stationary or at high speed. A propeller must be able to absorb the power it receives from the engine; to do this it must balance the engine torque by a torque resistance. If such is not the case then the propeller will race and both the propeller and the engine will be inefficient. The power absorption of a propeller can be improved by introducing any of the following measures either individually or in any combination at the design stage:

a. increased blade camber;
b. increased blade chord;
c. increased blade angle and blade angle of attack;
d. increased blade length;
e. increased number of blades;
f. increased RPM.

Propellers have a similar aerofoil cross-section to that of a low-drag wing and as such perform poorly when at any angle of attack other than the optimum. Adjustment of the helix angle to match the TAS can improve the performance of a propeller. A very low pitch and helix angle produce good performance against torque resistance but generate very little thrust, whereas larger angles generate a lot more thrust but perform poorly against torque resistance.

16.4   Airspeed

16.4.1   Fixed-Pitch Propellers

The blade angle of a fixed-pitch propeller is constant but the helix angle and the angle of attack will vary according to its forward speed. To ensure its efficient operation the blade angle of attack of a fixed-pitch propeller must be close to the optimum and is that which produces the greatest excess of horizontal lift (thrust) over drag. However, to operate the propeller in this manner will restrict the range of speeds within which the aeroplane can function. Thus, the blade angle selected during the design of the aeroplane is dependent on the operational requirements for that aeroplane.

A large blade angle or coarse pitch is suitable for high-speed flight. The higher the forward speed of the aeroplane the smaller is the blade angle of attack and the thrust generated. The optimum angle of attack for the blades of a fixed-pitch propeller designed for high-speed flight is attained when the aeroplane is stabilised in the cruise.

A small blade angle or fine pitch suits a slow-flying aeroplane because it produces a large amount of torque and is ideal for short-field take-offs or landings and crop spraying. During the take-off run, the thrust of a fixed-pitch propeller decreases slightly as the aeroplane accelerates because the movement of the aeroplane transforms the thrust from static thrust to net thrust.

The angle of attack of a fixed-pitch propeller increases with increased RPM and/or decreased TAS. Maximum propeller efficiency for any given RPM occurs at only one TAS and the blade angle of attack for that RPM increases with decreased TAS. For a given TAS the blade angle of attack will increase with increased RPM. During a climb with a constant RPM, the blade angle of attack increases because the TAS decreases. These details are summarised in Table 16.1.

Table 16.1   Factors Affecting Fixed-Pitch Angle of Attack.

TAS	RPM	Blade AoA
Constant	Increase	Increase
Decrease	Constant	Increase
Climb	Constant	Increase
Descent	Constant	Decrease

Figure 16.5 shows a fixed-pitch propeller at a constant RPM at three forward speeds for comparison.

a. **Low Forward Speed.** In Figure 16.5 the vector for the forward speed is short therefore the helix angle for (**a**) is small but the angle of attack for (**a**) is large. This produces a large amount of thrust but at the same time it creates a large amount of drag. To maintain a constant RPM it is essential to overcome the drag and that requires a high propeller torque. From the foregoing formula it can be determined that the propeller efficiency in these circumstances is low.

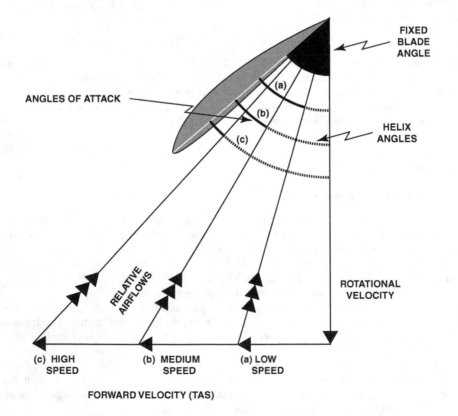

Figure 16.5   The Effect of Speed. (Fixed-Pitch Propeller)

b. **Medium Forward Speed.** In Figure 16.5 the forward speed vector (**b**) is for the optimum speed for the fixed-pitch propeller and is longer than (a). The helix angle (**b**) is larger than (a). Although the angle of attack of (**b**) is smaller than (a) and less thrust is produced the amount of drag induced is significantly less, this makes the propeller efficiency proportionately greater than (a). It is this speed that attains the greatest propeller efficiency.

c. **High Forward Speed.** In Figure 16.5 the longest forward speed vector is (**c**), which produces a very small angle of attack and a very large helix angle. This angle of attack generates very little forward lift, therefore the thrust element of the propeller efficiency formula is small, which indicates that the propeller is operating at a very inefficient angle of attack.

Because it is the relative length of the rotational velocity of the propeller vector and the vector length of the aeroplane's forward speed that determine the direction of the relative airflow (RAF) it is possible to vary the blade angle of attack of a fixed-pitch propeller by altering the RPM to match the forward speed of the aeroplane. However, this method of operation is very restricted because of the limitations imposed on the engine being used to drive the propeller. It is much easier to operate the aeroplane at a fixed RPM and vary the blade angle to produce a blade angle of attack to suit the forward speed of the aeroplane.

16.4.2 Variable-Pitch Propellers

The purpose of a variable-pitch propeller is to either manually or automatically maintain the optimum angle of attack to maximize the lift/drag ratio of the propeller as the aeroplane speed changes. This characteristic is the most significant advantage of the constant-speed (fixed RPM) propeller over the fixed-pitch propeller because it is able to maintain maximum efficiency over a range of aeroplane speeds. Originally there were only two pitch settings, fine for take-off and climb and coarse for cruise and high-speed flight. Subsequently, the fully variable pitch constant-speed propeller was introduced so that the optimum blade angle of attack could be maintained over the whole range of operating speeds for that aeroplane. This facility not only improves the propeller efficiency but also decreases the fuel consumption. *The pitch angle or blade angle of a constant-speed propeller (fixed RPM) increases (or coarsens) with increasing TAS.* During moderate horizontal turbulence the pitch angle varies very little because it has only a minor effect on the forward airspeed.

Constant-speed propellers enable the pilot to select the RPM for maximum power or maximum efficiency and a propeller governor varies the propeller pitch angle as required to maintain the selected RPM. Propeller efficiency improves with increased TAS; however, it is limited to that blade angle beyond which the propeller will no longer produce any thrust. The most practical value of the propeller blade angle is between 300 kt and 350 kt TAS, any increase of speed beyond this results in reduced thrust and diminished propeller efficiency.

16.5 Power Absorption

The ability of a propeller to convert engine power into forward thrust is its power-absorption capability. It is affected by:

a. the propeller-blade shape;
b. the number of propeller blades;
c. the propeller solidity.

16.5.1 Propeller-Blade Shape

16.5.1.1 Blade Length

The lift-producing capability of the propeller blades determines the amount of engine power that can be absorbed and converted to thrust. Similar to a wing, the aspect ratio of a propeller blade fixes the amount of lift generated; the greater the aspect ratio the greater the lift generated. The aspect ratio of a propeller blade is the length of the propeller blade divided by the blade mean chord. Therefore, increasing the propeller diameter increases the aspect ratio and hence the lift and thrust.

There is a finite aerodynamic limit to the maximum length of a propeller blade because of its tip speed. For any given RPM the speed reached by the blade tip is most often sonic, even at low RPM. Propeller efficiency is adversely affected by the compressibility created at sonic tip speeds reducing the thrust generated and increasing the torque required (the blade drag). Apart from the aerodynamic limitation the noise produced by the propeller at supersonic tip speeds is unacceptably high and could cause structural fatigue as well as environmental damage.

Sonic tip speed also limits the maximum TAS at mean sea level theoretically to approximately Mach 0.7. But because of the adverse effects, caused by the sudden increase of drag and torque resistance when the propeller blade becomes supersonic, conventional propeller aeroplanes fly at a maximum speed of Mach 0.6. Nevertheless, some military aeroplanes do operate at speeds in excess of this maximum.

16.5.1.2 Blade Chord

Increasing the blade chord has the advantages that blade interference is reduced; the blade area is increased to transmit the power available within a set diameter and the power absorption is improved. Its disadvantage is that it decreases the aspect ratio, and consequently the lift and thrust.

16.5.2 Propeller-Blade Number

The number of propeller blades is usually limited to six because of the complexity of the pitch control contained in the propeller hub and aerodynamic interference between propeller blades. Having a large number of blades has the advantage that it decreases propeller noise but its primary purpose is to improve the power absorption properties and efficiency of the propeller. Its disadvantage is that it diminishes propeller efficiency because it increases the total drag.

Power absorption can be improved by fitting contrarotating blades. Increasing the camber on propeller blades increases the thrust and the drag that the propeller develops, hence its power absorption is improved. Recent design developments have seen the introduction of propellers that have six thin blades that are swept back in the fashion of a 'scimitar' shape that allows the blades to have a large helix angle. This reduces the work done by each blade. These are often fitted to aeroplanes as contrarotating propellers to further improve their efficiency.

16.5.3 Solidity

To improve the power-absorption capabilities of a propeller it is necessary to increase the 'solidity' of the propeller disc. The solidity is the ratio of the propeller blade chord to the circumference of the propeller disc, at the radius selected for the measurement of the chord. Alternatively, it can be defined as the ratio of the total area of all the blades to that of the propeller disc. Normally, the radius selected for such calculations is 70% of the distance from the shaft to the tip. See Figure 16.6 The formula for calculating the solidity of a propeller is:

$$\text{Solidity} = \frac{\textbf{Number of propeller blades} \times \textbf{the blade chord at 70\% Tip Radius}}{\textbf{70\% Tip Radius Circumference}}$$

From the formula it can be seen that by either increasing the blade chord or increasing the number of propeller blades the solidity of the propeller will increase. Both methods have advantages and disadvantages.

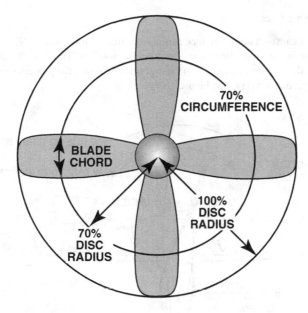

Figure 16.6   Propeller-Disc Solidity.

16.6 The Effects of a Propeller on Aeroplane Performance

16.6.1 Torque

The force rotating a propeller is the torque and according to Newton's third law of motion there is an equal and opposite reaction to this force. For a single-engined aeroplane there is, therefore, a force attempting to rotate the aeroplane around the propeller shaft but in the opposite direction to the propeller. This creates a rolling and a secondary yawing tendency especially at take-off when the propeller is at high RPM generating a considerable amount of torque and the forward speed is low.

On a single-engined aeroplane, if the propeller rotates clockwise, when viewed from behind, the torque reaction will tend to roll the aeroplane to the left. The wheels, being in contact with the ground, prevent the rolling motion and the load on the left wheel increases and supports most of the aeroplane mass. Consequently, the rolling resistance of the left wheel increases and the aeroplane will tend to swing left during the take-off ground roll. If this tendency is not adequately controlled particularly on a single-engined aeroplane fitted with a powerful engine, such as the Harvard, it can cause the aeroplane to make a complete ground loop.

There are several ways that may be used to counteract the effects of torque:

a. **Counterrotating Propellers.** By mounting two propellers in tandem that rotate in opposite directions mounted on concentric shafts but driven by separate engines the torque effect of one engine cancels out the torque effect of the other engine.

b. **Multiengined Aeroplane Counterrotating Propellers.** For multiengined aeroplanes with the engines mounted on the wings the engines on one wing are made to rotate in the opposite direction to those mounted on the other wing. The torques generated by the propellers on each wing will cancel each other out.

c. **Contrarotating Propellers.** This is similar to the counterrotating method except that the propellers are driven by one engine on concentric shafts through gearboxes that ensure that the propellers rotate in opposite directions. Again, the torque effects cancel each other out.

16.6.2 *Slipstream Effect*

The effect of the slipstream on a single-engined aeroplane is that the rotating propeller imparts a corkscrew rotation to the airflow in the same direction as the propeller rotation. This impacts the fin and rudder as an asymmetric force that induces a yaw. When viewed from behind if the propeller rotation is clockwise the yaw is to the left and vice versa. See Figure 16.7.

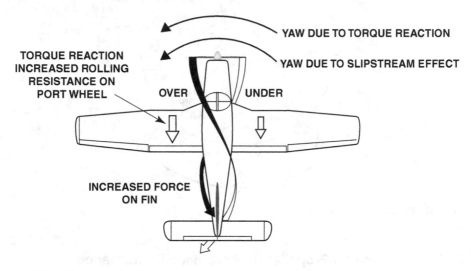

Figure 16.7   Slipstream Effect.

This effect can be compensated at the design stage by offsetting the fin and rudder at a slight angle to the longitudinal axis, into the asymmetric flow, and so reduce its effect. At high RPM especially during take-off, even with this built-in compensation, it is still necessary to apply rudder to cancel the asymmetric slipstream effect.

16.6.3 *Asymmetric Blade*

The asymmetric-blade effect is the result of different values of thrust being produced by the upgoing and downgoing propeller blades when the aeroplane's attitude is other than level. At any time that the propeller shaft is inclined at an angle to the horizontal plane, i.e. climbing, then the angle of attack of the upgoing and the downgoing propeller is different. This is caused by the vector addition of the horizontal speed of the aeroplane to the rotational velocity of the propeller in the plane of the disc. During a climb the downgoing blade has a greater angle of attack than the upgoing blade and therefore develops more thrust than the upgoing blade. Viewed from behind for a propeller rotating clockwise this will cause the aeroplane to yaw to the left because of the asymmetric-blade effect. See Figure 16.8.

In addition to this feature, the distance travelled per unit of time by the downgoing blade is greater than the upgoing blade, thus it has a greater speed relative to the airflow which will also produce more thrust for a given angle of attack. Together, the total thrust differences of the upgoing and downgoing blades supply an unbalanced thrust to the propeller disc.

When viewed from behind if the propeller rotation is clockwise the right side of the disc produces more thrust than the left side and this imbalance of thrust will cause the aeroplane to yaw to the left and vice versa if rotation is anticlockwise. A similar imbalance of thrust is produced between the top and bottom of the propeller disc if the aeroplane is yawed in level flight. This is because the upper and lower

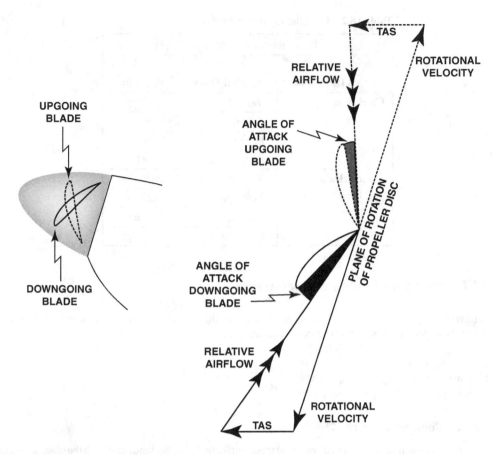

Figure 16.8 Asymmetric-Blade Effect.

blades will have different angles of attack and as a result will cause the nose of the aeroplane to pitch-up or -down depending on the direction of rotation and the direction of the yaw.

When viewed from behind if the propeller is rotating clockwise and the yaw is to the left then the lower blade has the greatest angle of attack and produces more lift which causes the nose to pitch up. If the yaw is to the right the nose will pitch-down. The asymmetric-blade effect increases when the angle between the propeller axis and the airflow through the disc increases or if the engine power increases.

16.6.4 Gyroscopic Effect

The characteristics of a propeller are similar to those of a gyroscope. RPM and the forces of inertia affect its rigidity and therefore its precession. When a force is imposed on the propeller disc it is precessed through 90° in the direction of rotation and is applied at that point. This means that if the aeroplane is yawed it is transposed as a pitching force; if the aeroplane is climbed or descended then the transposed force becomes a yaw. The gyroscopic effect occurs during aeroplane pitch changes and is most noticeable during low-speed flight with high propeller RPM.

When viewed from behind, if the propeller is rotating clockwise then if the aeroplane is yawed left then it is transposed as a nose-up pitch. Similarly, if the aeroplane is put into a climbing attitude it transposes as a yaw to the right. If the propeller rotates anticlockwise then the transposition will be in the opposite sense. A summary of the propeller gyroscopic effect is at Table 16.2.

Table 16.2 Propeller Gyroscopic Effect.

Propeller Gyroscopic Effect		
Propeller Rotation Viewed from behind	**Applied Force**	**Reaction**
Clockwise	Yaw left	Pitch up
	Yaw right	Pitch down
	Pitch up	Yaw right
	Pitch down	Yaw left
Anticlockwise	Yaw left	Pitch down
	Yaw right	Pitch up
	Pitch up	Yaw left
	Pitch down	Yaw right

16.7   Propeller Forces and Moments

A rotating propeller is subject to several different forces that stretch, bend or twist the propeller and as such it is subject to great stress. The most significant are:

a. centrifugal force (CF);
b. centrifugal twisting moment (CTM);
c. aerodynamic twisting moment (ATM).

16.7.1   Centrifugal Force (CF)

The propeller rotating at high speed creates a large centrifugal force that tends to stretch the blade outward away from the central hub or shaft. This same force affects the gearing mechanism housed within the spinner of the propeller. In addition to this, the resistance of the air applies a force in the opposite direction to the propeller rotation and tends to bend the propeller blade in that direction. Furthermore, thrust forces tend to bend the propeller blade forward of the plane of the propeller disc.

16.7.2   Centrifugal Twisting Moment (CTM)

The centrifugal force to which a rotating propeller is subject creates tensile stress at the blade root and a torque across the chord of the blade, about the pitch-change axis. In Figure 16.9 the centrifugal force (CF) acting along a radial from the propeller hub is divided into components parallel to the pitch-change axis and at right angles to that axis.

Because in normal flight the propeller blade has a positive angle of attack the blade leading edge is above the blade trailing edge, which creates a twisting moment about the pitch-change axis (the blade's longitudinal axis) attempting to bring the blade mass in line with the plane of rotation. The CTM tends to decrease the blade angle of attack, thereby 'fining' the pitch. Therefore, the fitted pitch-change mechanism requires little energy to reduce the blade angle but needs a lot more effort to increase the blade angle, i.e. coarsen the propeller pitch.

16.7.3   Aerodynamic Twisting Moment (ATM)

If the propeller blade CP is ahead of the pitch-change axis on the blade chordline the torque imposed on the blade is increased when the blade angle of attack is increased and vice versa this is aerodynamic

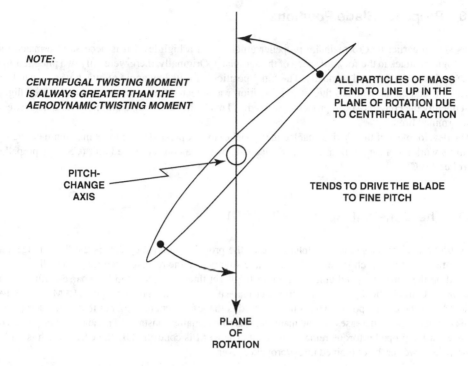

Figure 16.9   Centrifugal Twisting Moment.

twisting of the blade. Under normal circumstances the ATM tends to coarsen the pitch and partially counterbalances the CTM. See Figure 16.10(a).

For a windmilling propeller, as for instance in a steep dive, the direction of the ATM is reversed and acts in the same direction as the CTM. The total moments can affect the operation of the pitch-change mechanism. See Figure 16.10(b).

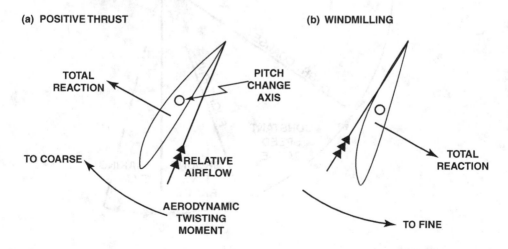

Figure 16.10   Aerodynamic Twisting Moment.

PROPELLERS

16.8 Propeller-Blade Positions

It was shown earlier that to maintain propeller efficiency at a high level it is necessary to match the blade angle of attack to the forward speed of the aeroplane. Originally, there were only two pitch settings and these had to be selected manually. The 'fine' position with its low angle of attack was selected for the take-off and climb, whereas the 'coarse' position was selected for the cruise and high-speed flight. Subsequently the constant-speed propeller evolved and made the blade angle infinitely variable between predetermined limits.

The development of this device enabled the propeller to be operated close to its maximum efficiency within its working range at all times. The device is known as a constant speed unit (CSU) or propeller control unit (PCU).

16.9 The Constant-Speed Unit (CSU)

For a turbopropeller aeroplane the alpha range of the propeller control system is the flight range and the beta range is that which is below the flight idle stop before the reverse pitch range. The limitations imposed on the constant-speed unit are those of the flight fine-pitch stop and the coarse-pitch stop. In flight the CSU automatically adjusts the propeller-blade pitch angle to maintain a given RPM irrespective of the TAS or the engine power. The constant speed is that of the engine and not that of the aeroplane. If the aeroplane speed increases but the manifold pressure remains constant then the propeller pitch will increase but the propeller torque remains constant. The RPM is controlled by the CSU, which is set by the propeller lever or the combined thrust/propeller lever.

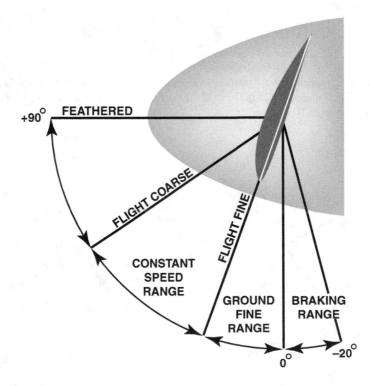

Figure 16.11   Range of Propeller-Blade Positions.

For feathering or reverse thrust, the CSU is overridden and the required pitch angle is set directly. The engine RPM, which is controlled by the fuel control unit adjusting the fuel flow through a governor, is constant for taxying and any required thrust changes are obtained by varying the propeller pitch. The range of pitch settings available is shown in Figure 16.11; the blade angle increasing with increased speed and the blade angle of a feathered propeller at 90°.

a. Feathering, that is turning the propeller blade through the coarse-pitch stop until its angle of attack produces zero lift.
b. Braking or reverse-thrust, that is rotating the propeller blades through the fine-pitch stop so as to use the lift of the propeller to assist in bringing the aeroplane to a halt during the landing ground run or abandoned take-off.

16.9.1 Propeller Windmilling

In the event of an engine failure both the engine and the propeller will slow down. The CSU will attempt to maintain the selected RPM by driving the propeller blades to a fine position that will continue until reaching the fine-pitch stop. Figure 16.12(a).

Despite the engine failing the aeroplane still has forward speed, for a single-engined aeroplane due to the glide descent or for a multi-engined aeroplane due to the thrust of the live engine(s). No matter which, the forward speed causes the propeller to have a negative angle of attack due to the direction of the relative airflow. This angle of attack generates sufficient torque to rotate the unpowered propeller in the normal direction of rotation. The component of the force at right angles to the plane of rotation now acts rearwards and is no longer thrust but is now drag and presents a flat disc to the direction of travel.

Because the propeller is being driven only by the relative airflow it is said to be windmilling. As such, it causes a considerable amount of drag; reduces the IAS and for multi-engined aeroplanes increases fuel flow on the live engine(s), decreases the maximum cruise altitude, decreases the maximum range and decreases the maximum endurance. The rotational drag, so caused, is greater than a stationary unfeathered propeller and tends to drive the engine that if allowed to continue will damage the engine. If the throttle is opened whilst the propeller blades are at this angle of attack it could seriously damage the engine. To prevent the likelihood of this damage happening the propeller must be feathered.

If, in flight, the propeller of a failed engine is not feathered but is windmilling it will cause a dramatic increase in the amount of profile drag. On a twin-engined aeroplane the asymmetric forces so produced could cause the aeroplane to become uncontrollable. The increased thrust required from the live engine, to prevent the aeroplane from stalling, could cause a yawing moment too large for the rudder to counteract, particularly at low speed on the approach to land.

16.9.2 Propeller Feathering

To prevent the propeller from 'windmilling' the propeller blade is rotated from the fine-pitch stop through the coarse-pitch stop to reach the angle of attack to the relative airflow that produces zero lift; and referred to as 'feathering' the propeller. This is shown in Figure 16.11 as 90° but usually it is approximately 85° because it is the best compromise for the whole blade length and accounts for the blade twist. This blade position stops the propeller from rotating because there is no lift force to produce a torque reaction; it reduces the drag to a minimum level and prevents damage being caused to the engine. In the past, the feathering angle was such that it caused a small backward rotation that locked the propeller and prevented any further damage to the engine. See Figure 16.12(b).

For a single-engined aeroplane feathering the propeller blades improves the glide performance because it does not require such a steep glide angle to maintain the required airspeed. As such, feathering the propeller increases the gliding range and endurance. For a multi-engined aeroplane, the decreased drag reduces the amount of thrust required to counteract the asymmetric force of the propeller of the failed engine and there is sufficient rudder authority remaining to enable the pilot to safely control the aeroplane.

(a) WINDMILLING

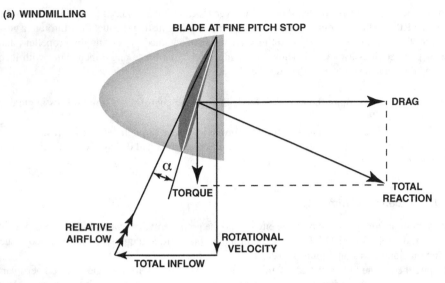

(b) FEATHERED POSITION

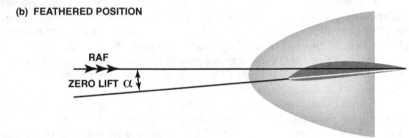

(c) REVERSE-THRUST POSITION

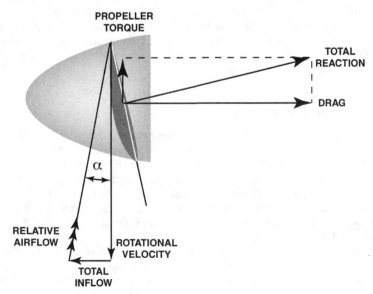

Figure 16.12   Propeller-Blade Angles.

16.9.3 Reverse Pitch

Some aeroplanes have a manual override to the constant-speed mechanism to enable the pilot to reverse the pitch angle and thereby reverse the thrust. Utilisation of this facility decreases the length of the ground run during a landing or an abandoned take-off and saves the wear on the brakes and tyres.

For the propeller to produce thrust opposite to the direction of travel the propeller blades have to be rotated through the fine-pitch stop to the 'braking' position. The forces acting on a propeller in this situation are shown in Figure 16.12(c). The propeller blades have a negative angle of attack and will produce the required negative thrust. The drag so produced will decrease the length of the landing ground run or in the event of an abandoned take-off decrease the distance required to bring the aeroplane to a halt.

When a propeller blade is in reverse pitch the maximum blade angle is restricted because the effect of blade twist is acting in the opposite sense to normal. The blade tip has a greater angle of attack than the blade root. Therefore, the blade generates the largest amount of thrust at the tip. In so doing it creates a significant bending moment that restricts the maximum reverse thrust that a blade can withstand. The usual maximum blade angle is approximately $-25°$.

16.10 The Effect of a Constant Speed Propeller On a Glide Descent

For a constant-speed-propeller aeroplane during a glide descent, maintaining a constant IAS with idle thrust and the RPM lever fully forward, the propeller pitch and lift/drag ratio will decrease but the rate of descent will increase because the TAS gradually increases during the descent.

With the throttle closed and the engine idling during a glide descent for an aeroplane with a constant-speed propeller, if the propeller pitch lever is pulled fully back the propeller pitch will increase to fully coarse, but the rate of descent, glide angle and the RPM will decrease. In a constant IAS descent if the propeller pitch is increased it will increase the lift/drag ratio and decrease the rate of descent. See Table 16.3.

Table 16.3 Glide-Descent Summary.

	Glide Descent		
	IAS	Constant	Constant
Setting	Throttle	Closed	Closed
	Thrust	Idle	Idle
	RPM Lever	Fully Forward	Fully Back
	RPM	Increase	Decrease
Result	Propeller Pitch	Decrease (Fine Pitch)	Increase (Coarse Pitch)
	Lift/Drag ratio	Decrease	Increase
	Rate of Descent	Increase	Decrease
	TAS	Increase	Increase

16.11 Engine Failure

The reaction of an aeroplane to an engine failure is dependent on where the engines are mounted on the airframe, whether they are counterrotating and whether they are propeller driven or turbojet. The advantage of having rear-mounted engines over those mounted on the wings is that any change of thrust has less affect on the longitudinal control of the aeroplane. Very few aeroplanes have rear-mounted propeller-driven engines; as a rule they are jet-propelled.

If the engines are wing-mounted, then the disadvantage of counterrotating propeller-driven engines over turbojet engines is that in the event of an engine failure the aeroplane has a tendency to roll in the direction of rotation of the dead engine. A jet-propelled aeroplane having suffered an engine failure is not prone to this effect.

Self-Assessment Exercise 16

Q16.1 Which of the following definitions of propeller parameters is correct?
  (a) Propeller angle of attack is the angle between the chordline and the propeller vertical plane.
  (b) Critical tip velocity is the propeller speed at which the risk of flow separation at some parts of a blade occurs.
  (c) Geometric propeller pitch is the theoretical forward distance a propeller blade travels in one revolution.
  (d) Blade angle is the angle between the blade chordline and the propeller axis.

Q16.2 The blade angle of attack of a fixed-pitch propeller increases when:
  (a) forward speed and RPM decrease
  (b) RPM increases and forward speed decreases
  (c) forward speed and RPM increase
  (d) forward speed increases and RPM decreases

Q16.3 Increasing the number of propeller blades will:
  (a) decrease the torque of the propeller shaft at maximum power
  (b) increase the maximum power absorption
  (c) not affect the propeller efficiency
  (d) increase the noise level at maximum power

Q16.4 Constant-speed propellers are more efficient than fixed-pitch propellers because they:
  (a) produce a higher maximum thrust than a fixed-pitch propeller
  (b) have more blades than a fixed-pitch propeller
  (c) produce an almost maximum efficiency over a greater speed range
  (d) have greater maximum efficiency than a fixed-pitch propeller

Q16.5 Why is a propeller blade twisted from root to tip?
  (a) To ensure that the root produces the greatest thrust.
  (b) To ensure that the tip produces the greatest thrust.
  (c) Because the blade angle of attack is dependent at any point on the ratio of the rotational speed to the TAS.
  (d) Because the blade angle of attack is dependent at any point on the ratio of the rotational speed to the propeller angular velocity.

Q16.6 One of the advantages of increasing the number of propeller blades is:
  (a) increased noise
  (b) less power is absorbed by the propeller
  (c) higher tip speed
  (d) improved solidity

Q16.7 How does the thrust of a fixed-pitch propeller vary during take-off run? The thrust
  (a) increases slightly while the aeroplane builds up speed
  (b) only varies with mass changes
  (c) is constant during the take-off and climb
  (d) decreases slightly while the aeroplane builds up speed

Q16.8 Seen from behind a propeller rotates clockwise, the torque effect during take-off will:
  (a) roll the aeroplane to the left
  (b) pitch the aeroplane nose-down
  (c) roll the aeroplane to the right
  (d) pitch he aeroplane nose-up

Q16.9 Which of the following statements regarding a constant-speed propeller is correct?
  (a) The blade angle increases with increasing speed.
  (b) The propeller system maintains the aeroplane at a constant speed.
  (c) The RPM decreases with increasing aeroplane speed.
  (d) The manifold pressure maintains the selected RPM.

Q16.10 During a glide descent of a constant-speed propeller aeroplane at idle thrust and a constant speed, if the RPM lever is pulled back the propeller pitch will:
  (a) increase and the rate of descent will decrease
  (b) increase and the rate of descent will increase
  (c) decrease and the rate of descent will decrease
  (d) decrease and the rate of descent will increase

Q16.11 During a glide descent of a constant-speed propeller aeroplane at idle thrust and a constant speed, if the RPM lever is pushed forward the propeller pitch will:
  (a) decrease and the rate of descent will decrease
  (b) decrease and the rate of descent will increase
  (c) increase and the rate of descent will decrease
  (d) increase and the rate of descent will increase

Q16.12 Viewed from behind a propeller rotates clockwise, the asymmetric-blade effect in the climb will:
  (a) roll the aeroplane to the right
  (b) yaw the aeroplane to the left
  (c) roll the aeroplane to the left
  (d) yaw the aeroplane to the right

Q16.13 When the blades of a propeller are in the feathered position:
  (a) the propeller produces an optimal windmilling RPM
  (b) the windmilling RPM is maximum
  (c) the RPM is just sufficient to lubricate the engine
  (d) the drag on the propeller is minimal

Q16.14 During medium horizontal turbulence, the pitch angle of a constant-speed propeller alters:
  (a) slightly
  (b) strongly
  (c) not at all
  (d) if the pitch is fully fine

Q16.15 The angle of attack for a propeller blade is the angle between the chordline and:
  (a) the local airspeed vector
  (b) the direction of the propeller axis
  (c) the aeroplane heading
  (d) the principal direction of the propeller blade

Q16.16 Gyroscopic precession of the propeller is induced by:
  (a) pitching and rolling
  (b) increasing RPM and yawing
  (c) increasing RPM and rolling
  (d) pitching and yawing

Q16.17 Propeller efficiency may be defined as the ratio between:
  (a) propeller power available and the shaft power
  (b) the thrust and the maximum thrust
  (c) the power available and the maximum power
  (d) the thermal power of fuel flow and the shaft power

Q16.18 For a fixed-pitch propeller designed for the cruise, the angle of attack of each blade, measured at the reference section is:
  (a) lower in the ground run than in flight with the same RPM
  (b) always positive during an idle descent
  (c) is optimum when the aeroplane is in a stabilised cruise
  (d) decreased when the aeroplane speed is decreased at the same RPM

Q16.19 In an idle-power glide at a constant IAS increasing the propeller pitch will (i) . . . . the L/D ratio and (ii) . . . . the rate of descent.
(a) (i) decrease; (ii) decrease
(b) (i) decrease; (ii) increase
(c) (i) increase; (ii) decrease
(d) (i) increase; (ii) increase

Q16.20 Which of the following statements regarding propeller drag is correct.
An engine failure can result in the drag of (i) a windmilling propeller and (ii) a nonrotating propeller.
(a) (i) is equal to (ii)
(b) (ii) is larger than (i)
(c) impossible to say which is the largest
(d) (i) is larger than (ii)

Q16.21 Decreasing the propeller pitch during an idle-power glide descent at a constant IAS will cause the L/D ratio to (i) and the rate of descent to (ii) . . . . :
(a) (i) increase; (ii) increase
(b) (i) decrease; (ii) increase
(c) (i) increase; (ii) decrease
(d) (i) decrease; (ii) increase

Q16.22 On a fixed-pitch propeller the angle of attack of the propeller is increased by (i) . . . . . . RPM and (ii) . . . . . . TAS:
(a) (i) increased; (ii) increased
(b) (i) increased; (ii) decreased
(c) (i) decreased; (ii) decreased
(d) (i) decreased; (ii) increased

Q16.23 During a glide descent with the throttle closed and the engine idling if the propeller lever is pulled back on an aeroplane with a constant-speed propeller it will cause the rate of descent to (i) . . . . and the RPM to . . . .
(a) (i) decrease; (ii) increase
(b) (i) decrease; (ii) decrease
(c) (i) increase; (ii) increase
(d) (i) increase; (ii) decrease

Q16.24 The purpose of increasing the number of blades on a propeller is:
(a) to improve the efficiency of the pitch control
(b) to increase the power absorption
(c) to reduce the size of the noise footprint
(d) to increase the maximum RPM

Q16.25 Viewed from the front, an anticlockwise rotating propeller during the take-off ground roll will produce a load (i) . . . . on the (ii) . . . . wheel due to (iii) . . . . . . :
(a) (i) increase; (ii) right; (iii) torque reaction
(b) (i) decrease; (ii) right; (iii) gyroscopic effect
(c) (i) increase; (ii) left; (iii) torque reaction
(d) (i) decrease; (ii) left; (iii) gyroscopic effect

Q16.26 During take-off the effect that the torque reaction of a clockwise rotating propeller (viewed from the behind) has on the left wheel load . . . . (i) and the right-wheel load . . . . (ii) is:
(a) (i) increased; (ii) constant
(b) (i) constant; (ii) increased
(c) (i) decreased; (ii) constant
(d) (i) increased; (ii) decreased

Q16.27 During a glide descent at idle thrust if the propeller pitch is decreased then the rate of descent will (i) . . . . and the lift/drag ratio will (ii) . . . . . .
  (a) (i) increase; (ii) increase
  (b) (i) increase; (ii) decrease
  (c) (i) decrease; (ii) increase
  (d) (i) decrease; (ii) decrease

Q16.28 The gyroscopic effect of the propeller is affected by:
  (a) roll and pitch
  (b) decreased blade angle
  (c) increased RPM
  (d) increased blade angle

Q16.29 Which of the following definitions is correct regarding the propeller parameters?
  (a) The propeller speed at which there is a risk of the airflow separating is the critical tip speed.
  (b) Blade angle of attack is the angle between the chordline and the propeller axis.
  (c) Blade angle is the angle between chordline and the propeller axis.
  (d) Geometric propeller pitch is the theoretical forward distance travelled in one rotation.

Q16.30 With regard to a constant-speed propeller:
  (a) pitch angle decreases with increasing TAS
  (b) pitch angle increases with increasing TAS
  (c) RPM decreases with increasing TAS
  (d) RPM increases with increasing TAS

Q16.31 Propeller efficiency is the ratio of:
  (a) Thrust HP to overall HP
  (b) Brake HP to maximum thrust
  (c) Thrust HP to Shaft HP
  (d) Shaft HP to overall HP

Q16.32 The advantages of increasing the number of blades on a propeller is:
  (a) decreased efficiency
  (b) increased noise
  (c) increased vibration
  (d) increased power absorption

Q16.33 The effect of feathering a propeller is to:
  (a) minimize the lift/drag ratio
  (b) maximize the windmilling speed
  (c) improve the thrust produced
  (d) reduce the drag on the propeller to minimum

Q16.34 During a glide with idle thrust at a constant IAS if the propeller pitch is decreased the lift/drag ratio will:
  (a) decrease
  (b) increase
  (c) remain the same
  (d) increase when the CSU activates

Q16.35 The effect of increasing the propeller pitch in a glide descent is that the L/D ratio (i) . . . . and the rate of descent (ii) . . . .
  (a) (i) increases; (ii) decreases
  (b) (i) increases; (ii) increases
  (c) (i) decreases; (ii) decreases
  (d) (i) decreases; (ii) increases

Q16.36 The asymmetric-blade effect of a clockwise rotating propeller (viewed from behind) during a climb is a:
   (a) yaw to the left
   (b) roll to the left
   (c) roll to the right
   (d) yaw to the right

Q16.37 A propeller blade is twisted from root to tip because:
   (a) maximum thrust is required from the tip
   (b) maximum thrust is required from the root
   (c) the local angle of attack is dependent on the TAS
   (d) the local angle of attack is dependent on the angular velocity

Q16.38 During a power-off descent if the propeller RPM lever is pushed forward the propeller pitch (i) ...... and the rate of descent (ii) ....
   (a) (i) decreases; (ii) decreases
   (b) (i) decreases; (ii) increases
   (c) (i) increases; (ii) increases
   (d) (i) increases; (ii) decreases

Q16.39 The advantage of a constant-speed propeller over a fixed-pitch propeller is:
   (a) increased maximum available thrust
   (b) increased maximum efficiency
   (c) it has a larger blade surface area
   (d) nearly maximum efficiency over a greater speed range

Q16.40 The drag of a windmilling propeller compared with that of a stationary propeller is that:
   (a) the windmilling propeller has less drag
   (b) the windmilling propeller has more drag
   (c) they are the same
   (d) it is impossible to determine

Q16.41 In a glide descent increasing the propeller pitch (i) .... the L/D ratio and (ii) .... rate of descent
   (a) (i) increases, (ii) increases
   (b) (i) decreases, (ii) decreases
   (c) (i) increases, (ii) decreases
   (d) (i) decreases, (ii) increases

Q16.42 The angle of attack of a fixed-pitch propeller for a general-purpose aeroplane:
   (a) is optimum during take-off
   (b) decreases with decreased speed at constant RPM
   (c) is optimum in stabilised cruising flight
   (d) is less during the take-off ground run than in level flight

Q16.43 The angle of attack of a propeller blade is the angle between the chord and the:
   (a) TAS vector
   (b) plane of the axis of the propeller
   (c) aeroplane's course
   (d) relative airflow

Q16.44 During a constant IAS climb, in a normal atmosphere, with a fixed-pitch propeller the angle of attack of the propeller:
   (a) decreases
   (b) remains the same
   (c) diminishes to zero
   (d) increases

Q16.45 In a glide descent increasing the propeller pitch (i) .... the L/D ratio and (ii) ....
... rate of descent
  (a) (i) increases; (ii) increases
  (b) (i) decreases; (ii) decreases
  (c) (i) decreases; (ii) increases
  (d) (i) increases; (ii) decreases

Q16.46 The advantage of a constant-speed propeller over a fixed-pitch propeller is:
  (a) nearly maximum efficiency over a greater speed range
  (b) increased maximum available thrust
  (c) increased maximum efficiency
  (d) it has a larger blade surface area

Q16.47 The torque of a clockwise turning propeller (when viewed from behind) produces a:
  (a) nose-up pitching moment
  (b) rolling moment to the left
  (c) nose-down pitching moment
  (d) rolling moment to the right

17 Operational Considerations

Any operational condition that adversely affects the efficiency of an aeroplane's performance is described in this chapter and includes surface contamination, aeroplane contamination and windshear.

17.1 Runway-Surface Contamination

Contamination of the runway surface will affect the take-off and landing performance of an aeroplane. A brief description of each of the runway-surface contaminant types is given in the following paragraphs. The recommended maximum depth for continued operations is shown in Table 17.1.

17.1.1 Surface Contaminants

17.1.1.1 Standing Water

Visible puddles, usually of rain, of a depth greater than 3 mm standing on the surface causing paved surfaces to glisten when the temperature is above 0 °C is standing water. On a natural surface it is assumed that more than 3 mm of water exists if under firm foot pressure the water rises to the surface. The specific gravity of water is 1.0. The maximum permissible depth of water permissible for take-off and landing is 15 mm (0.6 in). A depth of less than 3 mm is not considered to be significant. *CS 25 Book 2 AMC 25.1591 paragraph 4.1.*

17.1.1.2 Slush

Partly melted snow or ice with a high water content, and from which water can readily flow, is referred to as slush. It is displaced with a splatter when a heel-and-toe slapping motion is made on the ground. Slush is normally a transient condition found only at temperatures close to 0 °C and has an assumed specific gravity of 0.85. The maximum depth permissible for take-off and landing is 15 mm (0.6 in). *CS 25 Book 2 AMC 25.1591 paragraph 4.2.*

17.1.1.3 Wet Snow

Loose snow taking the form of large flakes, which if compacted by hand will stick together to form a snowball, but will not readily allow water to flow when squeezed. It forms a white covering on all surfaces, which when stamped on does not splash up. The temperature for this type of snow is between minus 5 °C and minus 1 °C, with an assumed specific gravity (SG) of 0.5. For take-off and landing the maximum permissible depth is 15 mm (0.6 in). *CS 25 Book 2 AMC 25.1591 paragraph 4.3.*

17.1.1.4   Dry Snow

Loose hard snow usually in the form of dry pellets that can be blown, or if compacted by hand will fall apart on release, is dry snow. For this contaminant to be present the temperature must be below minus 5 °C (and to have not risen since the snow fell). Its specific gravity is assumed to be 0.2, which is not applicable to snow that has been subjected to the natural ageing process. The maximum permissible depth for take-off and landing is 60 mm (2.4 in) on any part of the runway, measured by ruler. *CS 25 Book 2 AMC 25.1591 paragraph 4.4.*

17.1.1.5   Very Dry Snow

Granular snow, usually in the form of powder that can be easily blown, is very dry snow. For this contaminant to be present the temperature must be well below minus 5 °C. Its specific gravity is assumed to be less than 0.2. The maximum permissible depth for take-off and landing is 80 mm (2.4 in).

17.1.1.6   Compacted Snow

Snow that has been compressed into a solid mass and resists further compression is compacted snow. It will hold together or break into lumps if picked up and is such that aeroplane wheels, at representative operating pressures and loadings, will run on the surface without causing significant rutting. This type of covering is normally caused by the transit of vehicles over the surface when snow is falling. Its specific gravity is 0.5 and over. *CS 25 Book 2 AMC 25.1591 paragraph 4.5.*

17.1.1.7   Ice

Ice is a frozen layer of surface moisture, including the condition where compacted snow transitions to a polished ice surface. The thickness of ice varies and produces a poor coefficient of friction according to the condition of the surface. *CS 25 Book 2 AMC 25.1591 paragraph 4.6.*

17.1.1.8   Specially Prepared Winter Runway

A runway, with a dry frozen surface of compacted snow and/or ice that has been treated with sand or grit or has been mechanically or chemically treated to improve the runway-surface friction. The runway-surface friction is measured and reported on a regular basis in accordance with national procedures. *CS 25 Book 2 AMC 25.1591 paragraph 4.7.*

17.1.1.9   Mixtures

Mixtures of ice, snow and/or standing water may, especially when rain, sleet or snow is falling, produce a substance having an SG above 0.8. This substance is transparent at high SGs, and is easily distinguished from slush, which is cloudy.

Table 17.1 Surface Contaminants.

Contaminant	SG	Maximum Depth	Contaminant Drag	Surface Friction Reduced	Maximum WED
Standing water	1.0	15 mm	YES	YES	15 mm
Slush	0.5 to 0.85	15 mm	YES	YES	12.75 mm
Very dry snow	<0.35	80 mm	YES	YES	<28 mm
Dry snow	0.20	60 mm	YES	YES	12 mm
Wet snow	0.35 to 0.5	15 mm	YES	YES	7.5 mm
Compacted snow	-	0 mm*	NO	YES	-
Ice	-	0 mm*	NO	YES	-
Winter Runway	-	0 mm*	NO	YES	-

*Aeroplane assumed to be rolling on the surface.

17.1.1.10   Contaminant Drag

Contaminant drag has two components that are:

a. drag due to fluid displacement by the tyres and
b. spray impingement drag from the tyres onto the airframe.

17.1.1.11   Water-Equivalent Depth

The limitations and corrections given in most Flight Manuals are calculated for a uniform layer of contaminant at the maximum possible depth and the specific gravity quoted in Table 17.1. Flight Manuals that do not contain this information express the correction in terms of water-equivalent depth (WED); that is the contaminant depth multiplied by its specific gravity. Estimated data are not valid for WEDs exceeding 15 mm.

17.2   The Effect of Runway Contamination

When calculating the field-length-limited TOM or the field-length-limited landing mass the runway-surface condition must be accounted. Most AFMs contain performance data for contaminated runway surfaces for which it is assumed that all engines are operating. The usual safety margins in the event of an engine failure are not generally provided by the data, but the AFM will contain advice on the procedure to be adopted in such circumstances.

17.2.1   Take-off

Depths of greater than 3 mm of water, slush or wet snow, or 10 mm of dry snow, are considered significant and likely to affect the take-off and landing performance of aeroplanes. The main effects during take-off are:

a. Additional drag – retardation on the wheels, spray impingement and increased skin friction.
b. The possibility of thrust loss or system malfunction due to spray ingestion or impingement.
c. Reduced wheel braking performance – reduced wheel to runway-surface friction with the possibility of aquaplaning, which significantly increases stopping distances and reduces directional control.

d. Directional control problems.
e. Possibility of structural damage.

A water depth or water-equivalent depth of less than 3 mm is not considered significant and does not require corrections to the take-off performance calculations other than the allowance where applicable for the effect of a wet runway. However, on such a runway where the water depth is less than 3 mm and where the performance effect of spray impingement and retardation of the wheels is insignificant, isolated patches of standing water or slush in excess of 15 mm located at the latter end of the TOR may still lead to ingestion and temporary thrust fluctuations, which could impair safety. Some aircraft types are susceptible to thrust fluctuations at depths greater than 9 mm and the limitations quoted in the AFM should be checked. For contaminant depths greater than 9 mm specific mention will be made in the surface conditions report as to the location, extent and depth of water patches.

The most important area when assessing the likely effect of runway contamination on aircraft performance is the upwind half of the runway. If the estimated braking action is poor or if the Mu-meter reading is below 0.40 on this portion of the runway, it is possible in the event of an abandoned take-off at V1, the aircraft may not be brought to a halt safely before reaching the end of the stopway. If the runway-surface conditions were such, it would be prudent to delay take-off until the depth of contaminant has drained below the significant level of 3 mm. When there is any doubt about the depth of the contaminant or about the area of standing water on the runway surface DELAY TAKE-OFF.

Because of the difficulty in measuring the coefficient of friction on a contaminated runway surface and the inability of the manufacturers to predict the effect it will have on the scheduled performance of the aeroplane, in the event of an engine failure during take-off there may be a time interval during which the aeroplane is unable to continue with the take-off or to stop within the distance available without risk of accident. The duration of the 'risk period' is impossible to forecast because of the indeterminate nature of the acceleration and deceleration of the aeroplane on such a surface. The best way to minimize the 'risk period' is to keep the TOM as low as possible by ensuring that unnecessary fuel is not carried: definitely do not tanker fuel.

The lengths of the TOR, ASD and TOD will all be greater than the lengths for the same conditions on a dry or wet runway because of the spray impingement drag and the decreased surface friction. The LD will be increased because of the decreased brake efficiency. As a result, of these factors both the FLL TOM and the FLL LM are less than would be possible on a dry runway.

For aeroplanes that have an approved procedure for take-off on runways that have a 'significant' depth of contamination the AFM will state the latest point at which a decision to abandon take-off may be made and the earliest point at which a decision to continue the take-off with one engine inoperative may be made.

In such instances the ASDR will not exceed the ASDA because the manufacturers have, in their performance calculations, to allow for 150% of the actual contaminant depth. Information is also given of the penalties incurred if less than full reverse thrust is used to abandon take-off.

If the runway-surface braking coefficient is 0.50 or less, the runway is considered to be icy or very slippery. The V1 to be used for take-off is that applicable to wet runway. The use of this decision speed is to cover the effect of icy surfaces or aquaplaning, and ensures that ASDR does not exceed ASDA. Some manufacturers schedule the effect of aquaplaning separately where more accurate assumptions can be made.

Most AFMs contain advisory information for abandoning take-off on such a surface and refer to VSTOP. This is the maximum speed from which the aeroplane can be safely stopped, with all engines operating on a contaminated runway, by the end of the stopway and is sometimes referred to as 'maximum abandonment' speed. It is based on the use of all available means of retardation, although the data scheduled uses only 50% of full reverse thrust on all engines.

'Maximum abandonment speed' is *not* a V1 because it does not imply an ability that take-off can be safely continued after an engine failure. In the event of a power unit failure it is the maximum speed from which the aeroplane should be able to stop within the length of the stopway, it is not guaranteed. Unlike V1, this speed can be less than VMCG.

17.3   Aeroplane Contamination

Contamination of an aeroplane surface may be caused by ice, frost, rain, dust or damage. The effect of these contaminants on the various parts of an aeroplane is considered here.

17.3.1   The Effect of Heavy Rain

The effects that heavy rain has on aeroplane performance are:

a. The impingement and rapid accumulation of water sufficiently distorts the shape of the upper wing surface and diminishes the total lift by up to 30% of its normal value. Thus, C_{Lmax} is greatly decreased.
b. The impact of the rain on an aeroplane in the landing configuration decreases the aircraft's forward speed and increases the total drag.
c. When operating in such conditions a jet engine is slow to respond to rapid demands for power.
d. The mass of the rain can momentarily increase the aeroplane mass.

The total effect of these factors is to produce a much lower level of performance for any aeroplane experiencing such weather conditions. The weight increase raises the stalling speed, whilst the reduced lift and increased drag together reduce the forward speed achieved. For an aeroplane in the take-off or landing configuration, this situation is *extremely dangerous* because of the close proximity of the normal operating speed to the stalling speed. There is, therefore, a very high risk of the aeroplane stalling when it is close to the ground.

If there is a possibility of encountering *heavy rain* during take-off, it is advisable to *delay* the *departure*. If the danger were during the landing phase, it would be prudent to divert to an alternative aerodrome or to hold off until the rain has cleared the area. It could prevent a disaster.

17.3.2   The Effect of Propeller Icing

Ice formation on the surface of a propeller blade will alter the aerodynamic shape of the blade and results in:

a. Reduced thrust output because the changed profile of the blade diminishes its ability to convert all of the power available from the engine.
b. Increased loading of the propeller caused by the increased mass.
c. Increased total drag due to the increased surface friction caused by the roughness of the ice.
d. Damage to the airframe and/or the engines as a result of the ice breaking off of the propeller.
e. Reduced efficiency of the pitch control due to ice formation on the mechanism.

Usually, the propeller blades are deiced by boots attached to the leading edge of the blades. The boots are usually electrically heated and melt the ice. However, should a single blade deicing boot fail, it leads to an exceptionally dangerous asymmetric blade situation. The result is an extremely unpleasant vibration of the whole aeroplane, which can lead to structural damage if the propeller is not feathered as soon as possible.

The only way to minimize the effect of asymmetric blade deicing is to shut the engine down and to feather the propeller. This prevents serious structural damage to the airframe and engine(s).

17.3.3   The Effect of Airframe Icing

A deposit of ice or frost on the airframe surface increases the skin friction and destroys the boundary-layer energy and encourages boundary-layer separation. It increases the mass of the aeroplane and changes the shape of the lifting surfaces that not only decreases the total lift generated and C_{Lmax} but also increases the total drag and the stalling speed. Thus, the performance of the aeroplane is significantly diminished the most serious effect of which is the reduction of C_{Lmax}, and inevitably decreases the lift/drag ratio.

The effect of ice formation on the leading edge of the wings is most critical during the take-off phase at the last part of the rotation when the main wheels are about to lift off the runway surface. This is because the lift generated by a deformed wing surface may not be enough to counteract the increased aeroplane mass caused by the ice. This will increase the length of the take-off ground run beyond that which was expected and could cause the aeroplane:

a. to overrun the stopway end if the take-off is abandoned or
b. to come perilously close to any obstacles that may be encountered after lift-off or
c. clear any obstacles in the take-off climb path obstacle accountability area by less than the statutory minimum vertical interval or to possibly impact the obstacle.

The effectiveness of any high-lift devices is adversely affected by ice formation. Such deposits on the flap guide tracks may prevent the use of the flaps, which seriously affects the aeroplane performance during the landing phase because it could induce a stall during the approach.

Ice formation on the flaps may seriously adversely affect their performance and if they are slotted flaps could restrict the flow of re-energising airflow through the slots to the boundary layer with the consequent loss of lift and increase of drag.

Any reduction of the boundary-layer energy increases the speed at which boundary-layer separation is caused by the adverse pressure gradient. This increases the stalling speed and decreases the stalling angle. Consequently, at low airspeeds and high angles of attack the use of the ailerons could lead to a loss of lateral control because of tip stalling. A similar situation can occur during a steep turn.

The formation of ice or frost on the control surfaces will alter the aerodynamic moment of that control, and as a consequence the input force required from the pilot's controls to move the surfaces to balance the moment is significantly greater than normal.

17.3.4   The Effect of Airframe-Surface Damage

Similar to the effect of ice and frost, the effect of any factor that adversely affects the surface friction destroys the boundary layer and encourages its separation from the surface. The factors that may cause this are flaking or scratched paintwork, surface indentations, careless repairs or modifications and minor abrasions. All of these factors individually have a trivial effect on the aeroplane performance. However, the cumulative effect of these individual factors over the life of an aeroplane can seriously affect the performance of the aeroplane and may significantly alter it to such an extent that it no longer matches the manufacturer's published specifications.

17.3.5   The Effect of Turbulence

Unlike a gust that has only a short duration, turbulence is a disturbance in the atmosphere caused by frictional influences of the air resulting in the same changes of air velocity but over a longer period of time. Turbulence is categorized as slight, moderate or severe. Throughout its duration an aeroplane may experience sudden changes from positive 'g' to negative 'g' and the 'bumps' or jolts will vary in severity and will cause stress to the airframe.

A swept-back wing aeroplane is least affected by turbulence provided that it is flying at the recommended speed for flying in turbulence, V_{RA}. It is also advised that the aeroplane should be flown in the clean configuration for attitude and not altitude because the turbulence is often associated with a standing wave, which can cause a height loss or gain of up to 8000 ft. Although extending flaps will decrease the stalling speed it will also increase the load factor, which diminishes the speed margin to the structural limitations and is not recommended.

17.4   Windshear

Windshear describes a change of wind direction or speed associated with a change of height or horizontal distance. Usually, the change is gradual and is of no significance to aviation. However, it is the sudden change of direction or speed that causes control problems to pilots experiencing it because it displaces the aeroplane from its intended position unexpectedly and requires a large control input to correct the change it causes.

17.4.1   The Effect of Windshear

17.4.1.1   Energy Loss

The momentum of an aeroplane is proportional to its size and will sustain the groundspeed on encountering windshear. *If the strength of a headwind is suddenly decreased or the strength of a tailwind is suddenly increased, the aeroplane will lose energy and the indicated airspeed of the aeroplane will decrease.* As a consequence, the aeroplane loses lift and if corrective action is not taken it will suddenly lose height; if it is descending the rate of descent will increase and if it is climbing it will experience a reduced rate of climb. See Figure 17.1.

Figure 17.1   Effect of Energy Loss During a Descent.

17.4.1.2   Energy Gain

If the aeroplane experiences a sudden increase in the headwind component or a sudden decrease of a tailwind component then the aeroplane gains energy and the indicated airspeed increases. If the aeroplane is descending the rate of descent will decrease and if it is climbing it will experience a sudden increase in the rate of climb. See Figure 17.2.

During take-off or approach and landing changes to the energy of an aeroplane are critical because of the close proximity of the ground. In the event of lost energy it takes on average 7 s from opening the throttle for an engine to spool up to go-around thrust in the event of an aborted landing or to maximum take-off thrust during the take-off. The reaction time can be crucial.

The energy change experienced may be caused by vertical windshear or horizontal windshear. The resulting change of wind component will alter the airspeed and not the groundspeed of the aeroplane in the short term.

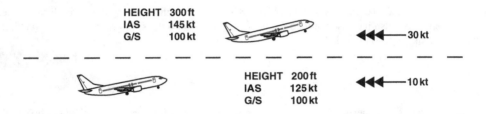

HEIGHT	300 ft
IAS	145 kt
G/S	100 kt

───── 30 kt

HEIGHT	200 ft
IAS	125 kt
G/S	100 kt

───── 10 kt

Figure 17.2   Effect of Energy Gain During a Climb.

17.4.2 Downdraught

17.4.2.1 Take-off

Often, a downdraught is descending cold air from a thunderstorm sometimes as a microburst. During take-off when close to the ground the energy change experienced is critical. The location of the thunderstorm relative to the runway will determine the effect that it will have on the take-off performance.

When taking off towards a thunderstorm into a headwind component from its downdraught as the climb progresses towards the thunderstorm the wind component changes to become a tailwind component causing the aeroplane to lose energy. Because of the momentum of the aeroplane the groundspeed remains the same but the airspeed decreases with the consequent loss of lift. As a result, the aeroplane descends and may even strike the ground. See Figure 17.3.

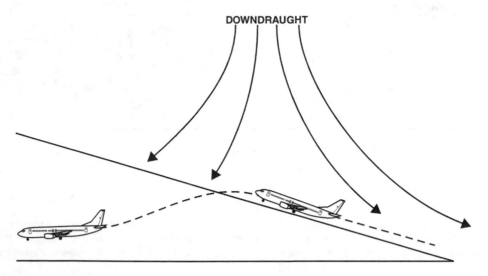

DOWNDRAUGHT

Figure 17.3   Take-off in a Downdraught.

17.4.2.2 Landing

When approaching a thunderstorm during a stabilised approach to land the wind component experienced is a headwind from the downdraught, Position A in Figure 17.4. The effect as before is that the groundspeed will remain unchanged due to the inertia of the aeroplane but the airspeed will increase. As a result, the aeroplane will start to climb, position B in Figure 17.4.

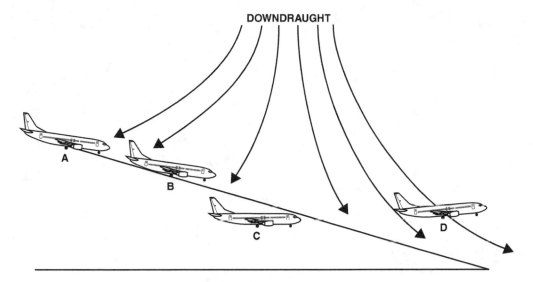

Figure 17.4   Landing in a Downdraught

The reaction of the pilot is to depress the nose of the aeroplane and reduce the thrust in an attempt to regain the glidepath. As the effect of the original momentum of the aeroplane is overcome the groundspeed decreases and the aeroplane drops below the glideslope. See Position C in Figure 17.4. It is now rapidly descending in an underpowered condition. Therefore, the thrust must be increased and the rate of descent reduced by raising the nose of the aeroplane and a go-around initiated.

17.4.3   Countering Windshear

During the approach to land if windshear is likely to be encountered then use an airspeed that is higher than normal but remember that the landing distance and the landing ground run will be longer than normal. If an unexpected loss of airspeed occurs on the approach to land then take the following actions:

a. Increase the thrust to the full go-around thrust.
b. Raise the nose to arrest the descent.
c. Coordinate thrust and pitch to counter the loss of airspeed.
d. Prepare to initiate a missed-approach procedure.

Should a downburst from a thunderstorm occur during the approach or take-off phase of flight then take the following actions:

a. If the encounter is severe then disengage the autopilot, autothrottles and/or flight directors and revert to manual handling.
b. Increase thrust to maximum take-off thrust.
c. Increase the pitch angle to 15° and maintain it.
d. If the stick shaker operates then ease the control column forward until it stops.
e. Maintain the new pitch angle.

Self-Assessment Exercise 17

Q17.1 In which phase of the take-off is the aerodynamic effect of ice located on the wing leading edge most critical?
  (a) During the climb with all engines operating.
  (b) All phases of the take-off are equally critical.
  (c) The last part of the rotation.
  (d) The take-off run.

Q17.2 In level flight with the port engine inoperative and wings level the correct turn and slip indication is:
  (a) turn indicator neutral and slip indicator neutral
  (b) turn indicator displaced left and slip indicator displaced left
  (c) turn indicator displaced left and slip indicator displaced right
  (d) turn indicator displaced right and slip indicator displaced left

Q17.3 The flight characteristic obtained at point B in the diagram below is:

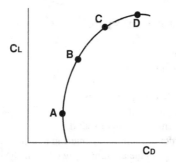

  (a) Maximum gliding range
  (b) Maximum gliding angle
  (c) Maximum gliding speed
  (d) Maximum coefficient of lift

Q17.4 The flight characteristic obtained at point C in the diagram below is:

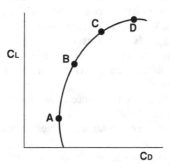

  (a) Maximum lift/drag ratio
  (b) Minimum sink rate
  (c) C_{Lmax}
  (d) V_{MO}

Q17.5 The stalling speed is represented as Point .... in the diagram below.

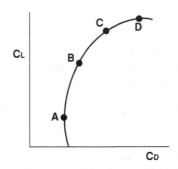

(a) A
(b) B
(c) C
(d) D

Q17.6 The wing shape least affected by turbulence is:
(a) straight wing
(b) swept-back wing
(c) swept-forward wing
(d) elliptical wing

Q17.7 If the speed exceeds V_A by a few knots:
(a) the aeroplane cannot stall
(b) a high-speed warning horn will sound
(c) there will be structural failure on entering a turn
(d) permanent damage to the airframe may occur if the elevators are fully deflected

Q17.8 V_{MCL} may be limited by (i) engine failure (ii) rate of roll.
(a) (i) correct; (ii) correct
(b) (i) correct; (ii) incorrect
(c) (i) incorrect; (ii) correct
(d) (i) incorrect; (ii) incorrect

Q17.9 The relationship of V_{MO} to M_{MO} in a climb and descent is:
(a) when climbing at V_{MO} the Mach number is decreasing
(b) M_{MO} can be inadvertently exceeded when climbing at V_{MO}
(c) when climbing at V_{MO} the TAS is decreasing
(d) when descending at M_{MO}, V_{MO} cannot be exceeded

Q17.10 Which of the following statements regarding the use of flaps is correct?
(a) Extending flaps in turbulence decreases the stall speed but reduces the margin to the structural limitations.
(b) Extending flaps in severe turbulence causes the CP to move aft, increasing the margin to the structural limits.
(c) Extending flaps in severe turbulence enables flight at lower speed and increases the margin to the structural limits.
(d) Extending flaps in severe turbulence decreases the stall speed at the risk of exceeding the structural limits.

Q17.11 With ice accretion on the leading edge of the wing the most dangerous phase of the take-off and climb is considered to be:
(a) at V_1
(b) just after V_R
(c) during the climb
(d) all phases are equally dangerous

Q17.12 The speed at which to fly in turbulence is:

(a) V_{RA}

(b) V_{MO}

(c) V_B

(a) V_C

Q17.13 The effect of heavy rain on C_{Lmax} (i) . . . . and on drag (ii) . . . .

(a) (i) increase; (ii) decrease

(b) (i) decrease; (ii) increase

(c) (i) increase; (ii) increase

(d) (i) decrease; (ii) decrease

Q17.14 The most serious effect of airframe ice accretion is:

(a) increased drag

(b) increased mass

(c) increased stick force

(d) reduced C_{Lmax}

Q17.15 The advantage of rear-mounted engines over wing-mounted engines is:

(a) the wing is less susceptible to flutter

(b) lighter wing construction reducing the mass

(c) thrust changes have less affect on longitudinal control

(d) easier maintenance access

Q17.16 Comparing the reaction of a multiengined turbojet aeroplane to that of a multiengined counter-rotating turboprop aeroplane to an engine failure:

(a) the turbojet aeroplane has the greater tendency to roll

(b) both have the same tendency to roll

(c) both have the same yawing tendency

(d) the turboprop aeroplane has the greater tendency to roll

Q17.17 Twin-engined variants of the same aeroplane have jet engines on one version and counterrotating propellers on the other version. If one engine fails the engine torque effect is jet (i) . . . . propeller (ii) . . . .

(a) (i) unchanged; (ii) roll in the direction of rotation of the dead engine

(b) (i) roll in the direction of the dead engine; (ii) roll in the direction of rotation of the live engine

(c) (i) unchanged; (ii) roll in the direction of rotation of the live engine

(d) (i) roll in the direction of the live engine; (ii) roll in the direction of rotation of the live engine

Part 7
Conclusion

18 Summary

18.1 Aerofoil-Profile Definitions

a. **Camber**. The curvature of the profile view of an aerofoil is its camber.
b. **Mean Camber Line**. The line joining points that are equidistant from the upper and lower surfaces of an aerofoil is the mean camber line.
c. **Chordline**. The straight line joining the centre of curvature of the leading-edge radius and the trailing edge of an aerofoil is the chordline.
d. **Chord**. The distance between the leading edge and the trailing edge of an aerofoil measured along the chordline is the chord.
e. **Leading Edge Radius**. The radius of a circle, centred on a line tangential to the curve of the leading edge of an aerofoil, and joining the curvatures of the upper and lower surfaces of the aerofoil is the leading edge or nose radius of an aerofoil.
f. **Maximum Thickness**. The maximum depth between the upper and lower surfaces of an aerofoil is its maximum thickness.
g. **Thickness/Chord Ratio**. The ratio of the maximum thickness of an aerofoil to the chord length expressed as a percentage is the thickness/chord ratio.
h. **Washout**. A reduction in the angle of incidence of an aerofoil from the wing root to the wing tip is the washout. This is also known as the aerodynamic twist of an aerofoil. See Figure 2.1(b).

18.2 Aerofoil-Attitude Definitions

a. **Angle of Incidence**. The angle subtended between the chordline and the longitudinal axis is the angle of incidence.
b. **Angle of Attack.** The angle subtended between the chordline of an aerofoil and the oncoming airflow is the angle of attack.
c. **Pitch Angle**. The angle subtended between the chordline of an aerofoil and the horizontal plane is the pitch angle.
d. **Climb or Descent Angle**. The angle subtended between the flight path of an aeroplane and the horizontal plane is the climb or descent angle.
e. **Pitching Moment.** The length of the arm multiplied by the pitching force is referred to as a pitching moment. A pitching moment is positive if it moves the aircraft nose-upward in flight and negative if it moves it downward.

The Principles of Flight for Pilots   P. J. Swatton
© 2011 John Wiley & Sons, Ltd

f. **Longitudinal Dihedral**. The difference between the angle of incidence of the mainplane and the tailplane is the longitudinal dihedral.

18.3 Wing-Shape Definitions

a. **Aspect Ratio.** The ratio of the span of an aerofoil to the mean chord is the aspect ratio, which is sometimes expressed as the square of the span divided by the wing area:

$$\text{Aspect Ratio} = \frac{\text{span}}{\text{chord}} = \frac{\text{span}^2}{\text{wing area}}$$

b. **Mean Aerodynamic Chord.** The chordline passing through the geometric centre (the centroid) of a wing is the mean aerodynamic chord (MAC).

c. **Quarter Chordline.** A line joining the quarter chord points along the length of a wing is the quarter chordline.

d. **Root Chord.** The line joining the leading edge and trailing edge of a wing at the centreline of the wing is the root chord.

e. **Sweep Angle.** The angle subtended between the quarter chordline and the lateral axis of an aeroplane is the sweep angle.

f. **Swept Wing.** Any wing of which the quarter chordline is not parallel to the lateral axis is a swept wing.

g. **Tapered Wing.** Any wing on which the root chord is longer than the tip chord is a tapered wing.

h. **Taper Ratio.** The ratio of the length of the tip chord expressed as a percentage of the length of the root chord is the taper ratio.

i. **Tip Chord.** The length of the wing chord at the wing tip is the tip chord.

j. **Wing Area.** The surface area of the planform of a wing is the wing area.

k. **Wing Centroid.** The geometric centre on a wing plan area is the wing centroid.

l. **Wing Loading**. The mass per unit area of a wing is the wing loading:

$$\text{Wing Loading} = \frac{\text{mass}}{\text{wing area}} = \frac{M}{S}$$

m. **Wingspan.** The shortest distance measured between the wing tips of an aeroplane is its wingspan.

n. **Load Factor (n)**. The total lift of an aerofoil divided by the total mass is the load factor, i.e.:

$$n = \frac{\text{total lift}}{\text{mass}}$$

o. **Centre of Pressure (CP)**. The point on the chordline through which the resultant of all of the aerodynamic forces acting on the wing is considered to act is the centre of pressure.

p. **Coefficient**. A numerical measure of a physical property that is constant for a system under specified conditions.

q. **Dihedral/Anhedral.** The angle of inclination of the front view of a wing subtended between the lateral horizontal axis is referred to as dihedral if the inclination is upward and anhedral if the inclination is downward.

18.4 High-Speed Definitions

a. **Critical Mach number (M_{CRIT}).** As M_{FS} increases so do the local Mach numbers. M_{CRIT} is that M_{FS} at any M_L has reached unity. It is the lowest speed in the transonic range.

b. **Critical Drag Rise Mach number (MCDR).** This is the MFS at which because of shockwaves, the CD for a given angle of attack increases significantly.

c. **Detachment Mach number (MDET).** This is the speed at which the bow shockwave of an accelerating aeroplane attaches to the leading edge of the wing or detaches if the aeroplane is decelerating.

d. **Indicated Mach number.** TAS is the difference between pitot pressure and static pressure. LSS is a function of static pressure and air density. Because air density is common to both TAS and LSS, both can be expressed as pressure ratios; this is what the machmeter measures. Indicated Mach number is therefore the ratio of the dynamic pressure to the static pressure.

e. **Free-Stream Mach number. (MFS)** The Mach number of the air at a point unaffected by the passage of the aeroplane but measured relative to the speed of the aeroplane is the Free Stream Mach number. (MFS). If the speed of the aeroplane is above the critical Mach number (MCRIT) a shockwave may well form, even with MFS below Mach 1.

f. **Local Mach number (ML).** This is the ratio of the speed of the airflow at a point on the aeroplane to the speed of sound at the same point.

g. **Shock Stall**. The airflow over an aeroplane's wings is disturbed when flying at or near the critical Mach number. This causes the separation of the boundary layer from the upper surface of the wing behind the shockwave and is the shock stall, which is described in greater detail later in this chapter.

h. **The Speed of Sound (a).** The speed of propagation of a very small pressure disturbance in a fluid in specified conditions. The local speed of sound through the air is equal to 38.94 multiplied by the square root of the absolute temperature (A). Therefore, as air temperature decreases so also does the local speed of sound (LSS).

$$LSS = 38.94\sqrt{A} \text{ in kt}$$

i. **True Mach number (M).** The ratio of the speed of an object to the value of the speed of sound in the same environmental conditions is the Mach number; it has no units of measurement because it is a ratio.

$$\text{Mach number} = TAS \div LSS \text{ in kt.}$$
$$\text{Therefore, Mach number} = TAS \div 38.94\sqrt{A}$$

18.5 Propeller Definitions

a. The **axis of rotation** is the shaft of the propeller.

b. The **propeller disc** is the area swept by a rotating propeller when viewed from the front.

c. The **leading edge** is the same as that of a wing and is the forward edge of the profile of a propeller blade.

d. The **trailing edge** is the same as that of a wing and is the rearward edge of the propeller profile.

e. The **blade face** is the forward surface of a propeller blade equivalent to the upper surface of a wing.

f. The **blade back** is the rearward surface of a propeller blade equivalent to the lower surface of a wing.

g. The **blade angle** sometimes called the geometric pitch angle is the angle subtended between the plane of rotation and the chordline of the propeller blade. It comprises two parts, the angle of attack and the helix angle, when the aeroplane is moving but when it is stationary it is equal to the angle of attack.

h. The **blade angle of attack** is the angle subtended between the chordline of the propeller blade and the relative airflow.

i. The **helix angle** sometimes called the angle of advance is the angle subtended between relative airflow and the plane of rotation of the propeller.

j. The **effective pitch** is the distance that a propeller **actually** advances in one revolution. Effective Pitch = Geometric Pitch minus propeller slip.

k. The **geometric pitch** the distance that a propeller ***should*** theoretically advance in one revolution. Geometric Pitch = Effective Pitch plus Propeller Slip.

l. The **propeller slip** is the difference between the effective pitch and the geometric pitch and is the result of the air not being a perfect medium for propulsion. Propeller Slip = Geometric Pitch minus Effective Pitch

18.6 V Speeds

The following is a summary of all of the 'V' speeds that must be known for the Principles of Flight syllabus and listed in alphabetical order:

V_A is the design manoeuvring speed. This is the highest speed at which the aeroplane will stall before exceeding the maximum load factor and does not need to exceed V_C. It is not less than the stalling speed with the flaps retracted multiplied by the square root of the limiting load factor at a speed of V_C. *CS 25.335(c)*.

V_B is the design speed for maximum gust intensity. At altitudes where V_C is limited by a Mach number V_B is chosen by the manufacturer to provide the optimum margin between the low-speed buffet and the high-speed buffet and need not be greater than V_C. *CS 25.335(d)*.

V_C is the design cruising speed. The minimum value of V_C must exceed V_B by a sufficient margin to allow for inadvertent speed increases caused by severe turbulence. V_C may not be less than V_B + (1.32 × the reference gust velocity EAS). It need not exceed V_C or the maximum speed in level flight at the maximum continuous thrust setting appropriate to the altitude. V_C is limited to a specific Mach number at altitudes at which a Mach number limits V_D. *CS 25.335(a)*.

V_{CLmax} is the calibrated airspeed when the C_L is maximised (i.e. maximum lift is attained) and is obtained with:

a. The engines developing zero thrust at the stall.
b. The pitch controls (if applicable) in the take-off position.
c. The aeroplane in the configuration being considered.
d. The aircraft trimmed for level flight at not less than $1.13V_{SR}$ and not greater than $1.3V_{SR}$. *AMC 25.103(b)*.

V_D is the design diving speed, which at high altitude may be limited to a specific Mach number. *CS 25.335(b)*.

V_{DD} is the design drag devices speed. For each drag device the design speed must exceed the recommended operating speed by a sufficient margin to allow for probable variations in speed control. For drag devices intended for use during high-speed descents V_{DD} may not exceed V_D. *CS 25.335(f)*.

V_F is the design flap speed. For each wing-flap position it must sufficiently exceed the recommended operating speed for that stage of flight to allow for probable variations in control airspeed and for the transition from one flap position to another. *CS 25.335(e)(1)*. It may not be less than:

a. $1.6V_{S1}$ with take-off flap at the maximum take-off mass;
b. $1.8V_{S1}$ with approach flap at the maximum landing mass;
c. $1.8V_{S1}$ with land flap at the maximum landing mass. *CS 25.335(e)(3)*.

V_{FE} is the maximum speed at which it is safe to fly with the flaps in a prescribed extended position. It must not exceed V_F the design flap speed. *CS 25.1511. CS Definitions Page 19*.

V_{FO} is the maximum speed at which the flaps may be operated either extended or retracted.

V_{LE} is the maximum speed at which an aeroplane may be safely flown with the undercarriage (landing gear) extended. *CS Definitions Page 20; CS 25.1515(b)*.

V_{LO} is the maximum speed at which the undercarriage (landing gear) may be safely operated, either extended or retracted. *CS Definitions Page 20; CS 25.1515(a).*

V_{MC} is the lowest CAS, at which, in the event of the critical power unit suddenly becoming inoperative when airborne, it is possible to maintain control of the aeroplane with that engine inoperative, and to maintain straight flight using no more than 5° of bank without a change of heading greater than 20°. V_{MC} may not exceed:

a. Class 'B' aeroplanes – $1.2V_{S1}$ at the maximum take-off mass. *CS 23.149(b).*
b. Class 'A' aeroplanes – $1.13V_{SR}$ at the maximum take-off mass at MSL. *CS 25.149(c).*

V_{MC} must be equal to or less than V_{LOF}, because if V_1 is equal to V_{MCG} the aeroplane must be able to continue the take-off safely. The lateral control may be used only to keep the wings level. *CS 25.149(e).*

V_{MCG} is the minimum control speed on the ground. It is the CAS, at which, when the critical engine of a multiengined aeroplane fails during the take-off run and with its propeller, if applicable, in the position it automatically takes, it is possible to maintain control with the use of the primary aerodynamic controls alone *(without the use of nose-wheel steering)* to enable the take-off to be safely continued using normal piloting skill.

The path of any multi-engined Class 'A' aeroplane from the point of engine failure to the point at which recovery to a direction parallel to the centreline of the runway is attained may not deviate by more than 30 ft laterally from the centreline at any point. *CS 25.149(e)*

V_{MCL} is the minimum control speed during the approach and landing with all engines operating. In the event of an engine failure, it is possible to maintain control:

a. With the critical engine inoperative, and the propeller of the failed engine feathered, if applicable, and the operating engine(s) set to go-around power or thrust:
   (1) To maintain straight flight using no more than 5° of bank. *CS 25.149(f).*
   (2) Roll through 20° from straight flight away from the operative engine in five seconds. *CS 25.149(h).*
b. Assuming the engine fails while at the power or thrust set to maintain a 3° approach path. *CS 25.149(f).*

$V_{MCL(1out)}$ is the one-engine-inoperative landing minimum control speed that may be provided by the manufacturer for use instead of V_{MCL}. AMC 25.149(f).
It is determined for the conditions appropriate to the approach and landing with one engine having failed *before* the start of the approach. The propeller of the inoperative engine, if applicable, may be feathered throughout. *AMC 25.149(f).*

V_{MCL-2} is the minimum control speed during the approach to land with the critical engine inoperative; it is the lowest CAS at which it is possible when a second critical engine becomes inoperative, and its propeller feathered, if applicable and the remaining engine(s) producing go-around power or thrust to:

a. Maintain straight flight using no more than 5° of bank.
b. Roll through 20° from straight flight, away from the operative engine in five seconds. *AMC 25.149(g).*
c. Assuming the power or thrust set on the operating engine(s) is that necessary to maintain a 3° approach path when one critical engine is inoperative. *CS 25.149(g)(5).*
d. The power or thrust on the operating engine(s) is rapidly changed, immediately the second critical engine fails, to the go-around power or thrust setting. *CS 25.149(g)(7).*

V_{MO}/M_{MO} is the maximum operating IAS (or Mach number, whichever is critical at a particular altitude), which must not be deliberately exceeded in any flight condition, is referred to as V_{MO}/M_{MO}. It must not exceed V_C. *CS 25.1505.*

V_{MS} is the lowest possible stalling speed, V_S, for any combination of AUM and atmospheric conditions with power off, at which a large, not immediately controllable, pitching or rolling motion is encountered.

V_{MS0} is the lowest possible stalling speed, V_{S0} (or if no stall is obtainable, the minimum steady-flight speed), in the landing configuration, for any combination of AUM and atmospheric conditions.

V_{MS1} is the lowest possible stalling speed, V_{S1} (or if no stall is obtainable, the minimum steady-flight speed), with the aeroplane in the configuration appropriate to the case under consideration, for any combination of AUM and atmospheric conditions.

V_{NE} is the maximum IAS that must never be exceeded. *CS Definitions Page 20.*

V_{NO} is the maximum normal operating speed is V_{NO}.

V_S is the stalling speed or minimum steady-flight speed at which the aeroplane is controllable. The stalling speed is the greater of:

a. The minimum CAS obtained when the aeroplane is stalled (or the minimum steady-flight speed at which the aeroplane is controllable with the longitudinal control on its stop).
b. A CAS equal to 94% of the one-g stall speed (V_{S1g}). *CS Definitions page 20.*

V_{S0} is the stalling speed or the minimum steady-flight speed for an aeroplane in the landing configuration. V_{S0} at the maximum mass must not exceed 61 kt for a Class 'B' single-engined aeroplane or for a Class 'B' twin-engined aeroplane of 2 722 kg or less. *CS 23.49(c). CS Definitions page 20.*

V_{S1} is the stalling speed or the minimum steady-flight speed for an aeroplane in the configuration under consideration, e.g. Flaps extended. *JAR 1 page 1-15.*

V_{S1g} is the 'one-g stalling speed' that is the minimum CAS at which the aeroplane can develop a lift force (normal to the flight path) equal to its mass whilst at an angle of attack not greater than that which the stall is identified. *CS 25.103(c).*

V_{SR} is the reference stalling speed, which is a calibrated airspeed used as the basis for the calculation of other speeds. Reference stalling speed (V_{SR}) is approximately 6% greater than Vs and is:

a. Never less than the one-g stalling speed. *CS 25.103(a).*
b. Equal to or greater than the calibrated airspeed obtained when the maximum load-factor-corrected lift coefficient, (V_{CLmax}), is divided by the square root of the load factor normal to the flight path. *CS 25.103(a).*
c. A speed reduction using the pitch control not exceeding 1 kt per second. *CS 25.103(c).*
d. Never less than the greater of 2 kt or 2% above the stick shaker speed, if a stick shaker is installed. *CS 25.103(d).*

V_{SR0} is the reference stalling speed in the landing configuration.
V_{SR1} is the reference stalling speed for the configuration under consideration.

18.7   PoF Formulae

7. The following is a list the formulae used in the topics expounded in this manual.

$$\text{Total Lift} = C_L{}^1\!/_2\rho V^2 S$$

where $^1\!/_2\rho V^2$ = dynamic pressure 'q'

C_L = lift coefficient = the ratio of lift pressure to dynamic pressure

The airspeed indicator measures 'q' but is square law compensated.
e.g. $2 \times$ IAS = $4 \times$ 'q'; $3 \times$ IAS = $9 \times$ 'q'.

8. If the total lift remains constant, then if the speed changes so also must the coefficient of lift for the formula to remain in balance. Here are three examples of past questions that have been derived from the lift formula.

Example 18.1

If the speed doubles then V^2 will increase by a factor of 4 so that C_L must change to a $\frac{1}{4}$ of its original value, i.e. $100 \div 4 = 25\%$ of its original value.

Example 18.2

If the speed triples then V^2 will increase by a factor of 9 so that C_L must change to a 1/9 of its original value, i.e. $100 \div 9 = 11.1\%$ of its original value.

Example 18.3

If the speed increases by 1.3 of its original value then V^2 will increase by a factor of $1.3 \times 1.3 = 1.69$ so that C_L must change to a 1/1.69 of its original value, i.e. $100 \div 1.69 = 59.2\%$ of its original value.

---

9. To express C_L as a percentage of C_{Lmax} divide 1 by the square of the Vs multiplicand and multiply by 100.

$$[\text{C}_L/\text{C}_{Lmax}]\% = [1/\textbf{Multiplicand}^2] \times 100$$

18.7.1 Drag

Profile Drag $(D_P) = C_D{}^1/_2\rho V^2 S$; $D_P \propto V^2$
C_D = drag coefficient = the ratio of drag pressure to dynamic pressure

Induced Drag (D_I); $C_{DI} = C_L{}^2/\pi A$; $D_I \propto 1/V^2$

Where C_{DI} = Induced drag coefficient; C_L = lift coefficient; A = Aspect ratio.

18.7.2 Wing Loading/Load Factor

Wing Loading = Mass/Wing Area = $C_L{}^1/_2\rho V^2$

Load Factor (n) = Lift/Mass.

Load Factor in a Turn (n) = 1/ cosine angle of bank = secant of the bank angle.

Gust Load Factor = (New C_L in gust)/(C_L before the gust)

Note: Increase of load factor = Gust Load Factor − 1

Climb Load Factor = cosine climb angle.
The climb load factor is always less than 1.

Note: To convert a climb gradient to a climb angle on the calculator then convert the gradient to a decimal and use the inv tan function, e.g. $20\% = 0.2$ inv tan $= 11.3°$. Then climb load factor $= \cos 11.3° = 0.98$.

18.7.3 Stalling Speed Calculations

18.7.3.1 Mass Change

Vs (New Mass) = Vs (Original Mass) $\times$ $\sqrt{}$(New Mass)/ (Original Mass)

18.7.3.2 Load Factor

Vs(n) = Vs1g $\times$ $\sqrt{n}$, where n = new load factor.

18.7.3.3 Turn

Vs(turn) = Vs1g $\times$ $\sqrt{}$1/cosine angle of bank.

Changed Angle of Bank:

Vs (new AoB) = Vs (original AoB) $\times$ $\sqrt{}$(cosine original AoB/cosine new AoB)

18.7.4 Design Manoeuvre Speed (V$_A$)

V$_A$ = Vs1g $\sqrt{}$limiting load factor

V$_A$ (New Mass) = V$_A$ (Original Mass) $\times$ $\sqrt{}$(New Mass)/ (Original Mass)

18.7.5 Turn Details

18.7.5.1 Radius of Turn

Radius of turn = V^2/g tangent ø

where ø = the angle of bank; V = speed in mps; g = 9.81 m/s^2; kt $\times$ 0.515 = mps

18.7.5.2 Rate of Turn

Rate of turn = g tangent ø/V radians per second

1 radian = 57.2958°.

18.7.6 Climb Calculations

Climb Gradient % = [(Thrust - Drag) $\div$ Mass] $\times$ 100.

Assuming Lift = Mass and g = 10 m/s^2.

Rate of Climb fpm = [(Thrust - Drag) kg $\div$ Mass kg] $\times$ TAS in fpm.

18.7.7 Descent Calculations

Descent Gradient = [(Drag - Thrust) kg $\div$ Mass kg] $\times$ 100

Rate of Descent = [(Drag - Thrust) kg $\div$ Mass kg] $\times$ TAS fpm

Max. Still-Air Glide-Descent Distance = Max. Lift/Drag Ratio × Height Difference.

18.7.8   Mach Angle (μ) Calculation

Sine μ = 1/Mach Number, e.g. Mach 1.5 μ = 41.8°

18.8   Key Facts

The following lists show all of the significant factors that affect the subject matter detailed in each subsection.

18.9   Stalling

18.9.1   The Maximum Coefficient of Lift (C_{Lmax})

Increased by:	Decreased by:
a. Large leading-edge radius	a. Small leading-edge radius
b. Highly cambered aerofoil	b. Symmetrical aerofoil
c. High aspect ratio	c. Low aspect ratio
d. Straight wing	d. Swept wing
e. Smooth leading edge	e. Rough/Iced leading edge
f. High Reynolds number	f. Low Reynolds number
g. Flap/slat combination	g. Rough/Iced wing surface
h. Trailing-edge flaps	h. Heavy rain

18.9.2   The Critical Angle

Increased by:	Decreased by:
a. Leading-edge slats	a. Trailing-edge flaps
b. A small amount of thrust	b. Ice accretion
c. Low aspect ratio	c. Heavy rain
d. Swept wing	d. High aspect ratio
e. Large leading-edge radius	e. Straight wing
f. High Reynolds number	f. Low Reynolds number
g. Symmetrical aerofoil	g. Highly cambered aerofoil

18.9.3   The Stalling Speed

Increased by:	Decreased by:
a. Increased mass	a. Decreased mass
b. Decreased flap angle	b. Increased flap angle
c. Forward CG position	c. Deployment of slats
d. Increased sweep angle	d. Decreased sweep angle
e. Increased load factor	e. Decreased load factor
f. Turning or pulling out of a dive	f. Aft CG position
g. Turbulence penetration	g. High aspect ratio
h. Ice accretion	
i. Heavy rain	

18.10   Stability

18.10.1   Static Stability

Positive Static Stability – an aeroplane having had its attitude disturbed by an external force returns to its predisturbed state, once the disturbance ceases, without assistance.

Neutral Static Stability – an aeroplane having had its attitude disturbed by an external force remains in the attitude that it attained when the disturbance ceases.

Negative Static Stability – an aeroplane having had its attitude disturbed by an external force continues to diverge from its original attitude after the disturbance ceases.

18.10.2   Dynamic Stability

Positive Dynamic Stability – the amplitude of any oscillation induced by an outside disturbance decreases with the passage of time.

'Dead Beat' Positive Dynamic Stability – any oscillation induced by an outside disturbance ceases immediately because of heavy damping.

Negative Dynamic Stability – the amplitude of any oscillation induced by an outside disturbance increases with the passage of time.

'Divergent' Negative Dynamic Stability – any oscillation induced by an outside disturbance ceases because the motion completely diverges from the original motion.

Neutral Dynamic Stability – the amplitude of any oscillation induced by an outside disturbance remains constant with the passage of time.

Spiral Instability has strong directional stability but weak lateral stability. Dihedral diminishes this effect. Tendency decreases with increased altitude.

Dutch Roll has weak directional stability and strong lateral stability. Tendency increases with increased altitude. Tendency increased by increased static lateral stability with a constant directional static stability.

General Static Stability	
Positive or Increased Stability by:	**Negative or Decreased Stability by:**
(1) Decreased Altitude.	(1) Increased Altitude.
(2) Increased IAS.	(2) Decreased IAS.
(3) Increased Air Density.	(3) Decreased Air Density.
(4) Decreased Aeroplane Mass.	(4) Increased Aeroplane Mass.
(5) Forward CG Position.	(5) Aft CG Position.
(6) Aft CP Position.	(6) Forward CP Position.
Directional Static Stability	
Positive or Increased Stability by:	**Negative or Decreased Stability by:**
(1) CG ahead of CP.	(1) CP ahead of CG.
(2) A long moment arm.	(2) Short moment arm.
(3) A forward movement of the CG.	(3) An aft movement of the CG.
(4) An aft movement of the CP.	(4) A forward movement of the CP.
(5) Wing sweepback.	(5) Forward wing sweep.
(6) A dorsal fin increases stability.	(6) Dihedral.
(7) A ventral fin and/or strakes.	(7) High aspect ratio fin.
(8) Large fin and rudder area.	(8) Small fin and rudder area.
(9) The vertical stabiliser design.	(9) Long Fuselage ahead of CG.
Lateral Static Stability	
Positive or Increased Stability by:	**Negative or Decreased Stability by:**
(1) Dihedral.	(1) Anhedral.
(2) Sweepback.	(2) Forward wing sweep.
(3) High-wing mounting.	(3) Low-wing mounting.
(4) Large, high vertical fin.	(4) Ventral Fin.
(5) Low CG.	(5) Extending inboard flaps.
Longitudinal Static Stability	
Positive or Increased Stability by:	**Negative or Decreased Stability by:**
(1) CG ahead of CP.	(1) CP ahead of CG.
(2) CG on forward limit.	(2) CG on aft limit.
(3) A forward movement of the CG.	(3) An aft movement of the CG.
(4) An aft movement of the CP.	(4) A forward movement of the CP.
(5) Longitudinal dihedral.	(5) High angle of attack.
(6) Long moment arm from tail CP.	(6) Extension of trailing-edge flaps.
(7) Large tailplane area.	(7) Increased wing downwash.
(8) Fuselage and engine nacelles aft of the wing.	(8) Fuselage and engine nacelles ahead of the wing.
(9) Excess tail volume.	(9) Increased altitude.

Directional Dynamic Stability	
Increased by:	**Decreased by:**
a. Directional Static Stability	a. Increased Altitude
b. Weathercock Effect	b. Increased Speed
c. Yaw Damper	
Lateral Dynamic Stability	
Increased by:	**Decreased by:**
a. Lateral Static Stability	a. Dutch roll
b. Dihedral	b. Spiral instability
c. Yaw dampers	c. Increased Altitude
d. Roll Damper	d. Increased Speed
e. Decrease Altitude	e. Engine Failure
Longitudinal Dynamic Stability	
Increased by:	**Decreased by:**
a. Longitudinal Static Stability	a. Increased Altitude
b. Aerodynamic Pitch Damping	b. High Temperature
c. Pitch Moments of Inertia	c. Increased Speed
d. Low Angle of Pitch	d. High Pitch Angle
e. Low Rate of Pitch	e. High Rate of Pitch
f. Decreased Altitude	
g. Low Temperature	

18.10.3 The Stick Force

Stick Force per 'g'	
Increased by:	**Decreased by:**
a. Forward movement of CG	a. Aft movement of CG
b. Increased Load Factor	b. Decreased Load Factor
c. Decreased Altitude	c. Increased Altitude

18.10.4 The Gust Load Factor

Gust Load Factor	
Increased by:	**Decreased by:**
a. Decreased Altitude	**a.** Increased Altitude
b. Decreased Mass	**b.** Increased Mass
c. Increased EAS	**c.** Decreased EAS
d. Decreased wing Loading	**d.** Increased Wing Loading
e. Steep Lift v Angle of Attack curve	**e.** Shallow Lift v Angle of Attack curve
f. Vertical Updraught	**f.** Vertical Downdraught
g. Increased Aspect Ratio	**g.** Decreased Aspect Ratio

18.11 Propellers

18.11.1 Propeller Efficiency

Power Absorption	
Increased by:	**Decreased by:**
a. Increased Blade Camber	**a.** Decreased Blade Camber
b. Increased Blade Chord	**b.** Decreased Blade Chord
c. Increased Blade Angle	**c.** Decreased Blade Angle
d. Increased Blade Angle of Attack	**d.** Decreased Blade Angle of Attack
e. Increased Number of Blades	**e.** Decreased Number of Blades
f. Increased RPM	**f.** Decreased RPM

18.11.2 Fixed Pitch Angle of Attack

Fixed Pitch Angle of Attack	
Increased by:	**Decreased by:**
a. Constant TAS and increased RPM	a. Constant TAS and decreased RPM
b. Decreased TAS and constant RPM	b. Increased TAS and constant RPM
c. Climb TAS and constant RPM (Constant IAS climb)	c. Descent TAS and constant RPM (Constant IAS descent)

18.11.3 Propeller Gyroscopic Effect

Propeller Gyroscopic Effect		
Propeller Rotation Viewed from behind	**Applied Force**	**Reaction**
Clockwise	Yaw left	Pitch up
	Yaw right	Pitch down
	Pitch up	Yaw right
	Pitch down	Yaw left
Anticlockwise	Yaw left	Pitch down
	Yaw right	Pitch up
	Pitch up	Yaw left
	Pitch down	Yaw right

18.12   The Effect of the Variables on Performance

11. The manner in which each or the relevant variables affects the performance of an aeroplane is shown and listed below by what its cause and effect are on the aeroplane.

18.12.1 Airframe Surface

Airframe Surface Condition:		
Variable	**Maximum C_L**	**Critical AoA**
a. Smooth Surface	+1.65	14.8°
b. Rough or Frost-Covered Surface	+1.20	12.2°
c. Ice-Covered Surface	+0.91	10.4°

18.12.2 Airframe Surface

Airframe Icing and/or Damage Causes:	
Increased	**Decreased**
a. Total Drag	a. Total Lift
b. Stalling Speed	b. C_{Lmax}
c. Take-off Ground Run	c. Stalling Angle
d. Take-off Distance	d. Climb Gradient
e. Landing Ground Run	e. Rate of Climb
f. Landing Distance	f. Flap/Slat Effectiveness
g. Stick Force (icing only)	
h. Total Mass (icing only)	

18.12.3 Altitude

Increased Altitude Produces:	
Increased	**Decreased**
a. Gliding Range	a. Climb Gradient
b. Gliding Endurance	b. Rate of Climb
c. Power Required	c. Damping
d. TAS at Constant IAS	d. Dynamic Stability
e. Mach Number at Constant IAS	e. Power Available
f. Mach Number at Constant TAS	f. IAS at Constant TAS
g. Stalling TAS (IAS Constant)	g. IAS at Constant Mach Number
	h. TAS at Constant Mach Number
	i. Thrust Available (Required Constant)
	j. V_Y (V_X Constant)

18.12.4 Aspect Ratio

Increased Aspect Ratio Produces:	
Increased	**Decreased**
a. Total Lift	a. Wing-Tip Vortices
b. C_{Lmax}	b. Induced Drag
c. Lift/Drag Ratio	c. Stalling Angle
d. Turbulence Sensitivity	d. C_{DI}
e. Speed Stability	e. V_{IMD}

18.12.5 Camber

High Camber Produces:	
Increased	**Decreased**
a. C_{Lmax}	a. Stalling angle of attack
b. Shockwave	b. Zero-lift angle of attack

18.12.6   CG Position

A Forward CG Position Produces:	
Increased	**Decreased**
a. Longitudinal Stability	a. Restorative Pitching Moment
b. Elevator Deflection Required	b. Manoeuvrability
c. Nose-down Pitching Moment	c. Maximum Range
d. Stalling Speed	d. Maximum Endurance
e. Wing Loading	
f. Stick-Force Stability	
g. Manoeuvre Stability	
h. Fuel Flow	
i. Stick Force Required	
j. Trim Drag	
k. Download on Tailplane	

18.12.7   Flap

Extended Flap Causes:	
Increased	**Decreased**
a. Total drag	a. Climb gradient
b. Thrust required	b. Rate of climb
c. Parasite drag	c. Stalling speed
d. Total lift	d. Lateral static stability
e. Likelihood of Tip stalling	e. Lift/Drag ratio
f. Effective Angle of Attack	f. Forward speed
g. Nose-down trim	g. Stalling Angle
h. Leading-edge upwash	
i. Downwash	
j. Aft CP movement	
k. Longitudinal static stability	
l. Effective Camber	
m. $C_{L max}$	

18.12.8 Sweepback

Increased Sweepback Generates:	
Increased	**Decreased**
a. Mcrit (greatest benefit)	a. CLmax
b. Lateral Static Stability	b. Total lift
c. Directional Static Stability	c. Trailing-edge control effectiveness
d. Dihedral effect	d. Trailing-edge flap effectiveness
e. Greater Turbulence Tolerance	e. Onset of supersonic airflow
f. Nose-up tendency	f. Elevator effectiveness
g. Stalling angle	g. Drag
h. Tip-stalling likelihood	h. CD in direct proportion to sweep

18.12.9 Dihedral

Dihedral Generates:	
Increased	**Decreased**
a. Lateral Static Stability	a. Directional Static Stability
b. Lateral Dynamic Stability	b. Directional Dynamic Stability
c. Tendency to Dutch roll	c. Tendency to Spiral Dive

18.12.10 Mass

Increased Mass Causes:	
Increased	**Decreased**
a. Total Drag	a. Climb Gradient
b. Induced Drag	b. Rate of Climb
c. Thrust Required	c. Maximum Speed
d. Total Lift Required	d. Lateral Static Stability (low wing-mounted engines)
e. Likelihood of Tip Stalling (swept-wing aeroplanes)	e. Gliding Range (maintaining optimum glide)
f. Vimd	f. Gliding Endurance
g. Stalling Speed	g. Maximum Cruise Altitude
h. Vx and Vy	h. Stabilizing Altitude
i. Power Required	
j. Fuel Flow	
k. Rate of Descent	
l. Descent Groundspeed	

Self-Assessed Exercise 18

Fill in the blanks:

Turn Calculations								
Formula/Question		**Q1**	**Q2**	**Q3**	**Q4**	**Q5**	**Q6**	**Q7**
IAS in knots		100 kt	150 kt	180 kt	220 kt	240 kt	270 kt	300 kt
Speed V mps	× 0.515							
V^2	Calculator							
Stalling Speed	Given	64 kt	72 kt	75 kt	70 kt	72 kt	75 kt	80 kt
Bank Angle	Given	45°	40°	35°	30°	25°	20°	15°
Cos ∅	Calculator							
Tan ∅	Calculator							
Load Factor	1/cos ∅							
Stalling Speed	Vs × √1/cos ∅							
g tan ∅	Calculator							
Radius of Turn	V^2mps ÷ g tan ∅							
Rate of Turn radians.	g tan ∅ ÷ V mps = Radians							
Rate of Turn °/s	Radians × 57.3							

19 SOLUTIONS (with page references)

Self-assessment Exercise 1

Q	A	Ref	Q	A	Ref	Q	A	Ref
1.1	c	8	1.11	c	7	1.21	d	10
1.2	b	11	1.12	c	11	1.22	a	12
1.3	d	12	1.13	b	13	1.23	b	13
1.4	c	12	1.14	c	13	1.24	d	10
1.5	c	13	1.15	c	13	1.25	d	10
1.6	b	13	1.16	b	8	1.26	d	12
1.7	a	11	1.17	a	13	1.27	c	12
1.8	a	12	1.18	a	13	1.28	d	12
1.9	b	11	1.19	b	11	-	-	-
1.10	d	13	1.20	b	11	-	-	-

Self-assessment Exercise 2

Q	A	Ref	Q	A	Ref	Q	A	Ref	Q	A	Ref
2.1	d	20	2.9	c	20	2.17	d	27	2.25	c	22
2.2	c	19	2.10	d	27	2.18	b	27	2.26	a	20
2.3	b	20	2.11	d	27	2.19	d	19	2.27	c	20
2.4	c	27	2.12	b	22	2.20	c	23	2.28	d	27
2.5	a	20	2.13	d	20	2.21	a	27	2.29	b	27
2.6	d	23	2.14	d	27	2.22	b	23	2.30	a	19
2.7	b	27	2.15	a	25	2.23	d	19	2.31	a	27
2.8	c	20	2.16	c	20	2.24	a	20	-	-	-

Self-assessment Exercise 3

Q	A	Ref	Q	A	Ref	Q	A	Ref
3.1	d	35	3.9	d	39	3.17	c	39
3.2	b	44	3.10	b	38	3.18	c	40
3.3	a	39	3.11	b	38	3.19	a	39
3.4	b	35	3.12	b	45	3.20	d	39
3.5	d	45	3.13	c	39	3.21	b	38
3.6	c	38	3.14	b	45	3.22	c	35
3.7	a	42	3.15	c	35	3.23	a	35
3.8	c	47	3.16	b	39	-	-	-

Self-assessment Exercise 4

Q	A	Ref	Q	A	Ref	Q	A	Ref	Q	A	Ref	Q	A	Ref
4.1	a	61	4.9	d	67	4.17	c	58	4.25	b	64	4.33	b	61
4.2	d	66	4.10	a	66	4.18	a	61	4.26	c	62	4.34	c	58
4.3	c	68	4.11	b	61	4.19	a	67	4.27	a	58	4.35	b	62
4.4	a	69	4.12	a	68	4.20	a	55	4.28	d	68	4.36	c	61
4.5	c	58	4.13	c	67	4.21	c	58	4.29	b	68	-	-	-
4.6	a	69	4.14	a	58	4.22	c	61	4.30	c	68	-	-	-
4.7	c	68	4.15	a	61	4.23	a	62	4.31	c	58	-	-	-
4.8	c	64	4.16	a	58	4.24	a	58	4.32	d	58	-	-	-

Self-assessment Exercise 5

Q	A	Ref	Q	A	Ref	Q	A	Ref
5.1	c	81	5.9	a	88	5.17	d	82
5.2	d	83	5.10	a	82	5.18	a	87
5.3	c	82	5.11	b	82	5.19	b	88
5.4	b	83	5.12	b	83	5.20	b	88
5.5	d	81	5.13	d	82	5.21	b	81
5.6	a	84	5.14	c	82	5.22	a	84
5.7	c	83	5.15	d	81	5.23	c	93
5.8	d	87	5.16	c	83	5.24	d	83

Q5.6(a). $[1/1.3^2] \times 100 = [1/1.69] \times 100 = \mathbf{59.17\%}$

Q5.7(c). Original IAS $= V^2$. New IAS $= [2V]^2 = 4V^2$
Then $C_LV^2 = [C_L/4] \times 4V^2$. C_L has to be reduced to balance the formula.
New $C_L = \mathbf{25\%}$ **of original** C_L

Q5.24(d). $V^2 = 1.3 \times 1.3 = 1.69$
Revised $C_L = 100 \div 1.69 = \mathbf{59.17\%}$.

Self-assessment Exercise 6

Q	A	Ref	Q	A	Ref	Q	A	Ref
6.1	c	106	6.18	d	106	6.35	c	105/100
6.2	b	106	6.19	c	106	6.36	d	100
6.3	d	100	6.20	c	106	6.37	c	105
6.4	b	103	6.21	b	106	6.38	b	106
6.5	c	106	6.22	c	105	6.39	a	106
6.6	b	106	6.23	a	106	6.40	a	109
6.7	a	107	6.24	a	100	6.41	b	100
6.8	d	106	6.25	b	100/105	6.42	c	106
6.9	c	100	6.26	d	109	6.43	d	112
6.10	c	109	6.27	a	106	6.44	b	100/105
6.11	a	100	6.28	c	101	6.45	c	100
6.12	b	100	6.29	d	105	6.46	a	106
6.13	a	102	6.30	c	109	6.47	a	106
6.14	d	109	6.31	d	109	6.48	b	106
6.15	d	100	6.32	b	106	6.49	b	109
6.16	b	106	6.33	a	106	6.50	a	100
6.17	b	109	6.34	b	106	6.51	b	106

Self-assessment Exercise 7

Q	A	Ref	Q	A	Ref	Q	A	Ref	Q	A	Ref
7.1	d	136	7.21	b	136	7.41	a	139	7.61	c	125
7.2	b	121	7.22	b	131	7.42	b	125	7.62	d	130
7.3	a	127	7.23	b	124	7.43	b	127	7.63	b	127
7.4	d	139	7.24	a	136	7.44	c	136	7.64	b	127
7.5	d	136	7.25	b	134	7.45	b	126	7.65	d	127
7.6	b	124	7.26	c	139	7.46	d	130	7.66	b	119
7.7	a	119	7.27	c	129	7.47	d	119	7.67	a	127
7.8	a	119	7.28	c	126	7.48	c	129	7.68	c	127
7.9	b	119	7.29	b	126	7.49	a	123	7.69	c	127
7.10	d	127	7.30	a	129	7.50	c	121	7.70	d	139
7.11	c	130	7.31	a	127	7.51	d	129	7.71	d	139
7.12	c	130	7.32	c	136	7.52	a	119	7.72	b	139
7.13	d	119	7.33	b	125	7.53	a	134	7.73	a	132
7.14	a	126	7.34	c	130	7.54	b	119	7.74	b	130
7.15	d	124	7.35	b	123	7.55	a	126	7.75	c	131
7.16	a	126	7.36	b	127	7.56	d	119	7.76	b	131
7.17	a	126	7.37	b	122	7.57	a	123	7.77	d	131
7.18	a	130	7.38	c	130	7.58	a	119	7.78	b	131
7.19	b	136	7.39	d	119	7.59	c	125	7.79	d	136
7.20	c	127	7.40	c	119	7.60	a	125	-	-	-

Q7.27(c) and Q7.51(d). The IAS has doubled from 80 kt to 160 kt. Therefore induced drag changes by $1/V^2 = 1/2^2 = \frac{1}{4}$.

The coefficient of induced drag $C_{DI} = \dfrac{kC_L{}^2}{\pi A} = \frac{1}{4} \times \frac{1}{4} = 1/16$

Q7.48(c). The IAS has halved. Therefore induced drag changes by $1/V^2 = 4$
The coefficient of induced drag $C_{DI} = 4 \times 4 = 16$

SOLUTIONS (WITH PAGE REFERENCES)

451

Self-assessment Exercise 8

Q	A	Ref	Q	A	Ref	Q	A	Ref	Q	A	Ref
8.1	d	160	8.19	a	174	8.37	d	170	8.55	b	171
8.2	c	168	8.20	a	171	8.38	b	174	8.56	c	156
8.3	b	173	8.21	c	174	8.39	c	174	8.57	d	156
8.4	b	156	8.22	d	160	8.40	a	174	8.58	c	167
8.5	a	172	8.23	a	168	8.41	b	174	8.59	b	166
8.6	c	156	8.24	b	179	8.42	c	176	8.60	d	172
8.7	c	177	8.25	c	178	8.43	b	177	8.61	d	162
8.8	d	166	8.26	c	177	8.44	d	156	8.62	a	162
8.9	b	166	8.27	b	160	8.45	c	159	8.63	c	174
8.10	d	174	8.28	b	162	8.46	b	160	8.64	c	174
8.11	d	156	8.29	a	174	8.47	a	178	8.65	b	177
8.12	c	179	8.30	d	160	8.48	c	159	8.66	a	162
8.13	d	167	8.31	a	171	8.49	b	160	8.67	c	162
8.14	a	172	8.32	d	168	8.50	d	156	8.68	c	162
8.15	a	166	8.33	b	174	8.51	a	154	8.69	b	171/3
8.16	b	159	8.34	c	173	8.52	d	160	8.70	d	173
8.17	a	153	8.35	b	173	8.53	a	172	8.71	b	160
8.18	b	167	8.36	a	173	8.54	b	177	8.72	b	177

Q8.3(b). $100 \times \sqrt{1.5} = $ **122 kt.**

Q8.10(d). $1 \times \sqrt{(1/\text{cosine } 45°)} = 1.19$
Increase $= $ **19%.**

Q8.19(a). Turn load factor $= (1/\text{cosine } 45°) = 1.41 = $ **41% increase.**

Q8.21(c). $100 \times \sqrt{2} = $ **141 kt.**

Q8.29(a). $100 \times \sqrt{(1/\text{cosine } 45°)} = $ **119 kt.**

Q8.32(d). Vs $+ 5\% = $ **1.05Vs.**

Q8.33(b). Load Factor for 15° banked turn $= 1/\text{cosine } 15° = 1.0352762$
Load Factor for 60° banked turn $= 1/\text{cosine } 60 = 2$
Vs at 60° turn $= 60 \times \sqrt{(2 \div 1.0352762)} = $ **83 kt.**

Q8.34(c). $2 \times 0.15 = 0.30$
New $C_L = 0.40 - 0.30 = 0.10$
New load factor $= 0.10 \div 0.40 = 0.25$
Change to load factor $= $ from $+ 1$ to $+ 0.25 = $ **−0.75**

Q8.35(b). $2 \times 0.15 = 0.30$
New $C_L = 0.40 - 0.30 = 0.10$
New load factor $= 0.10 \div 0.40 = $ **0.25**

Q8.36(a). $2 \times 0.15 = 0.30$
New $C_L = 0.40 + 0.30 = 0.70$
New load factor $= 0.70 \div 0.40 = \mathbf{1.75}$

Q8.37(d). $140 \times \sqrt{(350000/250000)} = \mathbf{165.7\,kt}$

Q8.38(b). $120 \times \sqrt{1.5} = \mathbf{147\,kt.}$

Q8.40(a). $1 \times \sqrt{(1/\text{cosine } 55°)} = 1.32$
Increase $= \mathbf{32\%}$

Q8.41(b). $1 \times (1/\text{cosine } 55°) = 1.74$
Increase $= \mathbf{74\%}$

Q8.63 & Q8.64(c). Turn stalling speed $=$ Unbanked stalling speed $\times \sqrt{(1/\text{Cos } ø)}$.
Factor $= \sqrt{(1/\text{Cos } ø)} = \sqrt{(1/\text{Cos } 60°)} = 1.412$

Q8.69(b). Vs @ 7500 kg $= 150 \sqrt{(7500 \div 10,000)} = 130\,kt.$
Changed load factor Vs $= 130 \div \sqrt{2} = 91.9\,kt.$
30° bank angle Vs $= 91.9 \sqrt{(1 \div \cos 30)} = \mathbf{98.8\,kt.}$

Q8.70(d). $3 \times 0.079 = 0.237$
New $C_L = 0.35 + 0.237 = 0.587$
New Load factor $= 0.587 \div 0.35 = 1.6771$
Load factor increase $= 1.68 - 1 = \mathbf{0.68}$

Self-assessment Exercise 9

Q	A	Ref	Q	A	Ref	Q	A	Ref	Q	A	Ref
9.1	a	193	9.9	b	198	9.17	c	198	9.25	a	193
9.2	c	198	9.10	a	200	9.18	b	198	9.26	c	198
9.3	d	191	9.11	a	190	9.19	c	200	9.27	b	190
9.4	a	190	9.12	b	190	9.20	a	198	9.28	b	190
9.5	a	198	9.13	b	190	9.21	a	195	9.29	c	190
9.6	a	198	9.14	c	191	9.22	c	193	9.30	d	195
9.7	a	190	9.15	c	190	9.23	a	193	9.31	b	200
9.8	a	198	9.16	a	191	9.24	a	190	9.32	b	198

Self-assessment Exercise 10

Q	A	Ref	Q	A	Ref	Q	A	Ref	Q	A	Ref
10.1	b	214	10.10	a	220	10.19	a	220	10.28	d	222
10.2	d	216	10.11	c	214	10.20	b	213	10.29	a	222
10.3	a	207	10.12	b	214	10.21	d	213	10.30	d	220
10.4	d	223	10.13	d	214	10.22	c	220	10.31	c	222
10.5	d	220	10.14	c	223	10.23	a	212	10.32	b	224
10.6	b	216	10.15	b	223	10.24	c	220	10.33	b	224
10.7	c	223	10.16	a	214	10.25	a	223	10.34	b	223
10.8	c	216	10.17	c	222	10.26	a	223	-	-	-
10.9	d	212	10.18	d	214	10.27	a	217	-	-	-

Self-assessment Exercise 11

Q	A	Ref	Q	A	Ref	Q	A	Ref	Q	A	Ref
11.1	d	251	11.21	d	251	11.41	b	233	11.61	a	246
11.2	a	259	11.22	a	260	11.42	c	248	11.62	b	239
11.3	b	250	11.23	a	248	11.43	c	234	11.63	c	260
11.4	d	256	11.24	a	252	11.44	a	251	11.64	d	251
11.5	a	252	11.25	d	252	11.45	a	252	11.65	a	234
11.6	b	251	11.26	c	259	11.46	b	259	11.66	c	242
11.7	a	265	11.27	b	255	11.47	d	242	11.67	b	242
11.8	b	235	11.28	c	242	11.48	b	252	11.68	c	233
11.9	a	264	11.29	d	249	11.49	c	246	11.69	a	255
11.10	c	263	11.30	b	252	11.50	c	242	11.70	b	265
11.11	d	256	11.31	a	264	11.51	a	252	11.71	b	250
11.12	c	249	11.32	c	249	11.52	a	255	11.72	c	251
11.13	a	235	11.33	c	233	11.53	d	247	11.73	b	243
11.14	d	259	11.34	b	235	11.54	d	251	11.74	b	243
11.15	c	250	11.35	a	252	11.55	b	234	11.75	b	252
11.16	a	242	11.36	d	252	11.56	c	242	11.76	a	253
11.17	b	242	11.37	d	252	11.57	a	245	11.77	c	245
11.18	c	233	11.38	a	252	11.58	d	234	11.78	b	246
11.19	b	233	11.39	c	261	11.59	b	234	11.79	d	250
11.20	c	239	11.40	b	242	11.60	c	242	-	-	-

Q11.10(c). $2.5 \times 250 = 375\,\text{N}$

Increase of stick force $= 375 - 150 = \mathbf{225\,N}$

Self-assessment Exercise 12

Q	A	Ref	Q	A	Ref	Q	A	Ref
12.1	c	282	12.9	c	280	12.17	b	280
12.2	b	282	12.10	c	280	12.18	c	282
12.3	b	279	12.11	d	279	12.19	c	282
12.4	d	277	12.12	b	277	12.20	a	282
12.5	b	282	12.13	a	277	12.21	b	283
12.6	b	277	12.14	d	281	-	-	-
12.7	a	280	12.15	a	282	-	-	-
12.8	a	280	12.16	d	281	-	-	-

Self-assessment Exercise 13

Q	A	Ref	Q	A	Ref	Q	A	Ref
13.1	b	298	13.22	b	299	13.43	a	298
13.2	d	298	13.23	a	295	13.44	b	295
13.3	c	294	13.24	a	305	13.45	c	294
13.4	b	303	13.25	c	298	13.46	d	298
13.5	d	300	13.26	d	298	13.47	c	294
13.6	b	294	13.27	a	298	13.48	a	295
13.7	b	304	13.28	c	300	13.49	a	294
13.8	d	294	13.29	b	299	13.50	d	300
13.9	d	305	13.30	d	296	13.51	b	294
13.10	a	304	13.31	d	303	13.52	d	294
13.11	d	294	13.32	d	303	13.53	c	299
13.12	d	299	13.33	b	303	13.54	a	299
13.13	a	295	13.34	c	303	13.55	d	299
13.14	a	298	13.35	a	304	13.56	d	299
13.15	b	294	13.36	b	304	13.57	b	297
13.16	d	296	13.37	d	303	13.58	c	296
13.17	c	298	13.38	c	304	13.59	a	300
13.18	c	303	13.39	b	303	13.60	a	303
13.19	c	303	13.40	b	303	13.61	b	298
13.20	d	303	13.41	d	305	13.62	c	298
13.21	a	294	13.42	c	304	13.63	d	306

Q13.1(b). Increase of $C_L = 2 \times 0.079 = 0.158$
New $C_L = 0.35 + 0.158 = 0.508$
New load factor $= 0.508 \div 0.35 = \mathbf{1.45143}$

Q13.9(d). Answer (a) Turn Load Factor $= 1/\text{Cosine } \phi$. A and B $= 1/\cos 20° = 1.064$. Turn load factors are equal.
Answer (b). Turn Radius $= V^2/g \tan \phi$; A $= 66.95^2/9.81 \tan 20° = 1255$ m. B $= 103^2/9.81 \tan 20° = 2971$ m. Turn radius of B greatest.
Answer (c). The C_L for the higher speed will be least. B's C_L is least.

Answer (d). Rate of turn $= \dfrac{g \times \tan \phi}{V}$ degrees per second.

$A = \dfrac{9.81 \times \tan 20°}{130} = 0.0275$ degrees per second;

$B = \dfrac{9.81 \times \tan 20°}{200} = 0.0179$ degrees per second

Rate of turn of A is greatest.

Q13.10(a). Turn Radius $= V^2/g \tan \phi$; A $= 154.5^2/10 \tan 45° = \mathbf{2387}$ m.

Q13.14(a). Increase of $C_L = 5 \times 0.09 = 0.45$
New $C_L = 0.4 + 0.45 = 0.85$
New load factor $= 0.85 \div 0.4 = \mathbf{2.125}$

Q13.19(c). Total Lift $=$ Total mass $\div$ Cosine $\phi = 50{,}000$ N $\div$ Cosine $45° = \mathbf{70{,}711}$ N

Q13.20(d). Turn Load Factor $= 1/\text{Cosine } \phi$;
Turn stalling speed $=$ Unbanked stalling speed $\times \sqrt{(1/\text{Cos } \phi)}$.

Q29.34(c). Lift required in a turn $=$ Mass $\div$ cosine $\phi =$ Mass $\div$ cosine $45° = 1/0.707 = 1.414$.
Increased lift $= \mathbf{41.4\%}$

Q13.36(b). Turn Radius $= V^2/g \tan \phi$; A $= 154.5^2/10 \tan 45° = 2387$m. $= \mathbf{7829\,ft}$.

Q13.62(c). Load Factor $=$ Lift/Mass. C_L at original load factor $= 0.42$.
Change to $C_L = 3 \times 0.1 = 0.3$
New $C_L = 0.42 + 0.3 = 0.72$
New Load Factor $=$ New $C_L \div$ Original $C_L = 0.72 \div 0.42 = \mathbf{1.71}$.

Self-assessment Exercise 14

Q	A	Ref	Q	A	Ref	Q	A	Ref
14.1	a	322	14.28	c	320	14.55	b	338
14.2	b	338	14.29	d	318	14.56	c	334
14.3	b	337	14.30	a	322	14.57	a	335
14.4	c	337	14.31	c	319	14.58	a	338
14.5	a	338	14.32	a	318	14.59	d	332
14.6	c	338	14.33	d	322	14.60	a	315
14.7	a	337	14.34	b	324	14.61 .	b	318
14.8	d	338	14.35	c	323	14.62	c	319
14.9	d	339	14.36	d	316	14.63	c	337
14.10	d	338	14.37	b	320	14.64	a	332
14.11	d	337	14.38	d	323	14.65	b	332
14.12	c	336	14.39	d	323	14.66	a	315
14.13	b	337	14.40	a	316	14.67	d	334
14.14	c	333	14.41	c	329	14.68	b	334
14.15	b	332	14.42	c	316	14.69	c	338
14.16	a	337	14.43	a	323	14.70	c	334
14.17	d	319	14.44	d	315	14.71	d	332
14.18	d	329	14.45	c	321	14.72	a	334
14.19	d	316	14.46	d	325	14.73	b	338
14.20	b	329	14.47	b	315	14.74	a	338
14.21	a	331	14.48	d	323	14.75	b	317
14.22	d	320	14.49	d	333	-	-	-
14.23	d	321	14.50	a	318	-	-	-
14.24	a	329	14.51	a	317	-	-	-
14.25	d	323	14.52	b	329	-	-	-
14.26	c	317	14.53	d	321			
14.27	d	331	14.54	d	338			

Q14.1(a). $\text{TAS} = 100\,\text{kt} = \dfrac{100 \times 6080}{60} = 10\,133.3\,\text{fpm}$

Vertical velocity $= V \sin \varnothing = 10\,133 \sin 3° = \mathbf{530.3\,fpm.}$

Q14.17(d). Given: (i) Thrust : **Mass** ratio $= 1{:}4$; (ii) Lift : Drag ratio $= 12{:}1$
Assuming Lift $=$ **Mass**; then ratio (i) expressed in terms of ratio (ii) becomes
Thrust : **Mass** ratio $= 3{:}12$.
Climb Gradient $= [(T - D)/M] \times 100 = [(3 - 1)/12] \times 100 = \mathbf{16.7\%}.$

Q14.28(c). $\text{TAS} = 400\,\text{kt}$. Groundspeed $= 450\,\text{kt}$.
Wind effective gradient $=$ Still-air gradient $\times$ TAS $\div$ groundspeed
Wind effective gradient $= 4 \times 400 \div 450 = 3.556\%$
Height gain $= 32\,000\,\text{ft} - 8\,000\,\text{ft} = 24\,000\,\text{ft}$

Ground distance travelled $= 24\,000\,\text{ft} \div 3.556 \times 100 = 674\,915.6\,\text{ft} = \textbf{111 nm}$

Q14.29(d). Climb Gradient = (Thrust - Drag) ÷ **Mass** × 100
Thrust $= 75\,000\,\text{N} = 75\,000 \div 10\,\text{kg} = 7\,500\,\text{kg}$
Lift = Mass $= 37\,500\,\text{kg}$
Drag $= 37\,500 \div 14 = 2\,678.6\,\text{kg}$
Climb Gradient $= (7\,500 - 2\,678.6) \div 37\,500 \times 100 = \textbf{12.86\%}$

Q14.30(a). SA Climb Gradient $= \dfrac{\text{ROC fpm}}{\text{TAS kt}} \times \dfrac{6\,000\%}{6\,080} = \dfrac{1\,000}{194} \times \dfrac{6\,000\%}{6\,080} = 5.09\%$

TAS $= 194\,\text{kt} = 194 \times 6\,080 \div 60\,\text{fpm} = 19\,658.7\,\text{fpm}$
Sine climb angle = opposite ÷ hypotenuse $= 1\,000 \div 19\,658.7 = 0.050868 = \textbf{2.92}°$

Q14.31(c). Climb mass $= 110\,000 \times (2.8/2.6) = \textbf{118\,462 kg}$.

Q14.32(a). Gradient $= [(\text{T} - \text{D})/\text{M}] \times 100$
$\text{T} = 2 \times 50\,000\,\text{N} = 100\,000 \div 10\,\text{kg} = 10\,000\,\text{kg}$; $\text{D} = 72\,069 \div 10 = 7\,206.9\,\text{kg}$
$2.75\% = [(10\,000 - 7\,206.9)/\text{M}] \times 100 = [2\,793.1\text{M}] \times 100$
$\text{M} = 279\,310 \div 2.75 = \textbf{101\,567.27 kg}$

Q14.33(d). SA Climb Gradient $= \dfrac{\text{ROC fpm}}{\text{TAS kt}} \times \dfrac{6\,000\%}{6\,080} = \dfrac{1\,000}{194} \times \dfrac{6\,000\%}{6\,080} = \textbf{5.09\%}$

or

Gradient = (Height difference / distance travelled) × 100
TAS $= 194\,\text{kt} = 194 \times 6\,080 \div 60\,\text{fpm} = 19\,658.7\,\text{fpm}$
Sine climb angle = opposite ÷ hypotenuse $= 1\,000 \div 19\,658.7 = 0.050868 = 2.92°$
Climb gradient = Tan climb angle × 100 = Tan 2.92° × 100 = **5.1\%**

Q14.37(b). ROC = still-air climb gradient × TAS
ROC $= 3.3 \times 100 = \textbf{330 fpm}$.

Q14.62(c). Gradient $= [(\text{T} - \text{D})/\text{M}] \times 100$
$\text{T} = 2 \times 60\,000\,\text{N} = 120\,000 \div 10\,\text{kg} = 12\,000\,\text{kg}$; $\text{L} = \text{M} = 50\,000\,\text{kg}$. $\text{D} = 50\,000 \div 12 = 4\,166.7\,\text{kg}$
Gradient $= [(12\,000 - 4\,166.7)/50\,000] \times 100 = [7\,833.3/50\,000] \times 100 = \textbf{15.67\%}$

14.0.1 *Vx & Vy Mathematical Proof*

The difference in speed between Vx and Vy in the following example is grossly exaggerated.

Assume that the lower speed is 100 kt TAS and the climb angle is 15°. Then the TAS $= 100 \times 6\,080\,\text{ft} = 608\,000\,\text{ft}$ per hour $= 608\,000 \div 60 = 10\,133.3\,\text{fpm}$. When resolved into its vertical and horizontal components becomes:
Vertical component $= 10\,133.3\,\text{sine}\,15° = 2\,622.7\,\text{fpm}$ = rate of climb
Horizontal component $= 10\,133.3\,\text{cosine}\,15° = 9\,788\,\text{fpm}$

Assume that the higher speed is 120 kt TAS and the climb angle is 15°. Then the TAS $= 120 \times 6\,080\,\text{ft} = 729\,600\,\text{ft}$ per hour $= 729\,600 \div 60 = 12\,160\,\text{fpm}$. When resolved into its vertical and horizontal components becomes:
Vertical component $= 12\,160\,\text{sine}\,15° = 3\,147.2\,\text{fpm}$ = rate of climb
Horizontal component $= 12\,160\,\text{cosine}\,15° = 11\,745.7\,\text{fpm}$
Therefore highest rate of climb is obtained at the higher speed, in this example Vy is 120 kt TAS.
Now assume that it is required to climb 2 000 ft in the distance travelled in one minute then the required gradient would be:

At 100 kt TAS $= 2\,000\,\text{ft} \div 9\,788 \times 100 = 20.43\%$
At 120 kt TAS $= 2\,000\,\text{ft} \div 12\,160 \times 100 = 16.45\%$

Therefore the highest gradient is obtained at the lower speed, in this example Vx is 100 kt TAS.

Self-assessment Exercise 15

Q	A	Ref	Q	A	Ref	Q	A	Ref	Q	A	Ref
15.1	a	359	15.34	a	357	15.67	b	371	15.100	c	352
15.2	a	367	15.35	a	359	15.68	b	359	15.101	a	360
15.3	d	361	15.36	d	369	15.69	c	351	15.102	a	351
15.4	d	353	15.37	c	354	15.70	c	358	15.103	b	371
15.5	c	359	15.38	a	352	15.71	d	358	15.104	a	358
15.6	a	352	15.39	d	371	15.72	d	357	15.105	b	367
15.7	b	369	15.40	c	365	15.73	a	358	15.106	a	359
15.8	a	351	15.41	a	352	15.74	a	360	15.107	d	369
15.9	a	359	15.42	c	369	15.75	b	360	15.108	c	354
15.10	a	358	15.43	a	369	15.76	c	371	15.109	a	359
15.11	a	360	15.44	b	359	15.77	b	369	15.110	b	367
15.12	d	357	15.45	b	367	15.78	d	360	15.111	b	369
15.13	c	367	15.46	c	354	15.79	c	360	15.112	c	367
15.14	d	360	15.47	d	358	15.80	d	369	15.113	d	366
15.15	d	367	15.48	a	359	15.81	a	369	15.114	b	358
15.16	a	368	15.49	a	352	15.82	b	369	15.115	c	357
15.17	b	360	15.50	a	370	15.83	c	359	15.116	d	358
15.18	d	359	15.51	b	370	15.84	a	359	15.117	c	360
15.19	a	367	15.52	a	360	15.85	c	367	15.118	a	369
15.20	b	369	15.53	b	357	15.86	a	359	15.119	a	352
15.21	a	367	15.54	c	360	15.87	b	359	15.120	b	369
15.22	c	358	15.55	b	367	15.88	a	357	15.121	a	369
15.23	c	362	15.56	a	359	15.89	b	353	15.122	b	359
15.24	b	352	15.57	a	359	15.90	d	367	15.123	d	367
15.25	c	359	15.58	c	371	15.91	b	360	15.124	b	367
15.26	c	367	15.59	d	358	15.92	c	357	15.125	c	368
15.27	d	360	15.60	c	371	15.93.	a	357	15.126	a	371
15.28	c	357	15.61	d	354	15.94	c	354	15.127	a	365
15.29	b	365	15.62	c	365	15.95	d	352	15.128	a	367
15.30	d	353	15.63	a	358	15.96	a	359	-	-	-
15.31	a	357	15.64	d	354	15.97	b	359	-	-	-
15.32	a	360	15.65	c	367	15.98	a	367	-	-	-
15.33	a	366	15.66	a	369	15.99	c	368	-	-	-

15.6(a). Mach number = TAS ÷ LSS

By transposition then LSS = TAS ÷ Mach number = $400 \div 0.8 = \mathbf{500\,kt}$.

Self-assessment Exercise 16

Q	A	Ref	Q	A	Ref	Q	A	Ref	Q	A	Ref
16.1	c	387	16.13	d	401	16.25	c	395	16.37	d	389
16.2	b	392	16.14	a	393	16.26	d	395	16.38	b	403
16.3	b	394	16.15	a	387	16.27	b	403	16.39	d	393
16.4	c	393	16.16	d	397	16.28	c	397	16.40	b	401
16.5	c	389	16.17	a	389	16.29	d	387	16.41	c	403
16.6	d	394	16.18	c	391	16.30	b	393	16.42	c	391
16.7	d	391	16.19	c	403	16.31	c	389	16.43	d	387
16.8	a	395	16.20	d	401	16.32	d	394	16.44	d	392
16.9	a	393	16.21	d	403	16.33	d	401	16.45	d	403
16.10	a	403	16.22	b	392	16.34	a	403	16.46	a	393
16.11	b	403	16.23	b	403	16.35	a	403	16.47	b	395
16.12	b	396	16.24	b	393	16.36	a	396	-	-	-

Self-assessment Exercise 17

Q	A	Ref	Q	A	Ref
17.1	c	416	17.9	b	317
17.2	a	306	17.10	a	417
17.3	a	136	17.11	b	416
17.4	b	136	17.12	a	416
17.5	d	136	17.13	b	415
17.6	b	416	17.14	d	416
17.7	d	294	17.15	c	403
17.8	c	45	17.16	d	404
			17.17	a	404

Self-assessment Exercise 18 Turn Calculations

Q	Formula	1	2	3	4	5	6	7
IAS	knots	100 kt	150 kt	180 kt	220 kt	240 kt	270 kt	300 kt
Speed V mps	× 0.515	51.5	77.25	92.7	113.3	123.6	139.05	154.5
V^2	Calculator	2652.25	5967.56	8593.3	12836.9	15277.0	19334.9	23870.25
Stalling Speed	Given	64 kt	72 kt	75 kt	70 kt	72 kt	75 kt	80 kt
Bank Angle	Given	45°	40°	35°	30°	25°	20°	15°
Cosine	Calculator	0.7071	0.7660	0.8192	0.8660	0.9063	0.9397	0.9659
Tangent	Calculator	1	0.8391	0.7002	0.5774	0.4663	0.3640	0.2680
Load Factor	1/cos ø	1.4142	1.3055	1.2207	1.1547	1.1034	1.0642	1.0353
Stalling Speed	VS × √1/cos ø	76 kt	82 kt	83 kt	75 kt	76 kt	77 kt	81 kt
g tan ø	Calculator	10	8.391	7.002	5.774	4.663	3.640	2.680
Radius of Turn	V^2 mps ÷ g tan ø	265.2 m	711.1 m	1227.3 m	2223.2 m	3276.2 m	5311.8 m	8906.8 m
Rate of Turn radians.	g tan ø ÷ V mps = Radians	0.19417	0.10862	0.07553	0.050962	0.037727	0.0261776	0.0173462
Rate of Turn °/sec	× 57.3	11.1	6.22	4.33	2.92	2.16	1.5	1.0

Index

The Principles of Flight for Pilots   P. J. Swatton
© 2011 John Wiley & Sons, Ltd